AF245502

Ecorestoration of the coalmine degraded lands

Subodh Kumar Maiti

Ecorestoration of the coalmine degraded lands

 Springer

Subodh Kumar Maiti
Indian School of Mines
Department of Environmental Science and Engineering
Centre for Mining Environment
Dhanbad, Jharkhand
India

ISBN: 978-81-322-0850-1   ISBN: 978-81-322-0851-8 (eBook)
DOI: 10.1007/978-81-322-0851-8
Springer New Delhi Heidelberg New York Dordrecht London

Library of Congress Control Number: 2012954478

Printed on acid-free paper

Springer is part of Springer Science+Business Media (www.springer.com)

Foreword

Indian economy has grown at a CAGR of 7.6% in the first decade of the twenty-first century, and the coal industry has played a very important role in sustaining this economic growth. Coal is the most important mineral mined in the country and accounts for 52% of its commercial energy consumption.

Indian government is giving a major thrust for development of the power sector as electrical energy is a key input for economic growth. The installed capacity for generation of electricity has gone up from 101.6 GW in 2000–2001 to 200 GW in 2011–2012 (CAGR 6.3%). In the XII Five-Year Plan, another 90 GW generation capacity is planned to be installed. As approximately 70% of the power generated in the country is coming from coal-fired power stations, such major expansion of the power sector requires commensurate expansion of the coal mining sector. Along with the power sector, the steel and cement sectors in the country are also undergoing expansion at a rapid rate adding to the increasing coal demand in the country. Indian coal production was 322.7 Mt in 2000–2001 and reached a level of 534.5 Mt in 2011–2012 registering a compounded annual growth rate of 4.7%. The XII Five-Year Plan now under formulation stage has projected an expansion of the coal mining sector at a higher rate of 8.2% for coal production to reach 795 Mt in 2016–2017. The demand for coal is so high that even with the projected rate of expansion of indigenous production, the country will need to import approximately 200 Mt of coal in 2016–2017.

Most of the coal deposits in India occur in forested tracts with many mine leaseholds containing large patches of forest land. These days, almost 90% of coal produced in India comes from opencast mines. Large mechanised opencast mines with high-stripping ratios exceeding 1 in 4 and in some cases even 1 in 6 or 1 in 7 and going to great depths inevitably cause large-scale deforestation and loss of soil cover over large areas and creation of huge external overburden dumps.

The total land requirement in India for mine operation, waste dumps and mine infrastructures is projected to increase from the level of 1,470 km^2 (including a forest area of 730 km^2) in 2006–2007 to 2,925 km^2 (including a forest area of 730 km^2 in 2025) as per 'Vision Coal-2025' document. The technology of opencast mining inevitably leads to complete degradation of land, destruction of forest ecosystem and fragmentation of wildlife habitat, and magnitude of devastation is so massive that entire landscape of the area is changed. The only saving grace is that unlike other industries, mining is a temporary user of land and with proper scientific restoration, a functioning ecosystem can be restored, and in some cases even a better landscape can be created. Several organisations are concerned about the problem and are actively involved in carrying out scientific restoration work in mined out areas. The universities and other organisations teaching mining course have incorporated the subject of mine closure and ecorestoration in the mining environment syllabus. One constraint faced by the students as well as practising engineers of the subject is the lack of availability of books dealing specifically with restoration of mine degraded land in India. Dr. Subodh Kumar Maiti has been working in the Centre of Mining Environment, Indian School of Mines, Dhanbad, for more than two decades and has been teaching and conducting research on the subject.

In this book, Dr. Maiti has discussed the scientific basis of ecorestoration, the different ways one can achieve the goals, and explains how the ecorestoration can be made self-sustainable. This book is presented in two parts: Part I dealing with different aspects of ecorestoration and Part II dealing with sampling, laboratory analysis and evaluation of overburden materials, soils and plants.

Public opposition to mining is increasing in most parts of the world. General public has become more aware of the importance of preserving the environment and ecorestoration of mine degraded lands. The social licence for mining, given by the Ministry of Environment in the form of environmental clearance for a project, generally requires the mine management to carry out systematic reclamation of degraded mined-out land. I hope this book by Dr. Maiti will be useful for the environmental professionals in the mining industry, students of mining engineering course, mine planners as well as the general public.

Former Professor and Head   S.P. Banerjee
Centre of Mining Environment
Indian School of Mines, Dhanbad

Preface

India is essentially a land-short economy. Its demand for usable land is growing at an exponential rate. Mining of minerals is put into the category of 'temporary use of land'. Centuries of mining have resulted in many mining land beyond any commercial, recreational or social use. It is now important to restore such land for fruitful use and also to plan the future mining in a way so that the used land can be reused in the future.

Coal mining industry in India plays a very important role in country's economy—more than 70% of the total power generated in the country is from coal. India's coal consumption ranks third in the world, and the country's demand for coal continues to grow much faster than the world average. Coal deposits are mainly found under the forest cover and are confined to eastern and south central parts of the country. Indian coal sector is poised to grow at a very fast rate in the near future due to steep increase in coal demand for providing power to all by 2012. Total indigenous coal production is expected to grow from the current level of around 540 Mt (2011–2012) to around 1,086 Mt by 2024 and 2,037 Mt by 2031–2032 as per the draft Coal Vision document, and more than 90% will be produced by opencast mining.

The quantum jump in coal production from opencast operation and consequent overburden removal will put significant stress on the environment due to total destruction of the vegetation and soil cover, formation of waste dumps, depletion of water tables, increase in dust pollution and deterioration of landscape and aesthetics of the area. The total land requirement for mine operation, waste dumps and mine infrastructures is projected to increase from the level of 1,470 km^2 (including a forest area of 730 km^2) in 2006–2007 to 2,925 km^2 (including a forest area of 730 km^2 in 2025) as per 'Vision Coal-2025' document.

In India, majority of the new and unmined mineral and coal deposits is under forest cover; thus, complete degradation of land, destruction of forest ecosystem and fragmentation of habitat are inevitable, and magnitude is so massive that entire landscape is changed. Fortunately, unlike other industries, mining is very temporary user of land, and proper scientific ecorestoration can restore the functioning of ecosystem and may bring better landscape. There is a growing concern to make the land useful yet again. Several organisations are concerned about these aspects and are actively involved to carry out restoration work, but no book as guide is available until date to deal with restoration of degraded land. After working more than two decades in Centre of Mining Environment, Indian School of Mines, Dhanbad, which is known as the Mining Capital of India, I feel a book in this line is overdue.

The purpose of this book is to discuss the scientific basis of ecorestoration and in what different ways one can achieve the goals and how to ensure that the ecorestoration is self-sustainable. This book is presented in two parts: Part I contains different aspects of ecorestoration, and Part II contains laboratory analysis and evaluation of overburden materials, soils and plants.

This book is presented as follows:

Part I

Chapter 1 provides an introduction to the importance of coal mining in India and relevant environmental issue, principles and components of ecorestoration, issues related to restoration, legal and statutory framework of ecorestoration.

Chapter 2 gives the basic concepts of ecology and functioning of ecosystem in mine degraded areas.

Chapter 3 discusses the importance of physical reclamation for ecorestoration process, estimation of sediment loss from bare areas, design of diversion ditches and sediment pond.

Chapter 4 reviews the important properties of minesoil that is going to affect the plant establishment and growth in mine degraded lands.

Chapter 5 stresses the importance of topsoil management, which includes removal, storage and redistribution of topsoil.

Chapter 6 discusses the methods of vegetation cover development, selection of plant species and case studies related to the existing tree covers in different dumps.

Chapter 7 introduces the importance of seeding for ecorestoration, its collection, preservation and breaking of dormancy.

Chapter 8 provides the details of nursery raising of forest tree species for the ecorestoration purposes.

Chapter 9 introduces the techniques of the establishment of grass and legume cover.

Chapter 10 introduces importance of mulching, geotextile and organic amendments.

Chapter 11 discusses the importance of bio-fertiliser technology for ecorestoration.

Chapter 12 introduces the importance of biodiversity, causes of biodiversity erosion and its methods of conservation.

Chapter 13 highlights the importance of monitoring and aftercare of ecorestored sites for the development of self-sustaining vegetation cover.

Chapter 14 discusses the criteria for the evaluation of ecorestoration success and indicators.

Chapter 15 provides the important forest and wildlife conservation acts.

Chapter 16 discusses the ecological impact assessment procedure for surface mining projects.

Chapter 17 highlights the importance of mine closure, objectives, issues, planning, activity, cost, mine closure guidelines in India and preparation of closure plan report.

Part II

Chapter 18 introduces the sampling and processing of soil and mine spoils for laboratory analysis, onsite test and description of parameters.

Chapter 19 provides detail laboratory analysis procedure of physical parameters like texture, bulk density, moisture content, infiltration test, rooting depth etc. and the interpretation of test results.

Chapter 20 provides detail analysis of chemical and nutritional parameters.

Chapter 21 contains laboratory procedure of analysis of soil microbiological parameters.

Chapter 22 provides detail of vegetation sampling procedure, plant nutrient analysis, pot and field trial experiments and analysis of plant growth parameters.

ISM, Dhanbad

Subodh Kumar Maiti
subodh_maiti@yahoo.com

About the Author

Dr. Subodh Kumar Maiti (b. 25-04-1960) is a professor in the Department of Environmental Science and Engineering, Centre of Mining Environment, Indian School of Mines, Dhanbad. He did B.Sc. (Hons) and M.Sc. in Botany from Calcutta University in 1984, M.Tech. (Environmental Science and Engineering) from IIT Mumbai in 1986 and worked as senior environmental engineer in Kirloskar Consultancy Division, Pune, until 1987. In 1988, he joined as lecturer in ISM, Dhanbad, and earned his Ph.D. in Environmental Science and Engineering in ecological aspects of reclamation of coal mine degraded lands. He underwent 3-month training on EIA and auditing at University of Bradford, UK, in 1996 and 1-month advanced training on mining and environment at University of Lulea, Sweden, under SIDA in 2001. He served as guest faculty in University of Technology, Papua New Guinea in the year 2008. He is teaching ecology, environmental microbiology and land reclamation in B.Tech. and M.Tech. levels and has guided several research students.

He has been working in the biological aspects of reclamation of mine degraded lands for more than two decades. He has published over 60 papers in international and national journals, - has presented over 120 papers in international and national seminars. He has attended International Conference organized by Society of Ecorestoration International (Perth, Western Australia) in 2009; 2nd International Conference on Constructed Environment (Chicago, USA) in 2011 and visited many more countries. He has also published two books" Handbook of Methods in Environmental Analysis Vol I

(Water and Wastewater analysis) and Vol II (Air, Noise, soil and overburden analysis). He has completed several R&D and consultancy projects on biological reclamation, biodiversity assessment and design and development of green belts in mining areas. He is life member of IAEM, IASWC, *Mycorrhiza News*, MGMI and *The Indian Mining & Engineering Journal* (IME). He was also selected as Fellow National Environmentalist Association (FNAE), Member SERI (Australia) and IPS (USA).

Contents

Part I

Ecorestoration of Coalmine Degraded Lands

Introduction

1

Contents

1.1 Importance of Coal Mining in India

Coal mining industry in India plays a very important role in the country's economy—more than 70% of the total power generated in the country is from coal, and considering the total energy requirement, coal contributes more than half. India's coal consumption ranks third in the world, and the country's demand for coal continues to grow much faster than the world average. The estimated recoverable reserve of coal and lignite is 101.9 billion tonnes (Bt), which is about 10% of the total world reserves (Table 1.1). As of 31.03.10, the estimated reserves of coal were around 277 Bt, an addition of 10 Bt over the last year. Coal deposits are mainly confined to eastern and south central parts of the country. The states of Jharkhand, Orissa, Chhattisgarh, West Bengal, Andhra Pradesh, Maharashtra and Madhya Pradesh account for more than 99% of the total coal reserves in the country. If India has to sustain an 8–10% economic growth rate, over the next 25 years, to eradicate poverty and meet its human development goals, an increase in its primary energy supply by 3–4 times compared to the level of 2003–2004 is required.

The Integrated Energy Policy (IEP 2006) document formulated by Planning Commission (August 2006) has presented several alternative scenarios of energy mix to sustain a GDP growth

S.K. Maiti, *Ecorestoration of the Coalmine Degraded Lands*,
DOI 10.1007/978-81-322-0851-8_1, © Springer India 2013

Table 1.1 World recoverable coal reserves (billion tonnes) (IEH 2011)

Region/country	Coal	Lignite	Total
World total	827.4	173.4	1,000.8
United States	234.7	36	270.7
Russia	161.6	11.5	173.1
China	89.1	20.5	109.6
India	99.3[a]	2.6	101.9

Source: Expert Committee on Coal Reforms
[a]Total proved 'in place' reserves instead of recoverable reserves relevant for other countries

Table 1.2 Coal demand projections (million tonnes)

Plan period	Power	Non-power	Total
XI 2011/2012	436	164	627
XII 2016/2017	603	221	824
XIII 2021/2022	832	299	1,131
XIV 2026/2027	1,109	408	1,517
XV 2031/2032	1,475	562	2,037

Source: Integrated Energy Policy (2006), Planning Commission

at 8% until 2031–2032. The requirement of coal demand has been projected to 2,037 Mt (2031–2032) against the 627 Mt projected for the end of 2011–2012 (XI plan) (Table 1.2). However, the requirement of coal-based energy has been projected to vary from 1,022 Mtoe (2,555 Mt) for a coal-dominant scenario to 632 Mtoe (1,540 Mt) in the scenario considering utilisation of full potential of nuclear, hydro and renewable resources along with all energy conservation measures. Therefore, coal will remain a dominant source of energy in India up to 2031–2032 and possibly beyond (Chaudhuri 2008).

Share of coal production from surface mines was increased from less than 20% in 1973–1974 to a level of more than 80% during 2001–2002. As of 31.12.2010, there are 717 mining projects of CIL and 140 mining projects in SCCL. Of these, a few mines are of large planned capacities—such as Gevra (20 Mt), Jayant (10 Mt), Nigahi (14 Mt), Dadhichua (10 Mt) and Rajmahal (10.5 Mt). Unfortunately, a large number of small-capacity surface mines (producing 0.1–0.3 Mt per annum)

are spread over in old coalfields posing a higher threat to the environment if proper corrective/controlling measures are not taken. However, for these smaller mines, effective environmental protection measures could not be adopted due to limited resources and other various reasons resulting in formation of large overburden (OB) dumps and huge voids at mining sites left as orphan land. Virtually all surface mining methods produce dramatic change in landscape due to large Indian coal sector poised to grow at a very fast rate in the near future due to steep increase in coal demand for the major reason of providing power to all by 2012. Total indigenous coal production is expected to grow from the current level of around 407 Mt (2005–2006) to around 1,086 Mt by 2024 as per the draft Coal Vision document.

The share of opencast production has increased from 26% (20.77 Mt) in 1974–1975 to 84.95% (345.79 Mt) in 2005–2006, whereas total underground production has declined from 74% (58.22 Mt) to 15.05% (61.25 Mt) in the same period. In the years 2006–2007, coal production was 430.83 Mt (terminal year of XI plan), out of which opencast contributed to around 373 Mt (87%) with an estimated overburden removal of 600 million m^3. A long-term perspective of coal production and OB removal presented in the 'Coal Vision 2025' document of Ministry of Coal indicates a coal production 1,086 Mt of which 900 Mt will be from opencast operation. The corresponding OB removal figure is estimated at 2,700 Mm^3. In 2009–2010, coal production was 533 Mt as compared to 493 Mt during 2008–2009, registering a growth of 8%. Even in the past 10 years, production figures indicate that OC production continues to rise year after year. It is estimated that at the end of terminal year of the 11th Five-Year Plan (2011–2012), the coal demand would be about 713 Mt, whereas the indigenous availability would be about 630 Mt. Therefore, there is likely to be a gap of 83 Mt, which is required to be met through imports (MOC 2011). Table 1.3 shows the technology-wise coal production since 1951.

Table 1.3 Technology-wise national coal production—past decade

Year	Opencast Production (Mt)	(%) Share	Underground Production (Mt)	(%) Share	Total production (Mt)
1951	4.8	13.7			35
1961	10.8	11.9			55.8
1970–1971	17.1	22.6			72.95
1980–1981	40.4	35.5			114.0
1982–1983	62.8	48.8			130.6
1983–1984	64.0	46.3			138.3
1984–1985	72.0	48.6			147.4
1985–1986	76.6	49.6			154.3
1986–1987	88.1	53.1			165.8
1987–1988	103.8	57.8			180.0
1988–1989	114.9	59.0			194.5
1989–1990	124.3	61.9			200.9
1990–1991	*139.6*	*66*			*214.7*
1991–1992	155.5	67.8			229.3
1992–1993	156.4	66			237.1
1996–1997	218.97	75.48	70.95	24.52	289.32
1997–1998	213.37	77.02	69.03	22.98	300.40
1998–1999	228.75	77.15	67.76	22.85	296.51
1999–2000	237.28	78.02	66.83	21.98	304.10
2000–2001	*247.63*	*78.94*	*66.07*	*21.06*	*313.70*
2001–2002	262.97	80.23	64.82	19.77	327.79
2002–2003	277.67	81.37	63.58	18.63	341.25
2003–2004	298.41	82.62	62.76	17.38	361.17
2004–2005	320.27	83.70	62.35	16.30	382.62
2005–2006	345.79	84.95	61.25	15.05	407.04
2006–2007	*373.1*	*86.59*	*57.8*	*13.41*	*430.83*
2007–2008	398.2	87.11	58.9	12.89	457.08
2008–2009	434	88.03	59	11.97	492.76
2009–2010	473.52	89	58.52	11	532.04
2010–2011	477.84	89.70	54.86	10.30	532.70
2011–2012	488.11	90.40	51.83	9.60	539.94

For the last 10 years (2002–03 to 2011–12), lowest growth rate of 0.1% recorded in 2010–11; 1.4% in 2011–12 and highest growth rate of 8% recorded in 2009–10.
Source: Provisional Coal Statics 2011–2012. Coal Controller's Organisation, MOC, GOI. http://www.coalcontroller.gov.in/coal_admin/files/Statistics/Statistic-9.pdf

1.2 Opencast Coal Mining and Environmental Issues

1. Land degradation due to mining, deterioration of air quality, water pollution and siltation, loss of vegetation and driving out fauna, noise and vibration, reduction in aesthetics and rehabilitation of person are some of the major issues where mining industry has to take care.

2. One of the foremost components of the environment that have been severely damaged due to surface mining is '*land*'. In comparison to the other user, mining industry uses less than 0.25% (0.7 m ha) of the total land (329 m ha). However, land degradation cannot be considered as insignificant because subsequent environmental impacts do not restrict within the boundaries of mining lease hold areas.

3. The quantum jump in coal production from opencast operation and consequent OB removal will put significant stress on the environment on account of total removal of the soil cover and formation of waste dumps (20–30% of OB excavated to be placed in external waste dump), depletion of water tables, etc., in the mining areas. The total land requirement for mine operation, waste dumps and mine infrastructures are projected to increase from the level of 1,470 km^2 (including a forest area of 730 km^2) in 2006–2007 to 2925 km^2 (including a forest area of 730 km^2 in 2025) as per 'Coal Vision 2025' document.

4. Next to land, it is *air quality*, which is severely affected by mining activities. Increase in respirable particulate matter (RPM or PM$_{10}$), suspended particulate matter (SPM) and settable dust (dust fall) is of great concern. Some of the impacts like damage to health (*bronchitis, asthma, pneumoconiosis*), soiling of material, aesthetic and loss of visibility need attention of the mining companies.

5. Deterioration in *water quality* particularly increase in suspended solids (SS), higher erosion from the barren areas during rain and deterioration of surface water sources due to dry deposition of dust are of major concern.

6. Near-total loss of vegetation and removal of topsoil result into creation of new habitat.

1.3 Land Degradation Due to Mining

In India, land degradation due to mining is inevitable as major coal deposits are under thick forest cover and more than 85% of coal is extracted by opencast method. Moreover, massive land degradation is unfortunately unavoidable due to localised deposit of minerals and geology of coal seams. In an old estimate (Business Line 2000), it has been reported that nearly 140,771 ha of land was covered under surface mining, and additionally, 57,000 ha of land was required of which 13,000 ha was under forest. The causes of land degradation during to mining are removal of vegetation cover and topsoil, excavation and dumping of overburden, subsidence, mine fire, etc. The overburden removal in Coal India alone increased from 500 million cubic metres (Mm^3) in 2003–2004 to 682 Mm^3 in 2009–2010 (CIL Report 2011).

Mine wasteland generally comprises the bare stripped area, loose soil piles, waste rock and overburden surfaces, subsided land areas, mine fire, etc. Overburden dumps created for the accommodation of mine waste have major effects on surrounding environment like deterioration of aesthetics and reduction in land productivity and complete destruction of landform (landscape) and habitat and act as continuous sources of dust pollution and water pollution and siltation.

Therefore, development of vegetation cover is essential on these dumps and other denuded areas to stabilise the dumps and minimise pollution and improve visual aesthetics to the surrounding population.

However, these newly created man-made habitats posses wide ranges of problems for establishing and maintaining vegetation cover, due to adverse physico-chemical and physico-mechanical properties. For example, in acidic dumps (which is common), along with elevated metal concentration, other adverse factors like high stone content, lack of moisture, higher compaction, shortage of soil-forming materials and organic matter also cause problems (Maiti 2003). Lastly, out of all alternatives, development of vegetation cover is the cheapest and easiest options, but one has to ensure that vegetation cover is self-sustained in long term. There are several process by which self-sustaining vegetation cover could be developed in mine-degraded land, starting from careful design of slope to selection of tree species (that spread and reproduce under severe conditions), human assistance (soil ameliorant, mulching, geo-netting of the area) and proper maintenance (Maiti 2010). In the impoverished site, sometimes exotic trees are chosen over native species for reclamation needs, so careful consideration is to be given before selection of exotic species. Hence, preference is always given for native species that are well adapted to the local environment they are most likely to fit into a fully functional ecosystem.

1.4 Current Ecorestoration Scenario in India

The ecorestoration of coal mine overburden (OB) dumps in developed countries is done by application of topsoil cover and liming. Topsoil is used to cover poor substrates and to improve growing condition of plants, whilst liming is used to ameliorate soil pH. In India, topsoil is scarce commodity, and in majority of the cases, it was generally not stored properly. Of late, in limited cases, topsoil is removed and concurrently used as cover material or stored and reused as required by legislative directives. In case topsoil is not available in mine site, it is borrowed from nearby areas.

As restoration is limited to planting tree species, it never spreads in the entire area of reclaimed site, rather it poured only in the plantation pit. Liming is yet to be practised for pH correction. It is recommended that ecorestoration of dump should be considered as a part of natural succession process, and it should be started with sowing of seeds of legumes, grasses, herbs and shrubs along with tree plantation. This concept has not become popular with coal mining industry because plantation is mostly done by State Forest Department and charges are calculated on the basis of trees that survived after 3 years. Sometimes OB dumps are acidic in nature (pH 4–5) which not only cause elevated metal concentration to the plant but also decrease microbial activity inhibiting soil organic matter decomposition and nitrogen mineralisation process. Acidic dumps are being restored by the principle of *phytostabilisation*. The normal practice is to choose drought-resistant, fast-growing trees which can grow in acidic, nutrient-deficient,

elevated metal-contaminated soils. The effort should be aimed at finding out restoration success of nutrient-poor acidic overburden dumps with liming, improvement of organic carbon, N and P status and reducing bioavailability of heavy metals due to elevated acidity.

1.5 Differences Between Natural Soil and Minesoil (Mine Spoil)

The spoil is characteristically different from soil in the following reasons:
- Undisturbed soils generally exhibit well-developed structure in the upper horizon, whilst mine spoils show little or no soil structure throughout their profile.
- Higher bulk density, lower porosity, lower permeability, sometimes higher clay contents and lower water-holding capacity.
- Unfavourable pH for plant growth.
- High electrical conductivity, high sodium and low potassium nutrients and sometimes saline.
- Coarse texture, high stoniness and low cation exchange capacity of mine spoil suggest that agronomic use is questionable without fertiliser addition.
- Low in organic matter, nitrogen, phosphorous and other nutrients and very low microbiological activity.

1.6 Ecorestoration

In mining context, reclamation often refers to the general process whereby the mined wasteland is returned to some forms of beneficial use (Cooke and Johnson 2002), whilst restoration refers to reinstatement of the pre-mining ecosystem in all its structural and functional aspects, rehabilitation means the progression towards the reinstatement of the original ecosystem, and replacement is the creation of an alternative ecosystem to the original (Bradshaw 2000).

Land reclamation is a broad term and is often used in varied sense in the literature of after-use possibilities of strip-mine spoil. According to Khoshoo (1988), '*It is the treatment of land creating conditions for bringing the land back to some beneficial use*'. Reclamation does not necessarily mean restoring the land to its original conditions. According to National Academy of Sciences study committee, USA (1974), '*Reclamation renders a site habitable to indigenous pre-mining condition organisms or organisms nearly so*'.

Reclamation is not synonymous to *restoration* or *rehabilitation*. Whilst *restoration* is the replication of site conditions prior to disturbance, *rehabilitation* implies returning the disturbed land to a form and productivity in conformity with a prior land use plan including a stable ecological state that does not contribute substantially to environmental deterioration and is consistent with surrounding aesthetic values. In UK, *the Commission of Mining and Environment* (Zuckerman Commission 1972) accepted the following terms:

- *Restoration*—recreating conditions suitable for previous use of area
- *Rehabilitation*—creating conditions for a new and substantially different use of the mining site
- *Reclamation*—returning a derelict site to some use

In the past 15 years, there have been major advances in '*restoration ecology*' as an academic discipline. The Society for Ecological Restoration (SER) has helped organise and formalise restoration ecology, providing a central authority and general guidelines for ecological restoration (SER 2004). Restoration ecology has also become an increasingly prominent topic in scientific publications, both in total articles published and as a percentage of all ecology publications.

Restoration-specific journals such as *Restoration Ecology* have blossomed into major scientific outlets, and restoration papers have had an increasing presence in top-tier applied ecology journals, including special issues dedicated to restoration for wide readership (e.g. *Restoration Ecology, Ecological Applications, Journal of Applied Ecology, Forest Ecology and Management, Science*). Numerous books that investigate scientific and practical facets of restoration have been published in this period.

1.6.1 Definition of Ecorestoration

There are several definitions of ecorestoration given by different ecologist, restoration ecologist and scientific societies (Bradshaw 1987, 1996; SER 2004; USDA Forest Service 2010) which are highlighted below:

Restoration—recreating conditions suitable for previous use of area 'often used to mean restoring the original land-use or vegetation or even the same land form', or 'it is the return of an ecosystem to an approximation of its structural and functional condition before damage occurred' or 'return of an ecosystem to a close approximation of its condition prior to disturbance'.

Ecological restoration is the process of renewing and maintaining ecosystem health. It requires understanding of not only the nature of the ecosystem itself but also the nature of the damage and how to repair it (Bradshaw 1987).

Ecological restoration 'is the process of assisting the recovery of an ecosystem that has been degraded, damaged, or destroyed' (SER 2004).

Rehabilitation—creating conditions for a new and substantially different use of the mining site.

Reclamation—returning a derelict site to some use. 'Process of creating a land use, which may be hard (industrial, commercial) or soft (agriculture, amenity) on a site where mining and quarrying operation have finished'.

Revegetation is the process of vegetation establishment and aftercare undertaken as part of reclamation, rehabilitation or restoration.

Recultivation—it generally applies to the agronomic and ecological aspects of reclamation, rehabilitation or restoration.

After use means a land use to which a site is returned, which should be beneficial although not necessarily economic.

Aftercare describes the crucial process of managing the soils and the vegetation systems after the initial revegetation or recultivation in order to ensure that the desired land use is attained within a reasonable time period. The process would involve soil amelioration and vegetation management that is more intensive than normally associated with land in that particular use.

The Society for Ecological Restoration (SER) defines ecological restoration as an 'intentional activity that initiates or accelerates the recovery of an ecosystem with respect to its health, integrity and sustainability'. The practice of ecological restoration includes wide scope of projects including erosion control, reforestation, the use of genetically local native species, removal of non-native species and weeds, revegetation of disturbed areas, reintroduction of native species, as well as habitat and range improvement for targeted species. The term 'ecological restoration' refers to the practice of the discipline of 'restoration ecology'.

Ecological restoration 'is the process of assisting the recovery of resilience and adaptive capacity of ecosystem that have been degraded, damaged, or destroyed. Restoration focuses on establishing the composition, structure, pattern, and ecological processes necessary to make terrestrial and aquatic ecosystems sustainable, resilient, and healthy under current and future conditions' (USDA forest service 2010).

1.6.2 Attributes of Restored Ecosystem

What is meant by '*recovery*' in ecological restoration? An ecosystem is '*recovered and restored*'—when it contains sufficient biotic and abiotic resources to continue its development without further assistance. It will be self-sustaining both structurally and functionally. It will demonstrate resilience to normal ranges of environmental stress and disturbance. The nine attributes listed below provide a basis for determining when restoration has been accomplished. Some attributes are readily measured. Others must be assessed indirectly (i.e. functional aspects of ecosystem) (SER 2004).

1. Comparison of characteristic of species of restored ecosystem, namely, the reference ecosystem.
2. The species composition in the restored ecosystem should be consisting of indigenous species to the greatest extent as practicable.
3. All functional groups necessary for the continued development and/or stability of the restored ecosystem are represented.
4. The physical environment of the restored ecosystem is capable of self-sustaining and reproducing species necessary for its continued stability or development along the desired trajectory.
5. The restored ecosystem apparently functions normally for its ecological stage of development, and signs of dysfunction are absent.
6. The restored ecosystem is suitably integrated into a larger ecological matrix or landscape, with which it interacts through abiotic and biotic flows and exchanges.
7. Potential threats to the health and integrity of the restored ecosystem from the surrounding landscape have been eliminated or reduced as much as possible.
8. The restored ecosystem is sufficiently resilient to endure the normal periodic stress events in the local environment that serve to maintain the integrity of the ecosystem.
9. The restored ecosystem is self-sustaining to the same degree as its reference ecosystem and has the potential to persist indefinitely under existing environmental conditions. Nevertheless, aspects of its biodiversity, structure and functioning may change as part of normal ecosystem development and may fluctuate in response to normal periodic stress and occasional disturbance events of greater

consequence. As in any intact ecosystem, the species composition and other attributes of a restored ecosystem may evolve as environmental conditions change.

1.6.3 Underlying Principles of Restoration

The term restoration normally implies return to an original state. In ecological restoration, it should be thought of as applying to whole ecosystems, and mine waste site rehabilitation or replacement is more practicable than restoration. The components of restoration are the chemical and physical aspects of the habitat and the species themselves. Each of these may require specific treatment, but natural restorative processes (succession) should not be neglected. The process of restoration being progressive, the criteria of success are not easy to define. The most important point is that ecosystem development should be on an unrestricted upward path (Bradshaw 1996). The details of this process are discussed below:

The word *ecosystem* includes the biotic and abiotic components occurring together in a particular area, and they are inclusive and closely interacting with each other through energy and material cycling. So when the restoration of *ecosystems* is being referred to, then the fundamental processes by which ecosystems work have to be restored. When we talked about *habitat restoration*, habitat just refers to a place where organisms live, therefore, more emphasis should be on the *restoration of place* than of important ecological functions.

1.6.4 Options in Restoration

There are many attributes to an ecosystem, but all are simplified into two main components, structural and functional components and in equilibrium in terms of exchange of matter and energy in the ecosystem. After degradation, some interventions are required to restore these components. In Fig. 1.1, a different option for improvement of degraded ecosystem is explained in terms of two major components of ecosystem—ecosystem structure (species composition and complexity) and ecosystem function (biomass and nutrient contents). When degradation occurs, both components are usually destroyed (*degraded ecosystem*). *Restoration* implies bring back the ecosystem to its original state in terms of both structure and function. There are other alternatives, *rehabilitation* in which this is not totally achieved and *replacement* of original with something new and may be better. All these general terms are covered under *reclamation. Mitigation* is a different consideration. Both components will have suffered and will have to be restored. *Rehabilitation*, in which progress has been made but the original state not achieved, and *reclamation*, is something different and that is similar to replacement as shown in same figure. In particular, it points to the fact that restoration may not be easy. It may be possible, perhaps, to restore the functions fairly completely, but to achieve the original structure may be more difficult. For example, in a forest ecosystem, full age structure may take 500 years, although biological function may be restored within 10 years.

So in many situations, true restoration may be unrealistic, and realistically, rehabilitation and replacement can be proper options. Replacement is a particularly interesting option since it may allow restoration of a component, such as productivity, to a higher level than existed previously.

The structural and functional ecosystem characteristics that are usually measured during ecorestoration process are given in Tables 1.4 and 1.5.

1.6.5 Components of Restoration

Once ecosystem is destroyed, factors essential for the redevelopment of the ecosystem should be considered. Naturally, this can involve many

Table 1.4 Ecosystem characteristics for consideration as ecological restoration objectives (Adapted from Cook and Jhonson 2002)

Sl no.	Ecosystem characteristics
1.	Composition of species presence and their relative abundance
2.	Structure: vertical arrangement of vegetation and soil components
3.	Pattern: horizontal arrangement of system components
4.	Heterogeneity: a variable composing of characteristics 1–3
5.	Function: performance of basic ecosystem processes (energy capture, water retention, nutrient cycling)
6.	Species interactions, for example, pollination and seed dispersal
7.	Dynamics and resilience: succession and state-transition processes, ability to recover from normal episodic disturbance events (e.g. drought, fire)

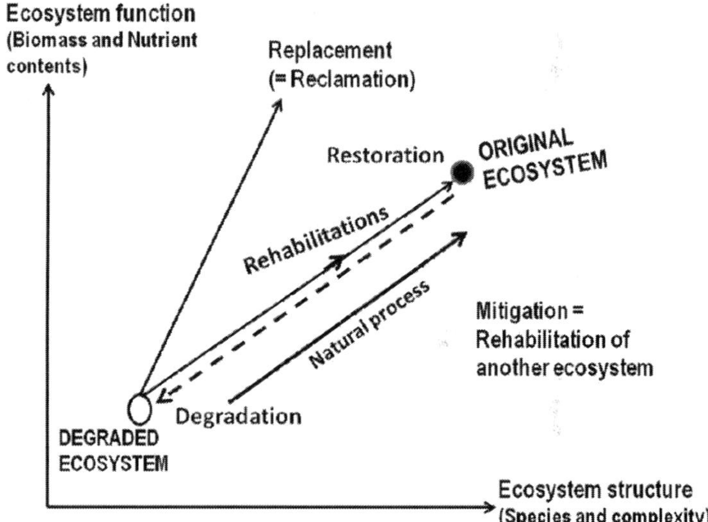

Fig. 1.1 Options for reconstruction of degraded ecosystem (After Bradshaw 1996)

different factors, depending on the nature of ecosystem and the magnitude of degradation. Essentially three factors will need attention: (1) remodelling the physical aspects of the habitat; (2) remodelling the chemical aspects, nutrients and toxicity; and (3) replacing missing species or removing undesirable exotic species and weeds. Attempts are now being made to rationalise the selection of tree species as well as maintain proper slope for revegetation. It is also essential to identify the controlling factors which are not going to ameliorate naturally should be addressed first.

- *Use of natural processes*: Many ecorestoration researches opined that natural process should be used wherever possible because (1) they cost nothing, (2) they are likely to be self-sustaining because they originate from nature and (3) they can be used on a large scale. Although it is clear that natural processes can eventually achieve full restoration, they take a long time and need to be assisted.

- *Physical problems*: Mine waste materials are compacted and have a high bulk density, yet changes take place naturally by the growth of vegetation, the incorporation of organic matter and the activities of soil fauna and flora. However, in some seriously compacted situations, natural recovery is so slow that mechanical treatment is necessary.

- *Nutrient problems*: One of the most common problems of degraded terrestrial environments is the lack of nutrients, particularly nitrogen. The soil nitrogen capital, normally

Table 1.5 Structural vegetation measurements commonly used to monitor restoration and the related functional characteristics that are implied (Adapted from Cook and Jhonson 2002)

The structural measurement	Functional characteristic needed
Biomass (g/m^2)	Productivity (g/m/a)
Species density	Species turnover (mortality, reproduction)
Species richness	Functional loss caused by missing species
Life-form spectra	Functional loss caused by absence of life forms
Indices of diversity/similarity	Species interactions which promote ecosystem functioning

at least 1,000 kg N/ha, can be rebuilt up by means of fertilisers, but this is expensive. It is much simpler to introduce legumes, such as white clover, *Stylosanthus*, *Trifolium*, which can accumulate nitrogen at the rate of 100 kg N/ha/year. However, calcium and phosphorus levels must be sufficient to maintain the growth of these species. Although N-fixing species can often be chosen that do not have as special nutrient requirements, nutrients such as phosphorus cannot be conjured out the air and will remain deficient if not added.

- *Species*: It is often presumed that those species are not introduced they will arrive on their own. However, natural invasion of species depends on surrounding seed pool and seed dispersal power of those species. For rapid establishment of diverse community, sowing of plant seeds is always required and very crucial to introduce the missing species.

1.6.6 Criteria of Ecorestoration Success

Now it is understood that intervention is required for ecorestoration process, and ecological professional may ask for everything to be restored completely, which is actually impossible (Bradshaw 1996). However, certain manipulations, such as retention and replacement of original topsoil and reintroduction of pre-existing species, may be the way to achieve end point. During ecorestoration process, most important is to set off succession in the right direction and then leave nature to continue itself. Nobody thinks wrong in planting small tree to re-establish a forest but to achieve satisfactory growth of tree; it may be necessary to ensure that soil is fertile or make necessary intervention to enhance fertility (i.e. increase nitrogen fertility by introducing legumes).

What Are the Criteria of Ecorestoration Success to Be Considered?

1. *Structure of ecosystem*: It should be based on structure of ecosystem and in particular the presence or absence of species and functional aspects, such as plant growth; sometimes, it should be a much simpler criterion, such as the amount of heavy metal emanating from a mine site, since water quality in the downstream will be directly related to the release of metals.

2. *Similarity of species composition*: If reliance is being placed on the progressive effects of natural processes, what level has to be achieved? Should species composition in the newly reconstructed site be 75 or 90% of the pre-existing species? There can be no fixed criterion for these, although SMCRA (US Surface Mining Control and Reclamation Act) expects 90% species similarity.

3. *Time*: Once targets have been set, when should they be achieved? Five years is taken as the period for bond release under SMCRA (USA). But for mine restoration process, such 5 years is not long enough to judge whether interventions are adequate. In UK conditions, time taken for build-up limiting nutrient stock of 700 kg N/ha in the soil for kaolin mine waste is taken as recovery time. By chronosequence study of coal mine overburden dumps, Mukhopadhyay and Maiti (2011) reported that 17 years is the minimum time period for

ecosystem recovery of coal mine overburden dumps on the basis of improvement in physico-chemical characteristics of mine spoils. The recovery time was calculated by comparing cation exchange (CEC) and base saturation (%), accumulation of organic matter, nitrogen, phosphorus, texture, porosity and moisture values with natural sal (*Shorea robusta*) forest. However, this recovery period will depend on geo-climatic conditions, types of tree species, nature of soil ameliorant, proximity of seed sources, aftercare and maintenance and magnitude of disturbance by anthropogenic activity.

Conclusive Remarks

- There is a close relationship between ecological understanding and successful restoration. If we do not understand the process of working in an ecosystem, then we are unlikely to be able to reconstruct it so that it works.
- Testing of different aspects of intervention in ecorestoration process has to be carried out in the field.
- It is essential that results are published in journals or reports that are readily available to everyone working in the field. Reviews, including failures as well as successes, are particularly important, and new findings and ideas can be promulgated and developed.
- Finally, the development of the science of restoration should go hand in hand with achievements of successful restoration itself. The legal and scientific framework for the restoration of drastically disturbed ecosystem is best described by Surface Mining Control and Reclamation Act of 1997 of USA (SMCRA). The environmental requirements established by the SMCRA are as follows:
- Restore mined lands to former or better use
- Backfill and grade the mined areas to their approximate original contour
- Control erosion and attendant air and water pollution
- Minimise disturbance to the hydrological balance—surface and groundwater
- Remove, separate and respread the topsoil (plus subsoil in case of prime farmland)

- Establish adequate vegetation on the mined lands

1.7 Relevant Issues of Dump Rehabilitation

1. Drainage, sedimentation and erosion control under different land use, namely, vegetation cover as well as geotechnical aspects of dumps. Quantification of sediment load using artificial rainfall simulator/in situ condition. Disposal/reuse of sediment. Design of sedimentation pond and garland drains
2. Seed bed ecology: natural invasion of seeds, plant succession, identification of constrains of seed germination in OB dumps, preparation of suitable seed mixtures and optimisation of green belt
3. Screening of suitable legumes: herbaceous forage and tree, nitrogen enrichment and nitrogen dynamics in dumps. Identification of physical, nutritional and microbiological constrains for dump reclamation
4. Ecorestoration technique of the forest area. Conservation of biodiversity and ecorestoration matching with surrounding landscape. Adoption of innovative approaches
5. Use leguminous forbs and grasses as pioneer colony in dumps stabilisation
6. Use of soil amendments/ameliorants
7. Identification of suitable mulch and mulching practices, in situ moisture conservation practices
8. Reclamation of erosion prone and/or steep slope areas: use of Geojute, Netlon, hydro-seeding and biological stabilisation measures
9. Transfer of heavy metals in food chain especially from metal mine dumps
10. Reclamation of mined-out orphan lands by using seed mixtures
11. Hydro-reclamation
12. Topsoil: feasibility of storing, reuse and quality assessment
13. Categorisation of dumps based on reclamation potential and regeneration of ecosystem

1.8   Aims of Biological Reclamation

The main aims of biological reclamation are

Short-Term Goals
- To control *erosion* quickly with fast-growing plants, those that could be acted as first coloniser in the derelict site may be grass–legume mixtures, for example.
- The criteria of selection of species having a good foliage cover, evergreen, and have good binding capacity of loose spoil materials, that is, preferably tuft fibrous root system.
- These plants should also have capacity to grow extreme environmental conditions—like low moisture, high temperature, less nutrients, no decomposers in spoil surface (without humus and organic matter) and no soil cover (may be arise in some cases).

Long-Term Goals
To create ecological equilibrium between the 'spoil, microflora and microfauna and plants' with surrounding environment. To achieve the ecological equilibrium, composition of plant species could be:
- Forest trees—mainly multipurpose trees (MPT)
- Fruit orchards
- Row crops (if possible and economically viable)
- Forage crops (forage legumes and grasses)
- Other long-term uses—eco-park and recreational areas
- Aesthetic beauty—shady, evergreen and flowering plants
- Dust control (by planting trees having dense branching, evergreen, closely arranged simple leaves and rough surface)

1.9   Philosophies of Revegetation

The approaches to revegetation can be described in terms of three different basic philosophies:
- Ameliorative
- Adaptive
- Forestry and agricultural

1. *The Ameliorative Approach*
   - It relies on achieving optimum condition for plant growth by improving physical, chemical and biological characteristics of waste dump by using amendments.
   - Most suitable plant species are grown based on edaphic properties.
   - This approach is commonly used in preference to the adaptive approach because it is quicker, requires less forward planning and is less labour intensive.
2. *The Adaptive Approach*
   - This approach emphasises selection of the most suitable species, subspecies, cultivars and ecotypes to meet the rigorous extreme conditions.
   - In addition, but not necessarily, the mine waste may be improved using amendments to achieve optimum establishment and loon-term growth.
   - This approach is simple but constrained by the availability of suitable plant species.
3. *The Forestry and Agricultural Approach*
   - This is used directly on less-toxic waste such as iron, solid waste from integrated steel plan, bauxite waste etc. The waste is covered with deep layer of topsoil. The crops or woodland and/or scrub species are established using conventional or specialised techniques.

1.10   Problems of Biological Reclamation

In great majority of the cases, the raw overburden material produced by mining activities does not possess any soil character. The heterogeneous overburden material usually possesses very poor physico-chemical characteristics and is devoid of any nutrients and organic matter. Natural soil formation starts only after the establishment of vegetation cover, which is the only source of nutrients on derelict sites. Due to acidic/alkaline nature of spoil material, natural plant succession process is delayed.

Secondly, due to lack of microbial activity on spoil dumps, nutrient cycling process does not start. Soil moisture also plays a vital role for initial plant establishment. For initial start-up of nutrient recycling in derelict sites, organic amendments have paramount significance. The soil generally is formed over long periods by weathering and disintegration of parent rocks. They have little nutrients to support plant life. For them to become productive, they must evolve through weathering, biological process and leaching.

Natural processes take a very long time to change the characteristic of mine spoil. Thus, artificial revegetation on mined land can accelerate the process. The soil slowly loses the characteristics of the parent body (overburden) and gets the nature of that of local environment. Soil thus establishes equilibrium with the environment, and mature soil is formed. Now-a-days, it is recommended that, for an economically attractive returns from derelict sites, raising food/fodder plants, particularly fodder grasses, legumes and fruit trees is necessary.

1.11 Dump Reclamation Practices in India

Case 1: External Dumps Created on Plain Ground. There is always a need to excavate the overlying waste materials to reach coal seam and dump OB materials outside the mining areas. These are known as *external OB dumps.* External OB dumps are necessary of any opencast mining, but once created, they seem to become external to the interest of mining industry despite posing a substantial treat to environment. The *MOEF stipulated that slope of the OB dumps should not exceed 28°.*

In India, those dumps are concurrently reclaimed once it is inactive (dead). Sometimes, OB materials are dumped on abandoned quarry (if it is available). The success of any biological reclamation depends on climatic conditions, nature of spoils, types of plant species, nature of dumps, proximity to seed banks (nearby vegetation) and types of amendments used. As all these factors are very much site specific and depends on geo-mining conditions, systematic bioremediation of these dumps and creation of database of that particular types of set-up will be used during mine closure planning process.

Case 2: In-Pit OB Dumps. Now-a-days, maximum efforts are given for in-pit dumping. The height of dumps is reduced, and excavated area is concurrently filled up, and there is an opportunity to restore the area to its original topography (i.e. popularly called as AOC, approximate original contour). As reclamation of in-pit dumps is also carried out simultaneously, collection of database like case 1 will be helpful for planning of 'closure'.

Case 3: Void Left at the Last Part of the Quarry. These large water bodies are sometimes useful for the community—for storage of irrigation water, pisciculture or used by day-to-day purpose. The banks must have gentle slope and afforested. In ECL, some of the water bodies are presently being used by local community for irrigation, washing, bathing and even sources of livelihood by catching fishes. Of course, in these water bodies, no planned or commercial pisciculture is practised due to problems of catching fishes, because of depth or may be leasing problems or protection from theft or the community commonly shares the water body.

Hydro-reclamation is also bioremediation process. Bearing all problems in mind, the best economic end use of water bodies is pisciculture or water sports, like boating. For pisciculture, important parameters to be monitored are shape, slope, depth (around 30 m preferable), quality of water whether suitable for pisciculture and productivity. The shape, slope and depth depend on geo-mining conditions; the quality of water depends on characteristics of strata; similarly food production for fishes depends on production of planktons (i.e. phytoplankton and zooplankton). Monitoring of water quality vis-à-vis enhancing of suitability for pisciculture activity should be taken care, and 'database' must be created by monitoring similar type of water-filled quarries. This database must be used for 'mine closure' plan.

Case 4: Shallow Voids. At mines where the stripping ratio is low, OB materials may not fill the void created due to the mining operation. In such case, *concurrent reclamation practices* may be adopted. The backfilling sequence has to be planned in details and inspected by mine management frequently to ensure compliance. This practice is carried out at Piparwar mine of NK area of CCL (Maiti 2006). Sometimes, it is used as dumping ground or could be used as disposal sites for fly ash. Once the voids are filled, these can be reclaimed.

- Fly ash, a waste from power generation plants, may be considered as filling materials (e.g. West Bokaro and Damoda of BCCL).
- OB materials from nearby quarry may be used as filled materials.
- If no filling materials are available from nearby sources, water body may be created and reclaimed (i.e. hydro-reclamation).

1.12 Biological Reclamation Planning

Before revegetation planning, types of vegetation cover requirement have to be planned, which will depend on characteristics of plant growth medium (i.e. nature of minesoil), quality of available of topsoil, proximity to the nearby seed sources (i.e. natural seed banks), climatic conditions (rainfall, temperature) and proneness to anthropogenic disturbance. A clear objective of reclamation and final land use after the end of mining operation should be clearly defined by mine authority along with the consultation of regulators and local bodies. The reclamation planning should be considered at an early stage of inception of project. All the planning of the mine degraded land should be based on the following basic principles:

1. *Community involvement*: Mining is a temporary activity, and as temporary occupiers of the land, mining company should conduct their business to facilitate post-mining land use. In proposals for redevelopment, community and environmental stewardship should be included in the planning and operational stages of the plan.
2. *Progressive reclamation*: The mining companies are encouraged to use progressive reclamation whenever possible.
3. *Visual impact assessment*: Surface mining has the potential to visually impact the natural landscape. A visual landscape assessment incorporated into reclamation planning should provide certainty that the final site design will be compatible with view sheds within surrounding natural landscapes.
4. *Compatible land use/land cover*: Reclamation and mine closure should provide land that is restored to a condition that matches surrounding land cover or accommodates another land use identified in the final reclamation plan. Final reclamation plans should ensure that subsequent land use/land cover objectives are clearly identified, described and are compatible with the surrounding land use and landscape.
5. *Topsoil management*: Topsoil management is vital to establishing a self-sustaining cover of vegetation in reclaimed areas. Topsoil should be preserved for reclamation wherever possible, and soil quality should be protected during moving and storage. Soil management should consider issues such as:
   - Quality assurance during stripping
   - Identifying stockpile locations to maintain soil quality
   - Temporary seeding
   - Permanent vegetation of stockpiles to control wind and water erosion
6. *Human intervention* is required to enhance moisture-holding capacities, drainage and use mulches and reduce compaction of the soil. Where little or no topsoil exists prior to mining, it may be necessary to amend or import soils depending on the final land use and site conditions.
7. *Revegetation planning*: A key element of successful reclamation projects is the

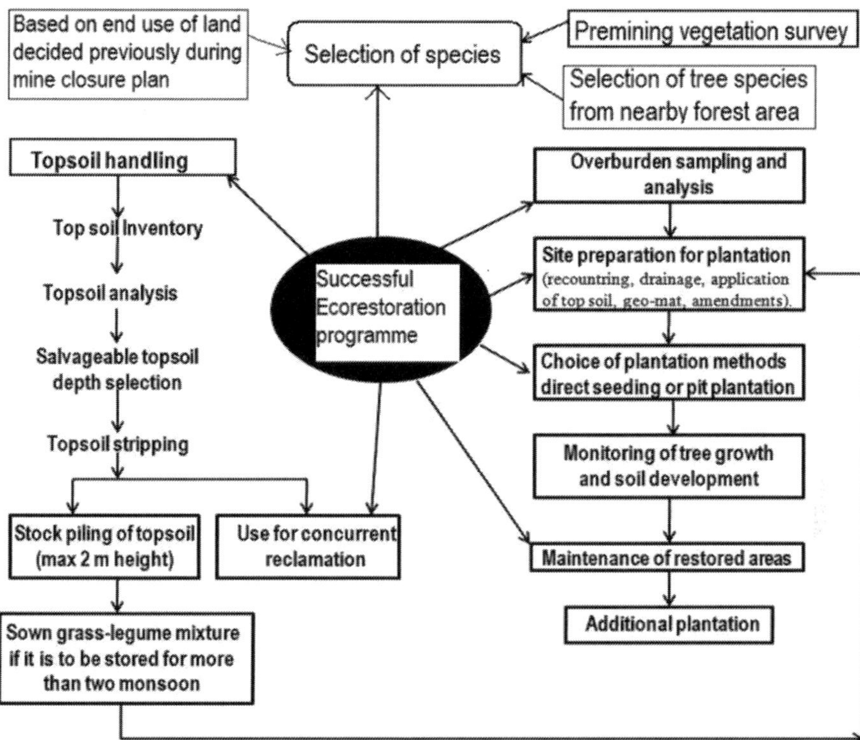

Fig. 1.2 Flow chart showing important activities required for revegetation of mine-degraded land (After Maiti 2010)

establishment of a self-sustaining, succession-based vegetation cover through the plantation and application of native grasses and legumes seed mixtures. This should include a 'revegetation plan' to establish the goals for vegetation at the beginning. Revegetation plans should mimic surrounding non-disturbed areas (control area, e.g. forest area) or encourage specific ecosystem establishment, incorporating strengths of both native and non-native species. This approach is referred to as successional reclamation, which refers to a multistaged process that relies on different treatments over a period of time.

8. *Ecological restoration using successional revegetation methods where appropriate*: This approach to reclamation planning is encouraged so that the eventual plant community promotes native species. Selection of native species is encouraged. Through this process, the initial use of non-native species will be succeeded by native species. This

cover is important as it is considered to be the bridge between initial colonisers and later developing vegetation.

- Protection of remaining patches of the original vegetation is encouraged to increase seed propagation and improve conditions for natural regeneration of native species. Development of floral banks is one of key activity for successful ecorestoration. The wild climbers, shrubs and small tree sapling may be preserved "habitat transplantation method". During transplantation, care should be taken like, soil should be moisten to protect the wire and tears of roots, operation should be carried out in rainy seasons, and transfer as much topsoil as possible along with the roots to the "Floral bank" site.

9. *Prevention of erosion and sedimentation*: Grasses and legumes may be required for temporary erosion control and soil rehabilitation. Reclamation plans that involve erosion

protection with revegetation must aim to encourage native vegetation and describe appropriate seed mixture, seeding and planting techniques.

10. *Site disturbance*: Issues from past mining operations such as old infrastructure, subsidence, underground workings, spontaneous combustion, underground mine fires and anthropogenic disturbances should be addressed in the reclamation plan.

Figure 1.2 shows the important activities that are required to be carried out carefully to achieve predefined vegetation cover development.

1.13　The Legal and Statutory Framework for Ecorestoration

Mineral deposits are governed by various statutes: The Mines and Minerals (Development and Regulation) Act (1957), The Mineral Concession Rules (1960), The Mineral Conservation and Development Rules (1988), The National Mineral Policy (1993) and The Granite Conservation and Development Rules (1999). These statutes also prescribe statutory guidelines for the restoration but either very vaguely or without reference to conservation needs. There are no legal provisions for key issues of monitoring amelioration practices.

1.13.1　The Mine and Mineral (Development and Regulation), MMDR Act 1957 (Amended on 1984 and 1994)

Under the Mines and Minerals (Development and Regulation) Act, 1957, the following sections are pertained to environment:

- Section 4A (1) and (2): Termination of PL and ML on environmental and other grounds
- Section 13(2): Rehabilitation of flora in leasehold area for major minerals
- Section 15 (1A): Rehabilitation of flora with respect to minor mineral licences
- Section 18(1): Rules making power on protection of environment in the mining areas

1.13.2　Mineral Conservation and Development Rules (MCDR), *1988* (Amended up to 25th Sep. 2000)

These have been enacted under the MMDR, 1957, for the conservation and development of minerals. *Chapter V, Rule 31 to 41* is entirely on environment.

Rule 31. Protection of Environment. Every holder of a prospecting licence or a mining lease shall take all possible precautions for the protection of environment and control of pollution whilst conducting prospecting, mining, beneficiation or metallurgical operations in the area.

Rule 32. Removal and Utilisation of Topsoil
1. Every holder of a prospecting licence or a mining lease shall, wherever topsoil exists and is to be excavated for prospecting or mining operations, remove it separately.
2. The topsoil so removed shall be utilised for restoration or rehabilitation of the land which is no longer required for prospecting or mining operations or for stabilising or landscaping the external dumps.
3. Whenever the top soil cannot be utilised concurrently, it shall be stored separately for future use.

Rule 33. Storage of Overburden, Waste Rock, Etc.
1. Every holder of a prospecting licence or a mining lease shall take steps so that the overburden, waste rock, rejects and fines generated during prospecting and mining operations or tailings, slimes and fines produced during sizing, sorting and beneficiation or metallurgical operations shall be stored in separate dumps.
2. The dumps shall be properly secured to prevent escape of material therefrom in harmful quantities which may cause degradation of environment and to prevent causation of floods.
3. The site for dumps, tailings or slimes shall be selected as for as possible on impervious

ground to ensure minimum leaching effects due to precipitations.

4. Wherever possible, the waste rock, overburden, etc., shall be backfilled into the mine excavations with a view to restoring the land to its original use as far as possible.

5. Wherever backfilling of waste rock in the area excavated during mining operations is not feasible, the waste dumps shall be suitably terraced and stabilised though vegetation or otherwise.

6. The fines, rejects or tailings from mine, beneficiation or metallurgical plants shall be deposited and disposed in a specially prepared tailings disposal area such that they are not allowed to flow away and cause land degradation or damage to agricultural field, pollution of surface water bodies and ground water or cause floods.

Rule 34. Reclamation and Rehabilitation of Lands. Every holder of prospecting licence or mining lease shall undertake the phased restoration, reclamation and rehabilitation of lands affected by prospecting or mining operations and shall complete this work before the conclusion of such operations and the abandonment of prospect or mine.

Comment: No clear definitions of the terms again.

Rule 35. Precaution Against Ground Vibrations. Whenever any damage to public buildings or monuments is apprehended due to their proximity to the mining lease area, scientific investigations shall be carried out by the holder of mining lease so as to keep the ground vibrations caused by blasting operations within safe limit.

Rule 36. Control of Surface Subsidence Stopping in underground mines shall be so carried out as to keep surface subsidence under control.

Rule 37. Precaution Against Air Pollution Air pollution due to fines, dust, smoke or gaseous emissions during prospecting, mining, beneficiation or metallurgical operations and related activities shall be controlled and kept within 'permissible limits' specified under various environmental laws of the country including the Air (Prevention and Control of Pollution) Act, 1981 (14 of 1981), and the Environment (Protection) Act, 1986 (29 of 1986), by the holder of prospecting licence or a mining lease.

Rule 38. Discharge of Toxic Liquid Every holder of prospecting licence or a mining lease shall take all possible precautions to prevent or reduce the discharge of toxic and objectionable liquid effluents from mine, workshop, beneficiation or metallurgical plants and tailing ponds, into surface water bodies, groundwater aquifer and useable lands, to a minimum. These effluents shall be suitably treated, if required, to conform to the standards laid down in this regard.

Rule 39. Precaution Against Noise Noise arising out of prospecting, mining, beneficiation or metallurgical operations shall be abated or controlled by the holder of prospecting licence or a mining lease at the source so as to keep it within the permissible limit.

Rule 40. Permissible Limits and Standards The standards and permissible limits of all pollutants, toxins and noise referred to in rules 37, 38 and 39 shall be those notified by the concerned authorities under the provisions of the relevant statutes from time to time.

Rule 41. Restoration of Flora
1. Every holder of prospecting licence or a mining lease shall carry out prospecting or mining operations, as the case may be, in such a manner so as to cause least damage to the flora of the area held under prospecting licence or mining lease and the nearby areas.

2. Every holder of prospecting licence or a mining lease shall:

    (a) Take immediate measures for planting in the same area or any other area selected by the Controller General or the authorised officer not less than twice the number of trees destroyed by reason of any prospecting or mining operations

(b) Look after them during the subsistence of the licence/lease after which these trees shall be handed over to the State Forest Department or any other authority as may be nominated by the Controller General or the authorised officer

(c) Restore, to the extent possible, other flora destroyed by prospecting or mining operations

1.13.2.1 Mineral Conservation and Development Rules, 1988

Rule 4(2) states that a scheme for prospecting shall include baseline information of prevailing environmental conditions before the beginning of the prospecting operations.

Rules 11, 12 and 13 relate to the submission of the mining plans by leases existing before the coming in of the MCDR, whereby a lessee has been asked to submit a plan within a year of the commencement of these rules and then work according to the plan.

- Rule 56 says that if the Controller General, Chief Controller of Mines or the Controller of Mines feel that a particular mine poses a grave and immediate threat to the environment, they may prohibit deployment of persons until the conditions specified by them are met.

References

Bradshaw AD (1987) The reclamation of derelict land and the ecology of ecosystems. In: Jordan WR (ed) Restoration ecology: a synthetic approach to ecological research. Cambridge University Press, Cambridge, UK, pp 53–74

Bradshaw AD (1996) Underlying principles of restoration. Can J Fish Aquat Sci 53(1):3–9

Bradshaw AD (2000) The use of natural processes in reclamation – advantages and difficulties. Landscape Urban Plann 51:89–100

Business Line (2000) Environmental issues in coal mining. The Hindu group of publications. http://www.thehindu-businessline.in/2000/07/05/stories/040567mp.htm

Chaudhuri S (2008) Environmental and social aspects of coal mining – an overview. In: Chaudhuri S, Singh G (eds) EDP course on environmental management in coal mining areas. ISM, Dhanbad

CIL Annual Report (2011). www.coalindia.in/.../ CIL_Annual

Cooke JA, Johnson MS (2002) Ecological restoration of land with particular reference to the mining of metals and industrial minerals: a review of theory and practice. Environ Rev 10:41–71

IEP (2006) Integrated energy policy, Planning Commission, Government of India, August 2006

India Energy Hand book (2011) http://www.psimedia. info/handbook/India_Energy_Handbook.pdf

Khoshoo TN (1988) Land reclamation in opencast mines, Ch.23. In: Environmental concerns and strategies. Aashis Publishing House, New Delhi, pp 213–237

Maiti SK (2006) MoEF report on "An assessment of overburden dump rehabilitation technologies adopted in CCL, NCL, MCL and SECL mines" (No. J-15012/ 38/98-IA II (M)), Delhi

Maiti SK (2010) Revegetation planning for the degraded soil and site aggregates in Dump sites. In: Bhattacharya J (ed) Project environmental clearance. Wide Publishing, Kolkata, pp 189–228

MOC (2011) The year 2010–11 at a Glance – Ministry of Coal coal.nic.in/annrep1011.pdf

Mukhopdhyay S, Maiti SK (2011) Minesoil reclamation by to tree plantation: a chronosequence study. Afr J Basic Appl Sci 3(5):210–218

National Academy of Science (NAS) Committee, USA (1974) http://www2.nas.edu/arc

SER (2004) The SER international primer on ecological restoration. www.ser.org and Society for Ecological Restoration International, Tucson

The Zuckerman Commission (1972) The Commission on Mining and the Environment. http://linkinghub.elsevier.com/retrieve/pii/0305048374900656

USDA Forest Service (2010) http://www.fs.fed.us/restoration/QandAs.shtml

Vision Coal - 2025 (2005). Ministry of Coal, Gov of India. www.pib.nic.in/release/release.asp?relid=13974

Ecology and Ecosystem in Mine-Degraded Land

2

Contents

2.1   Preamble

Surface mining completely destroyed the delicate plant–microbes–soil link during its operation. Unlike other use, mining uses the land very temporarily, and at the end of project or even during the project, the ecosystem is regenerated on degraded land by afforestation or left to the nature as it is. *The question arises now, what is the capacity of these lands to regenerate the ecosystem?* As per the laws of succession, the new ecological link will be established itself by nature. However, the natural process is slow; thus, artificial intervention is required; for example, planting is done to enhance the speedy recovery of ecosystem.

The most significant environmental damages due to opencast mining in India are the complete destruction of ecosystem, habitat fragmentation, alteration of land use pattern, deterioration of aesthetics and change in local drainage system due to inadequate landscape management during mining operation. Very often, degraded land and overburden dumps are reclaimed by planting fast-growing tree species, without considering the ecological principles like initiation of nutrient cycling, functioning of ecosystem attributes (stability, diversity and resiliency) and succession process which acts as delicate links between biotic and abiotic components. Therefore, in this chapter, basic knowledge of ecology and ecosystem is introduced very briefly so that during ecorestoration programme, importances of ecological principles should be considered.

2.2   Ecology

The basic two components of nature—the *organisms* and their *environment*—are not only much complex and dynamic but also interdependent, mutually reactive and interrelated which is studied under ecology. The word *ecology* (*Greek oikos* = home, habitat; *logos* = study) was first used by Ernst Haeckel over a century ago (1866) and described as to the study of 'household of nature' and 'the science of the relationship of the organism to the environment'. Simply, the study of interaction between biotic and abiotic components is known as *ecology*. The central theme of ecosystem concept is that at any place where an organism lives, there is a continuous interaction between the living and the nonliving components, that is, between plants, animals and their environment through flow of energy and cycling of materials (nutrient cycling). Ecology is studied at population, community and ecosystem level. Conventionally, ecology has been defined variously by different modern ecologists:

1. '*Ecology to be science of community*'
2. '*Ecology as the science of the relations of all organisms to all their environments*'.
3. According to Woodbury (1954), *ecology is the science which investigated organisms in relation to their environment and a philosophy in which world of life is interpreted in terms of natural process.*

Recently, some modern ecologists have provided somewhat more boarder definitions of ecology:

1. According to Southwick (1976), '*Ecology is the science of study of the relationship of living organisms with each other and their environment*'. *He also clarified that interaction among individuals, population and communities is also a science of ecosystem*—that is, interaction of biotic communities with their nonliving environment.

2. More recently, renowned American ecologist Eugene P. Odum (1975) defined ecology as '*The study of structural and function of ecosystem- or broadly the nature*'.
3. Ecology is the scientific study of the processes influencing the distribution and abundance of organisms, the interactions among organisms, and the interactions between organisms and the transformation and flux of energy and matter (Cary Institute of Ecosystem Studies 2012).

2.3   The Ecosystem

The term *ecosystem* (eco = environment and the system = a complex of coordinate units) was first coined by A.G. Tansley (1935) as a basis for the fundamental understanding of *ecosystem dynamics* and pointed out the relevance of *material interchanges between organisms and their abiotic environment:*

Later on, Odum (1971) appropriately defined as a 'Unit that includes all of the organisms i.e. the community in a given area interacting with the physical environment so that a flow of energy leads to clearly defined tropic structure, biotic diversity and material cycles (i.e exchange of material between living and nonliving parts) within the system'. In 1992, the Convention on Biological Diversity (CBD) formulated that an 'Ecosystem means a dynamic complex of plant, animal and microorganism communities and their non-living environment interacting as a functional unit' (CBD 1992).

2.3.1   Components of Ecosystem

As said above, the ecosystem consists of two major components: (a) biotic components comprising of all the living organisms and (b) abiotic components which include the physical (nonliving) environments

The *biotic components* are usually divided into *two categories:*

(a) The *autotrophs* which can produce their own food by taking inorganic carbon as CO_2 from atmosphere and use sunlight as source of energy. These are green plants (with chlorophyll) and certain bacteria (chemosynthetic and photosynthetic). Since these organisms produce food for all the other organisms, they are also known as *producer.*

(b) The *heterotrophs* which depend directly or indirectly upon the autotrophs for their food. These can again be divided into two groups based on the mode of nutrition: (1) *Phagotrophs* or the organisms which ingest food and digest it inside their bodies are called *consumers.* The consumers may be herbivores (plant eating), carnivores (animal eating) or omnivores (eating both plants and animals).

- *Decomposers* are osmotrophs which secrete digestive enzymes to break down dead organic matter into simper substances and then absorb the digested food. They are mostly parasitic and saprophytic bacteria and fungi.
- *Decomposition:* Decomposition is a natural process, but decomposers speed up the process of decomposition. Bacteria, fungi and actinomycetes are three main types of decomposers. *Bacteria* make up about 90% of all microorganisms and are the most abundant of decomposers. They can eat anything from dead trees, dead animals and oil slicks on the surface of the ocean. *Fungi and actinomycetes* work on harder substances like cellulose, bark, paper and woods. These decomposers usually only work to a certain stage in decomposition, then bacteria will finish the process, similar to primary and secondary succession. The heterotropic *decomposers*, mainly bacteria and fungi, decompose organic into inorganic matter (*mineralisation*).

The *abiotic components* consist of the solid mineral matter on the earth (the lithosphere); the water in the ocean, lakes, river, ice-caps, etc. (the hydrosphere); the gaseous mixture in the air (the atmosphere) and the radiant solar energy.

Climate, soils and rehabilitation strategy are important considerations in minimising impacts on native flora and fauna. Soil erosion can be minimised by a proper understanding of soil structure, conservative landform design, utilising complex drainage networks, incorporating runoff silt traps and settling ponds in the rehabilitated landform. Careful use of topsoil can promote vegetation cover if the topsoil material is structurally appropriate and contains propagules of native vegetation. Selection of native floral species is desirable in promoting a stable and robust vegetation cover. Where possible, species endemic to the area should be used, preferably those from the site itself.

2.3.2   Characteristics of Ecosystem

The ecosystem characteristics usually measured are those related to the composition, structure and pattern of the vegetation. There are *two* aspects of an ecosystem—the *structural* and *functional* aspects.

Structural aspects of ecosystem:
- The quantity and distribution of the nonliving materials, such as nutrients, water, temperature, light and meteorological conditions
- The composition of biological communities including species, number, biomass, life history and distribution in space
- The range or gradient of conditions of existence

Functional aspects of ecosystem:
- Energy flow
- Nutrients or biogeological cycles
- Food chain and food web
- Diversity pattern in time and space
- Developments and evolution
- Control and cybernetics

With the help of the following flow chart (Fig. 2.1), we can interpret the functional aspects of an ecosystem or the interaction between various components which involve the flow of energy and cycling of materials. The health of an ecosystem depends on the integration of structural and functional components (Fig. 2.2). The characteristics of ecosystem are shown in Box 2.1.

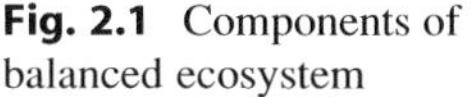

Fig. 2.1 Components of balanced ecosystem

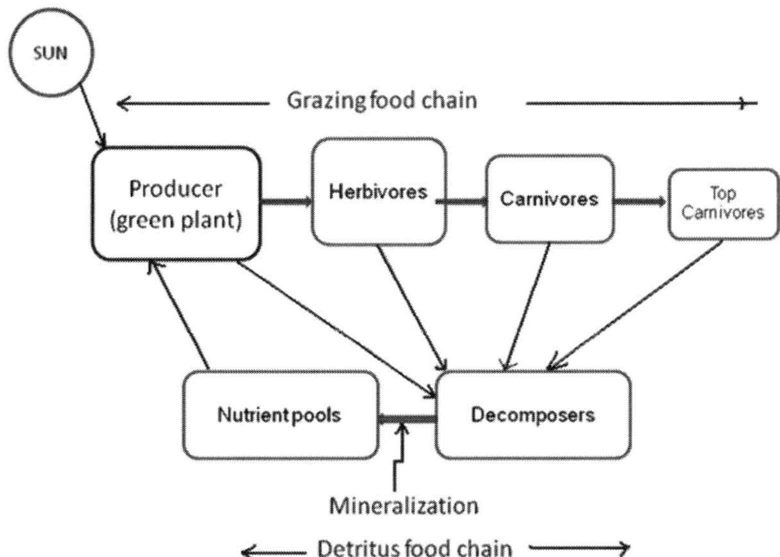

2.3.3 Food Chain, Food Web and Tropic Level

2.3.3.1 Food Chain and Energy Flow

The energy used for all life processes is derived from solar radiant energy. Solar energy is converted into chemical energy through photosynthesis by plants. These green plants are grazed subsequently by heterotrophs. Energy movement is unidirectional unlike the nutrients/materials in an ecosystem, that is, the initial energy trapped by autotrophs does not revert back to solar input.

Energy flow in an ecosystem is unidirectional, thus, the sequence of organisms through which the energy flows can be identified, and this sequence is known as *food chain*. Or simply, the process of eating and being eaten forms a food chain.

At each transfer 80–90% potential energy is lost as heat. Therefore, the number of steps linked in a sequence is limited, usually 4 or 5. The shorter the food chain, the greater the available energy.

In nature, *two types of food chain* have been distinguished:

1. *Grazing food chain*: This food chain begins from green plants at the base, and the primary consumers are herbivores. Ecosystem with such type of food chain directly depends on an influx of solar energy. Most of the ecosystem in nature follows this type of food chain. From energy point of view, these chains are very important.

2. *Detritus food chain*: This food chain starts from dead organic matter of decaying animals and plant bodies $\rightarrow$ microorganisms $\rightarrow$ predators. This type of food chain does not depend on solar energy.

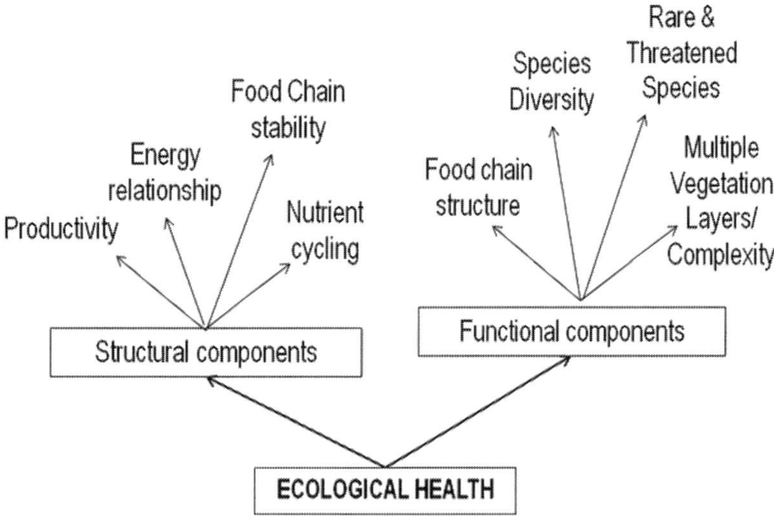

Fig. 2.2 Integrataion of structural and functional components

2.3.3.1.1 Grazing and Detritus Food Chains: The Y-Shaped Energy Flow Model

In nearly all ecosystems, some of the net production is consumed as living plant material and some is consumed later as dead plant material. We can conveniently designate these primary consumers that eat living plants as grazing herbivores, whether they may be large animals such as cattle or deer, or small animals such as zooplankton. The energy flow through grazers can be designated as the grazing food chain. Likewise, consumers of dead matter can be conveniently designated as detritus consumers and the energy flow along this route as the detritus food chain, as shown by the lower pathway in the Y-shaped diagrams (Fig. 2.3). Two types of organisms consume detritus: (1) small detritus-feeding animals, such as the soil mites or millipedes on land and various worms and molluscs in water, and (2) microbes (i.e., bacteria, fungi, actinomycetes etc).

2.3.3.2 Food Web

A *food web* is a graphical description of feeding relationships among species in an ecological community (i.e., who eats whom), and simply defined as 'the interlocking pattern of organisms' (Fig. 2.4). It is also a means of showing how energy and materials (e.g. carbon) flow through a community as a result of these feeding relationships. Typically, species are connected by lines or arrows called 'links', and the species are sometimes referred to as 'nodes' in food web diagrams.

In nature, simple food chain occurs very rarely. We obtained several food chains linked together and intersecting each other to form a network known as food web. The position of the organisms in the food chain is indicated by tropic levels. A tropic level may be defined as the number of links by which it is separated from the producer. R.H. MacArthur (1955) first suggested that the larger the number of species in food web, the more '*stable*' the community, because the greater number of alternate pathway of energy flows. If one species is vanished by stress, a predator had more feeding options in a species-rich food web rather than a species-poor one.

Tropic structure: The position of organisms in the food chain is indicated by tropic levels.

2.3.3.3 Ecological Pyramids

Ecological pyramids are the characteristic of an ecosystem, and it is defined as 'graphical representation of tropic structure'. The steps of tropic level can be expressed in a diagrammatic way which are referred to as *ecological pyramids*. The food producer forms the base of the pyramids

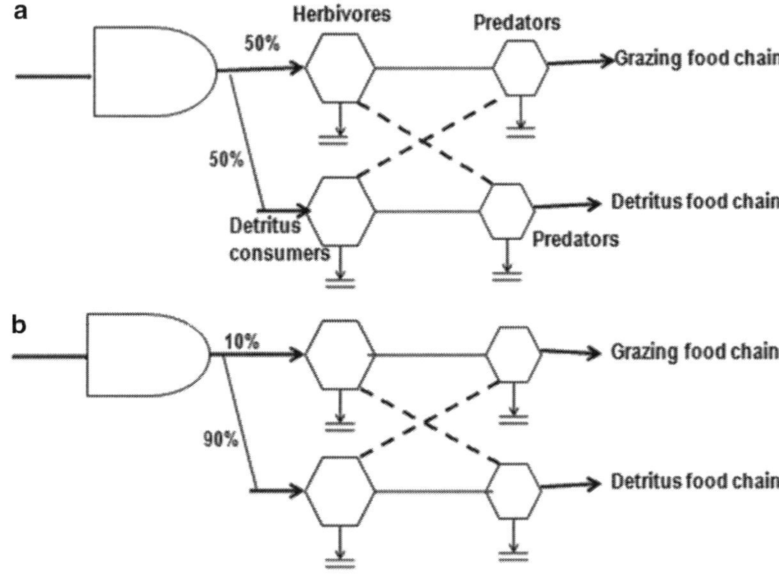

Fig. 2.3 The Y-shaped energy flow model of an ecosystem showing linkage of two major food chains: the grazing food chain and the detritus food chain. Diagram (**a**) represents an ecosystem, such as a grazed pasture, with a large proportion of energy (50%) flowing through the grazing pathway. Diagram (**b**) represents an ecosystem, such as mature forest, with most energy flowing along the detritus pathway (Odum 1975)

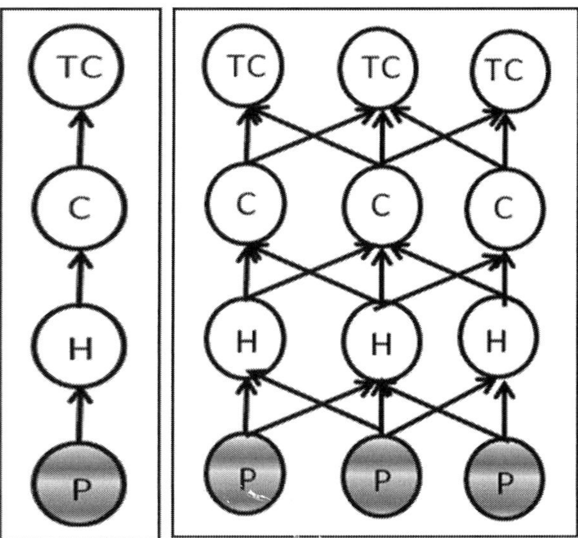

Fig. 2.4 Simple food chain and a food web. *P* producer, *H* herbivores, *C* carnivores, *TC* top carnivores

and the top carnivores form the tip. The ecological pyramids are of three categories:

- *Pyramids of number*: showing the number of individual organisms at each tropic level (Fig. 2.5)
- *Pyramids of biomass:* showing the total dry weight and other suitable measures of the total amount of the living matter

- *Pyramids of energy or productivity:* showing rate of energy flow and/or productivity at successive tropic level

The pyramids of biomass and number may be *upright or inverted*, depending upon the nature of the food chain in the particular ecosystem, whereas pyramid of energy are always *upright*.

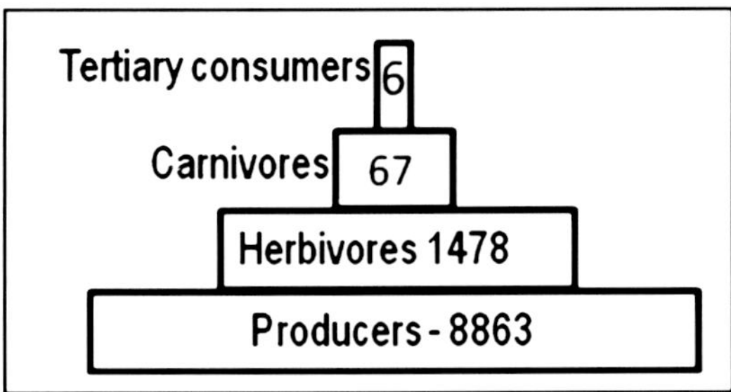

Fig. 2.5 Pyramids of number

would need to includes faunal component (e.g. invertebrates, birds, animals, reptiles) and decomposers. In restoration studies these must also be considered, but in order to achieve any ecological resilience, the vegetation and soil must be established first. The diversity–stability hypothesis predicts that as species diversity increases so too does ecosystem productivity and resilience. Nutrient cycling has often been used as a measure of stability.

2.4 Guild and Keystone Species

Species competing for the same resources in a similar fashion in an ecosystem are known as *guilds*. They are classified according to how they acquire their nutrients, their state of mobility and their mode of feeding. Some examples of guilds are shrubs, trees, vines etc. A guild is much more stable than a single species since more than one species can balance out the system.

A *keystone species* is a species having a disproportionate effect on the ecosystem. They provide stability to an ecosystem. Normally, they are not the dominant species but required for a community to have stability. Many times keystone species are predators that keep some type of herbivore from consuming all of the dominant plant species. One interesting aspect of keystone species is that since they normally feed on predators that consume small numbers of prey, they can effectively control a system without actually needing to have a large population size.

2.5 Ecosystem Stability

Ecosystem stability is the ability of an ecosystem to maintain its given trajectory in spite of stress; it denotes dynamic equilibrium rather than stasis. Stability is achieved in part on the basis of an ecosystem's capacity for resistance and resilience. The stability of an ecosystem is usually measured either by vegetation structure (biomass/cover of component species) or by a measure of ecosystem function (nutrient cycling). This is an oversimplification, because for any biotope, the vegetation and soils are only two components; hence, a complete assessment

2.6 Functioning of Ecosystem

Ecosystem too, in an *open system* it received energy from outside source (*sun*), fixes and utilises it and ultimately dissipates heats to space and has some sort of feedback system to make it self-regulating. Ecosystems are capable of self-maintaining and self-regulating as are their components and organisms. Ecosystem is a dynamic and self-regulating system, while, in nature there is always a disturbance and according to the nature and magnitude of disturbance, ecosystem responding in its own, and changes according. The dynamic property of ecosystem is explained by two terms: *inertia property* (resistance to change) and *resilience properties* (a return time to a stable state following a disturbance).

2.6.1 Ecosystem Inertia and Resilience

Inertia is the inherent property of ecosystem, which arises due to its stability property, and it is always resistant to change against perturbation. For example, force needed to stretch a coil over a given distance or amount of oil that must accumulate over a given area in a given time period to cause ecosystem to damage or, how much (energy) disturbance is needed to cause the ball to move (Fig. 2.6).

Resilience properties refer to ways in which a disturbed system responds. Resilience is the tendency of a system to return to a previous state after a perturbation. Species diversity is often cited as a key feature of ecosystem resilience. In an ecosystem, more the species diversity, higher is the resilience capacity. There are four components of resilience:

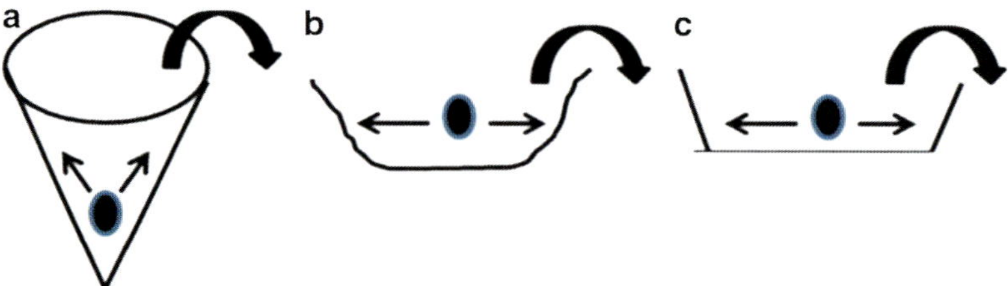

Fig. 2.6 Inertia and resilience analogised to a ball in a cup. Inertia: a > b > c: How much disturbance is needed to cause the ball to move?. Elasticity: a > b > c How fast will the ball return to its original position? Amplitude c > a > b: How much disturbance is needed to cause the ball to roll out of the cup?. Hysteresis: a = b < c : Will the ball roll back by the same route it took when initially displaced?. Malleability: b > a > c: How far away will the ball land when displaced from the cup with a given force of disturbance?

Elasticity and *amplitude* are measures of resilience.

(a) *Elasticity* is the speed with which a system returns.

(b) *Amplitude* is a measure of how far a system can be moved from the previous state and still return.

(c) *Hysteresis*: The extent to which the path of restoration is exactly reversal of path of degradation. For example, are there any differences in paths of alteration and recovery?

(d) *Malleability*: Degree to which stable state (old viz. new) established after disturbance differs from the original steady state. How closely do the species composition and equitability of new climax resembles the old?

2.6.2 Resistance and Resilience

Resistance is the term describing an ecosystem's ability to maintain its structural and functional attributes in the face of stress and disturbances. *Resilience* is the ability of an ecosystem to regain structural and functional attributes that have suffered harm from stress or disturbance.

2.6.3 Ecosystem Integrity and Ecosystem Health

The terms ecosystem integrity and ecosystem health are commonly used to describe the desired state of a restored ecosystem. Although some authors use the terms interchangeably, they are distinct in meaning. *Ecosystem integrity* is the state or condition of an ecosystem that displays the biodiversity characteristic of the reference, such as species composition and community structure, and is fully capable of sustaining normal ecosystem functioning.

Ecosystem health is the state or condition of an ecosystem in which its dynamic attributes are expressed within 'normal' ranges of activity relative to its ecological stage of development. A restored ecosystem will be called as healthy ecosystem, if it functions normally relative to its reference ecosystem or to an appropriate set of restored ecosystem. A state of ecosystem integrity suggests, but does not necessarily confirm, a concurrent state of ecosystem health and a suitable abiotic environment.

2.7 Species Diversity

Ecosystem is considered to be stable, if its structure and functions remain more or less same from year to year. In a mature and stable community, solar energy is fully utilised in the maintenance of living organisms; therefore, photosynthesis/respiration ratio is close to 1. In such community, species diversity is usually high. The simplest measurement of species diversity is to count the number of species in an area. In such count, we should include only residential species, no accidental or temporary immigrants be considered.

2.7.1 Species Richness Index (d)

The first and oldest concept of species diversity is species richness index and relates between number of species (S) and number of individuals of each species (N). Following species richness indices are used:

- Margalef's index (d1): $(D_{Mg}) = \frac{(S-1)}{\ln N}$
- Menhinik index (d2): $(D_{Mn}) = S/(\sqrt{N})$
- General index (d3): $S/1{,}000$.

Drawback: does not take species abundance into account, which is explain in Example-1.

Example 1: Imagine two hypothetical plant community A and B, both with 100 individuals

	Community A	Community B
Number of individuals of species 1	1	50
Number of individuals of species 2	99	50
Total plants	100	100

By applying three richness indices, both communities A and B show equal in species richness, but B has a higher equitability or evenness of the species richness:

$$d_1 = \frac{2-1}{\log_e 100} = \frac{1}{4.605} = 0.217;$$

$$d_2 = \frac{2}{\sqrt{100}} = \frac{2}{10} = 0.2;$$

$$d_3 = 20.$$

In both case the species diversity is same. The drawback of the species richness index is that it never considers the evenness of distribution. Hence, two indices are used:

Therefore, during the measurement of species diversity, both abundance (i.e., evenness of distribution of species) and species richness has to be considered. There are many indices, but we can divide them into two broad categories:

Dominance indices weighted towards the abundance of the most common species: (a) Berger and Parker index (1970) and (b) Simpson's index (Simpson 1949).

Information-based index also known as statistic indices—Shannon index (1948) and Brillouin index (1962).

2.7.2   Berger–Parker Index (D_{BP})

The simplest index based on dominance of species is Berger–Parker (D_{BP}) index defined as

$$D_{BP} = \frac{(N_{max})}{N}$$

where

N_{max} = total number of individuals in the most common species

N = total number of individuals in the community

Example 2: Calculate species diversity for community 1 and 2 by using for Berger–Parker index

Species	Individuals per species	
	Community 1	Community 2
a	10	5
b	10	5
c	10	5
d	10	5
e	10	5
f	10	5
g	10	5
h	10	5
i	10	5
j	10	55
Total number of individuals	100	100
Total number of species	10	10

D_{BP} for community 1 = $10/100 = 0.1$, while for community 2 = $55/100 = 0.55$.

To express greater diversity with a numerically greater value, we usually use a reciprocal form of the index. Thus, $D_{BP} = 1/D$, so that more diverse community actually has a higher index of diversity. In this case, for community 1, $D_{BP} = 1/0.1 = 10$ and community 2, $D_{BP} = 1/0.55 = 1.82$. A huge advantage of the Berger–Parker index over the others is that it is very easy to compute.

2.7.3   Simpson's Index (Ds)

A more widely used index than Berger–Parker's index is Simpson's index (Simpson 1949), which gives the probability that any two individuals drawn at random from an infinitely large community will belong to different species.

Simpson's index (Ds) = $1-C$, where 'C' is calculated as

1. For infinite sample,

$$C = \sum_{i=1}^{s} (ni/N)^2$$

2. For finite samples (i.e. which only a portion of the community has been measured),

$$C = \sum_{i=1}^{s} \frac{ni(ni-1)}{N(N-1)}$$

where

ni = total number of individuals in each species
N = total number of individuals in all species
S = total number of species

After C value is calculated, Simson's index (Ds) is calculated as:

Ds = 1 − C or
Ds = 1/C (Williams 1964)

Solution: Example 1

For community 1:
D = 1 − $\sum[(99/100)^2 + (1/100)^2] = 0.0198$.
For community 2:
D = 1 − $\sum[(50/100)^2 + (50/100)^2] = 0.50$.

Example 3: Hypothetical data sets for Simpson's index (finite sample)

Name of tree species	Number of individuals
a	100
b	50
c	30
d	20
e	1
Total (N)	201

Solution for Example 3:

$$Calculate \ \ C = \frac{(100 \times 99)}{(200 \times 201)} + \frac{(50 \times 49)}{(200 \times 201)}$$

$$+ \frac{(30 \times 29)}{(200 \times 201)} + \frac{(20 \times 19)}{(200 \times 201)}$$

$$+ \frac{(1 \times 0)}{(200 \times 201)} = 0.338,$$

$$D_s = \frac{1}{C} = \frac{1}{0.338} = 2.96$$

$$D_s = 1 - C = 1 - 0.338 = 0.662$$

The disadvantage of Simpson's index is that it is heavy weighted towards the most abundant species, as are all dominance indices. Thus, addition of many rare species of trees with one individual will fail to change the index value. As a result, Simpson's index is of limited value in conservation biology if an area has many species with just one individual.

2.7.4   Shannon Index (Hs)

The widely used Shannon index (H) is one of the best indices which is reasonable independent of sample size and also normally distributed (Odum 1971). The greater the value of H, the higher is the diversity. The value of H can be more than 1. Higher diversity value occurs when the number of species and the evenness component are large (low dominance). The Shannon index given by

$$\overline{H} = -\sum p_i \ln p_i$$

where pi = ni/N and 'ln' denotes the natural logarithms.

Solution: Example 1

For community − 1: $-\sum[0.99\,(\log_e 0.99) + 0.01$ $(\log_e 0.01)] = 0.056$.

For community − 2: $-\sum[0.5\,(\log_e 0.5) + 0.5$ $(\log_e 0.5)] = 0.69$.

The second plant community is much more diverse than the first community. The higher the value of H, the greater the diversity. Maximum value of H can be more than 1.

Example 4: Calculation of Shannon index of data given in Example 3

Name of species	Number of individuals (ni)	pi = (ni/N)	ln pi	pi ln pi
a	100	0.4975	−.6981	−.3473
b	50	0.2487	−1.3912	−0.3460
c	30	0.1495	−1.9021	−0.2844
d	20	0.0995	−2.3075	−0.2296
e	1	0.00497	−5.3033	−0.0263
				−1.2336

Note that rare species with one individual contributes some value to the Shannon index, so if an area has many rare species, their contributions would be accumulated. Values of Shannon index for real communities are often found to fall between 1.5 and 3.5.

Could these two indices be comparable? Answer is yes.

Indices	Community 1	Community 2
Simpson index (Ds)	0.0198	0.50
Shannon index (H)	0.056	0.69
For comparison (H/ln$_e$ S)	0.056/ ln$_e$2 = 0.0808	0.96/ ln$_e$S = 0.995

Table 2.1 Comparison of the effectiveness of different diversity indices Stiling (2002).

Index	Discriminant ability	Sensitivity to sample size	Biased towards rare species (R) or dominant (D) species	Calculation	Widely used
S (species richness)	Good	High	R	Simple	Yes
Shannon	Moderate	Moderate	R	Intermediate	Yes
Simpson	Moderate	Low	D	Intermediate	Yes
Berger–Parker	Poor	Low	D	Simple	No

For comparison, Shannon index value (H) is divided with $\ln_e$ of total S (species).

2.7.5 Shannon Equitability Index (E_H) or Evenness Index

For any information–statistic index, the maximum diversity of a community is found when all species are equally abundant. A comparison of the effectiveness of different diversity indices is shown in Table 2.1. We can compare a community actual diversity, (Hs) to the maximum possible diversity (Hmax), by using a measure called evenness:

$$Evenness\ (E_H) = \frac{\overline{H}}{\ln S},$$

Evenness index (E_H), for *Example 4*,

$$evenness\ (E_H) = \frac{1.2336}{\ln 5} = \frac{1.2336}{1.6094} = 0.621$$

In case of *Example 2* data sets,

$$evenness\ (E_H) = \frac{2.30}{\ln 10} = \frac{2.30}{2.30} = 1$$

E_H *is ranged between 0 and 1.*

2.7.6 Index of Similarity

Indices that compare the composition of pairs of sites are called 'resemblance functions' and calculated by 'indices of percentage similarity (S)'. The following indices are commonly used: Czekanowski index (1913) Jaccard index (1912) and the Sorensen index (1948). It is useful for a comparison of the species diversities between unpolluted and polluted sites. If the level of pollution is low, similarity index tends to 1, and if pollution level is high, the index value tends towards 0. The similarity index (S) between two samples is given by the following (Czekanowski index):

$$S = \frac{2C}{A + B} \times 100$$

where
S = index of similarity (in %)
A = number of species in sample A;
B = number of species in sample B,
C = number of species common to both A and B.

2.7.7 α-, β- and γ-Diversity

α-*Diversity*: Called intrabiotopic diversity or diversity within one components. It measures the number of species in single community. Diversity within a particular area, community or ecosystem is known *alpha diversity*. Number of taxa (species usually) present in an ecosystem is a measure of *alpha diversity*.

β-*Diversity*: Species diversity between two ecosystems is β-*diversity*. It compare the differences of population between two adjacent biotopes, that is, diversity between two habitats, for example, change in species composition along the environmental gradients. Comparing the number of taxa that are unique to each of the ecosystem is β-*diversity*.

γ-*Diversity*: It is also known as Macrodiversity, for example, mixed biotopes contained in geographical areas. Measure of overall diversity for different ecosystems within a region is γ-*diversity*.

2.7.8 Biotope and Ecoregion

The area that is uniform in environmental conditions and in its distribution of animal and plant life is called *biotope*. An area constituting a natural ecological community with characteristic flora, fauna and environmental conditions and bounded by natural borders is called *ecoregion*.

2.8 Biogeochemical Cycles (Nutrient Cycles)

There are two basic principles based on which entire ecosystem functions are governed, (a) energy flow cycle and (b) biogeochemical cycle, shown in Fig. 2.7. In this figure, both the cycles are superimposed and clearly be seen that it is the energy cycle which is unidirectional, acts as a force to run the biogeochemical cycle in a cyclic manner. Initially, energy is received by the autotrophs; therefore, without plant establishment, initiation of ecosystem cannot be started on degraded sites.

Mining operation totally destroys the nutrient cycle on the sites; however, once sites are left undisturbed, due to natural succession process, again nutrient cycles slowly regenerate. The process can be enhanced by the addition of topsoil and subsoil in ecorestoration sites along with plantation. The topsoil actually provides the microbial inoculum initially which will start decompositing the organic matter. Monitoring nutrient loss rates may provide a useful indication of ecosystem response to mining disturbance. The long-term sustainability of restored ecosystems is closely linked with the establishment of nutrient cycles. In most natural ecosystems, recycling rates of nutrients limit primary production and regulate biotic energy flow through the tropic structures.

The litter accumulation and ground vegetation enhance the infiltration of water, minimise erosion and contribute to diversity and activity of soil microbes including decomposers, thus stabilising the biogeochemical cycle in mine degraded sites. Therefore, the composition and types of ground vegetation cover significantly contribute the operation of biogeochemical cycles.

2.9 Ecological Succession

Ecological succession is a process by which an ecosystem is born. Ecosystem development and evolution is a dynamic process in the nature but culminates once ecosystem reached to its climax stage (forest stage). During the passage of time, an unstable ecosystem reaches to a greater stability. This change is due to variation in climate and activities of the species in community itself. Odum (1971) has defined ecological succession with respect to changes in three parameters:

(a) It is an orderly process of community development that involves changes in species structure and community process with time; it is reasonably directional and therefore predictable.

(b) It results from modification of the physical environment by the community.

(c) It is culminated in a stabilised ecosystem in which maximum biomass and symbiotic function between organisms are maintained/unit of available energy.

Ecological succession is a directional change in an ecological community and trends to proceed along with (a) continuous change in the plant and animal composition, (b) increasing trend in species diversity, (c) increase in the organic matter and biomass supported by the available energy flow and (d) decrease in net community production or annual yield. The course of ecological succession depends on initial environmental conditions; thus, based on the nature of substrate available, succession is divided into two types:

(a) *Primary Succession*: If succession begins on an area that has not been previously occupied by a community, the process is known as *primary succession*. It starts from a bare rock where there is no living matter, i.e., newly exposed rock or sand surface, new island, etc.

(b) *Secondary Succession*: Starts from built-up structure, that is, where community was removed (cutover forest, an abandoned cropland, etc.). The process is appropriately called secondary succession. The secondary succession is more rapid because some organic matter and propagules are present.

The succession occurs during the ecorestoration of coal mine derelict sites may be termed as *secondary succession* because overburden materials already contain soil-forming materials

Fig. 2.7 Biogeochemical cycle superimposed on energy flow cycle. *Pg* gross primary productivity; *Pn* net primary production, which may be consumed within the system by heterotrophs or exported; *P* secondary production and *R* respiration (Odum 1971)

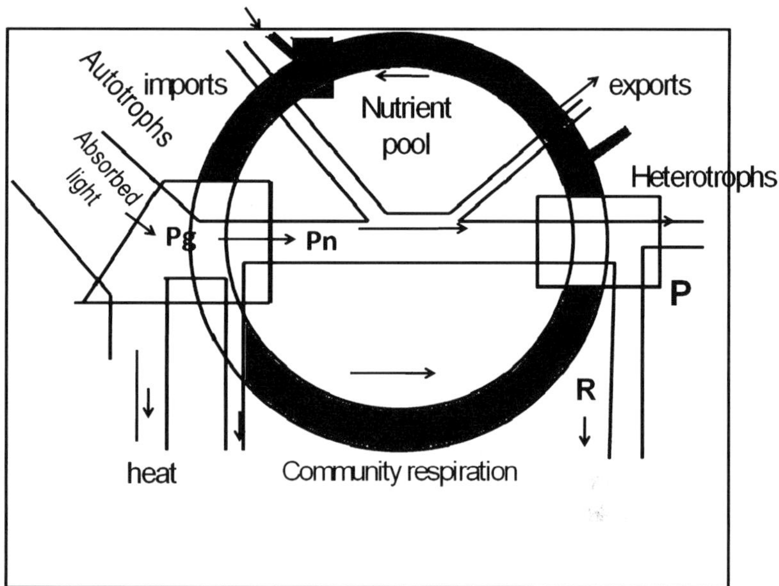

(soil particles and other nutrients) which foster succession process. At an early stage, succession rates of species turnover and soil changes tend to be relatively rapid, and gradually it becomes slower as the community matures. Eventually, a steady stable community is established. The terminal stabilised system is known as climax, which is self-perpetuating and in equilibrium with the physical habitat. Actually, there is no accumulation of organic matter. The important changes in community structure and function taken place during succession process are given in Table 2.2.

This unified concept of succession, the monoclimax hypothesis, implies the ability of organisms progressively to modify their environment until it can support the climatic climax community. Although plants and animals do sometimes ameliorate environmental conditions, evidence suggests overwhelmingly that succession has a variety of stable end points. This hypothesis, known as the polyclimax hypothesis, which suggests that the end point of a succession depends on a complex of environmental factors that characterise the site, such as parent material, topography, local climate and human influences.

2.9.1 Important Steps Involved in Succession

1. *Invasion*: It is the process of successful establishment of a species in a bare area. The species actually reaches this new site from another area. The whole process is completed in following three successive stages:
   - *Migration (Dispersal):* The seeds, spores or other propagules of the species reach the bare area. This process, known as migration, is generally brought about by air, water, etc.
   - *Ecesis (Establishment):* After reaching to new area, the process of successful establishment of the species, as a result of adjustment with the conditions prevailing there, is known as ecesis.
   - *Aggregation:* After ecesis, as a result of reproduction, the individuals of the species increase in number, and they come close to each other. This process is known as aggregation.
2. *Competition and Coactions*: After aggregation of a large number of individuals of the species at the limited place, there develops competition (inter- as well as intraspecific)

Table 2.2 Progressive changes during ecosystem development Modified after Odum (1971)

Community structure	
1. Species composition	Changes rapidly at first, then more gradually
2. Size of individuals	Tends to increase
3. Number species of autotrophs	Increases in primary and often early in secondary succession, may decline in older stages as size of individuals increases
4. Species diversity	Increases initially, then becomes stabilised
5. Total biomass	Increases
6. Nonliving organic matter	Increases
Energy flow (community metabolism)	
1. Gross production (p)	Increases during early phase of primary succession, little or no increase during secondary succession
2. Food chains	From linear chains to more complex food webs
3. P/R ratio	P > R to P = R
4. Community respiration(R)	Increases
Biogeochemical cycles	
1. Turnover time	Increases
2. Role of detritus	Increases
3. Nutrient conservation	Increases
Natural selection and regulation	
1. Growth form	From R-selection (rapid growth) to K-selection (feedback control)
2. Overall stability	Increases

mainly for space and nutrition. Thus, individuals of a species affect each other's life in various ways, and this is called *co-action*. The species, if unable to compete with other species gradually perished. To withstand competition, reproductive capacity, wide ecological amplitude, etc., are of much help to the species.

3. *Reaction*: This is the most important stage in succession. The mechanism of the modification of the environment through the influence of living organisms is known as *reaction*. Thus, as a result of reactions, changes take place in soil, water, light conditions, temperature, etc., of the environment. Due to all of these, the environment is modified, becoming unsuitable for the existing community, which sooner or later is replaced by another community (*seral community*).

4. *Stabilisation* (*Climax*): Finally, there occurs a stage in the process, when the final terminal community becomes more or less stabilised for a longer period of time, and it can maintain itself in the equilibrium with the climate of that area. This final community is not replaced and is known as *climax community* and the stage as *climax stage*.

2.9.2   Effects of Ecological Succession on Coal Mine Spoil

Plant succession is law of nature and improves the spoil physico-chemical properties by reducing bulk density, pH, conductivity and increase in organic matter, NPK. The main factors responsible for success of succession process are (1) nature of substrate, quality and depth of topsoil on restored sites, and nutrient availability, moisture status, acidity or alkalinity and microbial activity; (2) proximity of seed banks; (3) sources of invasion; and (4) faunal population (Clark 1985). Out of all these factors, soil moisture and organic matter are two most important factors for natural succession on mine spoil; however, in initial stage water availability alone is one of the limiting factors for initial plant establishment (Maiti 1995). Succession in coal mine degraded land happens to be both primary and secondary in nature; therefore, it ends up with a complex vegetation characteristics rather than a uniform successional trend which is due to heterogeneous nature of spoil material (Wali and Freeman 1973). Even though succession process very seldom ends up with undesirable species, it is essential for establishment of diverse microflora and -fauna for development of self-sustaining nutrient cycling (Wilson 1965). For topsoiled restored sites, the species were found in the different successional stages.

2.10 Terminology of Ecology

Each science has its own terms and concepts. Some terms frequently used in this subject of ecology are explained below for the convenience in understanding the subject matter:

1. *Species, Population and Community*
   - *species* is a natural biological unit tied together by sharing of a common gene pool. The species defined as 'a uniform population spread over time and space'. For example, tiger and mango species.
   - *population* is a group of individual organism of the same species in a given area (e.g. tiger population).
   - *community* is a group of population of different species in a given area. It, thus, includes all the populations in that area—all plants, all animals and microorganisms.
2. *Ecads and Ecotypes*
   The species acclimatised themselves in the environment, which depends on the plasticity of the genetic constituents of the species. The various forms of species, in order to meet the challenge of changed environment, may arise by virtue of somatic plasticity, the *ecads*, or the reorganisation of their genes during sexual reproduction, the *ecotypes*. Thus, species acclimatised to their surrounding environment by developing *ecads* and *ecotype*.

3. *Biome*
   Under similar climatic conditions, they may simultaneously develop more than one community, some may be on climax stage and other may be on successional stage. This complex of several communities in any area, represented by an assemblage of different kinds of plants and animals sharing a common climate, is called a *biome*.
4. *Functional Aspects of Ecosystem*
   Population interaction: There exists varying degrees of positive (+), negative (-) or even neutral interaction among organisms, at both inter- and intraspecific levels. Examples are competition and prey–predator,
   Energy flow in ecosystem: Sun energy is the driving force of all natural ecosystem. The autotrophic organisms (producers) trapped the energy and are transferred as organic molecules to the heterotropic organisms (consumers). This energy flow is unidirectional or non-cyclic.
   Biogeochemical cycles: The cycling of all essential elements of the protoplasms between environment to the organisms and back to the environment are known as the *biogeochemical cycles.*
   Limiting factors: Successful growth of the organisms is governed by limiting factors. The growth of any organism needs various essential elements from its environment. The success of a growth of organisms is not only limited by deficiency of elements but also by excess conditions. The minimal and maximal levels of tolerance for all ecological factors of a species vary seasonally, geographically and according to the age of the population.
   Managed and natural ecosystem: As a result of disruption by natural condition or exploitation by the activities of man, species diversity of an ecosystem is reduced or equilibrium is destroyed. Ecosystem which is substantially altered by human activities is called *managed ecosystem,* where those free from such disturbance are referred to as *natural ecosystem.*

5. Level of Study of Ecology
   - *Population ecology*—pure stands of individuals of a single species.
   - *Community ecology*—group of individuals belonging to different species.
   - *Biome ecology*—In nature, we generally find complex of more than one community, that is, some are climax, approaching climax and under early stages of succession. These are growing under more or less similar climatic conditions in an area.
   - *Landscape ecology*—is the science of studying and improving relationships between ecological processes in the environment and particular ecosystems. This is done within a variety of landscape scales, development spatial patterns and organisational levels of research and policy.
   - *Ecosystem ecology*—It has been established that biotic and abiotic components of the nature interact between each other. There is an interaction within the biotic and abiotic components itself.

2.11 Ecosystem Components in Restored Site

The components of ecosystem in an OB dumps are consisting of biotic, abiotic factors and their interaction. The components could be described as:

- *Biotic components*: consist of producer, functional aspects are enhanced by plantation and natural colonisation.
- *Abiotic components:* consist of substrate (nature and property of overburden materials), and surrounding climatic which naturally influence the selection the plant species.
- *Decomposer* acts as link between biotic and abiotic components; this is a very week but vital link.

The speedy recovery of decomposer could be achieved by addition of organic mulch, topsoil and afforestation which are practised in India. The main aim is to *enhance the microbial life on OB dump vis-a-vis establishment of decomposer cycling or nutrient cycling.* These decomposers are heterotroph and detritus feeder; thus, accumulation of litter layer is essential for self-sustaining ecosystem development. At the same time, optimal spoil conditions are also maintained so that microbes can survive and function. Based on this concept, the ecosystem recovery on OB dump could be assessed by studying the following:

Indicator Parameters of Ecosystem Recovery in Restored Site
- *Dump physico-chemical status* in relation to plant growth
- *Nutrient status*—organic matter, organic carbon, total and available nitrogen and plant available phosphorous and potassium
- *Ecological status*—standing biomass, litter fall and litter decomposition
- *Microbial status*—density of bacteria, nitrifiers, actinomycetes, fungi and mycorrhiza
- Density of *microfauna*

References

Cary Institute of Ecosystem Studies (2012) http://www.caryinstitute.org/discover-ecology/definition-ecology

Berger WH, Parker FL (1970) Diversity of planktonic foraminifera in deep sea sediments. Science 168:1345–1347

Brillouin L (1962) Science and Information Theory, 2nd edn. Academic press, New York, p 386

CBD (1992) Convention on biological diversity. Convention text, article 2 – use of terms. Online available at http://www.cbd.int/convention/articles.shtml?a=cbd-02

Clark WA (1985) Plant succession on abandoned mine lands in the Eastern united states. In: Schlosser et al (ed) National symposium on abandoned mined land reclamation, Northwood, England, pp 613–631

MacArthur RH (1955) Fluctuations of animal populations, and a measure of community stability. Ecology 36:533–536

Maiti SK (1995) Some experimental studies on Ecological aspects of reclamation in Jharia coalfield. Ph.D dissertation, Indian School of Mines, Dhanbad

Odum EP (1971) Fundaments of ecology, 3rd edn. W. B. Sunders and Co., Philadelphia

Odum EP (1975) Ecology, 2nd edn. Oxford/IBH Publications, Calcutta

Shannon CE (1948) A mathematical theory of communication. Bell Sys Tech J 27:379–423

Simpson EH (1949) Measurement of diversity. Nature 163:688

Southwick CH (1976) Ecology and the quality of our environment, 2nd edn. Van Nostrand, New York, p 426

Stiling P (2002) Ecology- theory and application, 4th edn. PHI, New Delhi, p 278

Tansley AG (1935) The use and abuse of vegetational concepts and terms. Ecology 16:284–307

Wali MK, Freeman PG (1973) Ecology of some mined areas in North Dakota. In: Wali MK (ed) Some environmental aspects of strip mining in North Dakota, vol 5, North Dakota, education series. North Dakota Geological Survey, Grand Forks, pp 25–47

William CB (1964) Patterns in the balance of nature and related problems in quantitative ecology. Academic press, New York

Wilson HA (1965) The microbiology of strip mined spoil. West Virginia Ag Stn Bull, 506T, p 44

Woodbury AM (1954) Principles of general ecology, 1st edn. McGraw-Hill, New York

Physical (Technical) Reclamation

3

Contents

3.1 Introduction

All landscape reclamation schemes has to be in *two* distinct stages: (a) *Physical, technical or engineering,* this is the high-cost, low-risk part of reclamation and takes over 60–90% of the total cost of reclamation, and (b) *Biological,* this is low-cost, high-risk part of reclamation work, which requires multidisciplinary inputs. The job of reclamation specialist is to bring these two parts together, along with other consideration of landscape, after use, etc.

Engineering reclamation is concerned with the creation of suitable landforms, compatible with the landscape as well as satisfying stability and other requirements. The development of the landscape plan has *three basic stages*:

- Initial survey
- Determination of ultimate landscape objectives including after use
- Preparation of working plans for each phase of the operation

Biological reclamation is concerned with establishing and maintaining vegetation cover on the new landform, which is compatible with surround landscape, stabile and fulfils after-use requirements. Vegetation establishment plays a key role in all aspects of the landscape plan. It has major functions, like screening, slope stabilisation and erosion control, improvement of soil conditions and structure, aesthetic value (visual improvement) and lastly the means for providing for some satisfactory return during after use.

S.K. Maiti, *Ecorestoration of the Coalmine Degraded Lands,*
DOI 10.1007/978-81-322-0851-8_3, © Springer India 2013

Methods of physical reclamation (progressive and final) for coal mines will greatly vary depending on geo-mining conditions of coal mines. In case of opencast coal mining, the surface land will be totally excavated, and external surface dumps will be created sterilising additional land outside the excavated area. The extent of backfilling and physical reclamation of mine site will depend primarily on factors like area of mine excavation, stripping ratio of the deposit, inclination of the strata, types of mechanisation adopted, etc. (Chaudhuri 2008).

In case of underground coal mining, the land damage is primary due to subsidence overextracted areas. However, the damage is comparatively less severe compared to opencast coal mining and will depend on factors like number of coal horizon(s), thickness(es) and depth(s) of coal seam(s), method of mining—Bord and Pillar or Longwall mining—and whether extraction will be with caving or stowing, etc. (Chaudhuri 2008).

There may be other cases where a combination of opencast and underground mining is envisaged. In such cases, the upper seams are extracted by opencast mining up to economic limit, and the balance part of these upper seams (if any) and the lower seams is extracted by underground mining. The sequence of such opencast and underground mining will have influence on the extent and timing of land damage, and the planning for physical reclamation (progressive and final) will have to be done accordingly (Chaudhuri 2008).

In opencast mining, the main pit design parameters that have influence on physical reclamation are volume of overburden to be accommodated in internal dump of the mine, volume of overburden to be placed in external dump and volume of residual void. In most cases, the physical reclamation will be facilitated if maximum volume out of total overburden is accommodated in internal dump, whilst the volume of external dump and of residual void is minimise.

Spoil or overburden waste should not be thought of as simply troublesome by-products to be disposed of as cheaply and simply as possible; they are a valuable resource with which to build a landscape. Bulk materials can be used to improve access roads, backfill around steep pit and quarry faces, as a cover material for land fill, construction of screening barrier, etc. In fact, without them, quarry landscaping would be very difficult.

3.2 Issues Related to Physical Reclamation

Technical reclamation consists of earthwork, construction of peripheral drainage and drains at the toe of the slope. In some cases, stability of dump has to be improved by re-profiling to give a flatter slope. In practice, the work is carried out by benching. If space permits, the slopes are buttressed by ring dumping with new material placed around the dump at a lower level.

3.2.1 Drainage

Water is detrimental to the stability of a dump. All attempts should be made to keep the dump as far as possible by providing peripheral drains at top and bottom, whereas longitudinal drains are required to convey and dispose the collected water in a planned manner. This keeps the dump materials in drained condition to eliminate erosion of slopes. Control of gullies formed due to surface erosion is important. Inspection of dumps is essential to pick up early signs of problems and to ensure that maintenance is carried out effectively.

3.2.2 Internal Dumps

These are the dumps within mine boundary and created generally through backfilling. When the ratio of OB/coal is high, the expected volume of OB will be far in excess than the void created by OB removal and coal extraction. In such situation external OB dumps are created and all necessary steps are to be taken to reclaim these dumps.

No external or internal dump should permit to be left as such. These dumps should be revegetated by applying a suitable layer of topsoil (15–20 cm) (previously stored or from outside) over it for its best economic use.

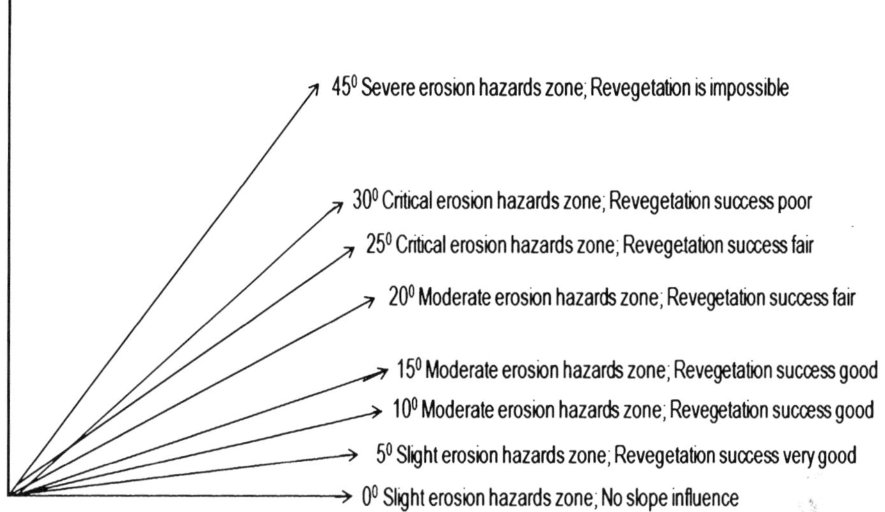

Fig. 3.1 Influence of slope on vegetation growth

3.2.3 External Waste Dumps

The most important aspect of physical reclamation is design of external dumps. The maximum allowable height and slope of external waste dumps should be determined by the slope stability study taking a factor of safety of 1.2 as the dumps will be permanent. Khandelwal and Mozumder (1987) stated that spoil dumps generally designed for factor of safety of 1.10 to 1.15 have only a minor risk of failure and less than 1.10 subject to greater risk. It depends on – dump geometry, water and material strength. (i) Geometry of dumps – height and slope, zonation within the dumps and potential failure surface, (ii) Water–water within a spoil pile is probably the single most important factor affecting stability. Water pressure can increase driving forces by filling tension cracks, developing seepage forces associated with ground water movement. (iii) Material strength depends on type of spoil, spoil placement method, zonation, grain size, compaction, Atterberg limits and dynamic forces. Usually, an overall dump slope of 28° to the horizontal and a maximum dump height of 60 m are stipulated at the time of approval of Environmental Management Plan (EMP). However, a slope stability study must be carried out during operating life of the opencast mine to confirm the safe slope angle and maximum height of dump that can be permissible with the physico-mechanical parameter of the waste rock obtained and competence of the basal ground over which the waste dump has been located. If necessary, the waste dump slope will have to be flattened involving sterilisation of additional land area.

3.2.4 Slope

One of the most important factors that must be considered is slope, and it is a major determining factor in after use of the land. Slope angle limits the establishment of the vegetation, in terms of vegetation cover and species richness. Steepness of the slope affects all after-use possibility. Spoils having more than 30% slopes pose severe limitation for establishment for vegetation cover. The overall slope is planned to have gradient of 1 in 2 or approximately 27°. For example, in Paraj OCP of Central Coalfield Limited (CCL), dumps are designed for 27°–28° and in Bina mine of Northern Coalfield Limited (NCL), 28°. Figure 3.1 shows the influence of vegetation establishment and after-use possibilities associated with various slope angles. It is clear that a slope of 1:4 to 1:5 is good for vegetation establishment.

3.2.5 Aspects of Slope

Slope type and aspect influence the amount and composition of the vegetation as well as the prevalence of the different water erosion processes

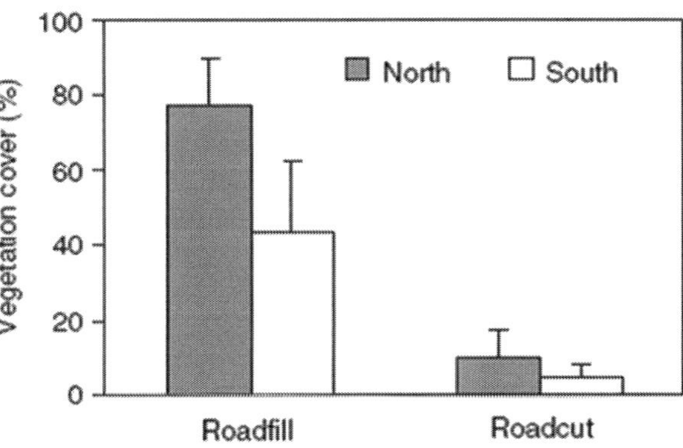

Fig. 3.2 Influence of aspects of slope ($<45°$) on mean vegetation cover (species number/100 m^2). Observe the difference of vegetation cover in north and south facing slope (Bochet and Garcı́a-Fayos 2004)

through the control of soil characteristics and soil moisture dynamics. Sunlight inputs are affected by aspect of slopes depending upon whether it is south facing or north facing. In the northern hemisphere, a south-facing slope will receive more sunlight and warm winds therefore generally be warmer and dryer due to higher levels of evapotranspiration than a north-facing slope (Jonathan et al. 2006). Therefore, effects of these factors have important role during the establishment of vegetation cover in overburden dumps. Coppin and Bradshaw (1982) also reported that in quarries, these effects can be very important. In West Virginia in USA, seed germination on northeast slopes exceeds that on a southeast slope by 50%. Similarly, tree density was found higher in the south-facing slope (880 plants/ha) compared to the north-facing slope (550 plants/ha) (Sidari et al. 2008). Figure 3.2 shows the influence of aspects of slope on vegetation cover development in a road fill slope. The aspect of slope has not been used in practice in designing of coal mine spoil dumps in India.

3.3   Site Preparation

Broadly the following are involved in site preparation:

1. Construction of waste dumps: Ensure that *all objectionable/toxic strata* are buried carefully whilst during construction and backfilling waste dumps.
2. Backfill and regrade the waste materials to the surface configuration.
3. *Spread and regrade* the material from B horizon (subsoil), over the graded waste materials in a planned thickness to form the subsurface layer of the reconstructed site.
4. *Spread and regrade A* horizon (topsoil) over the subsurface horizon in a uniform thickness.
5. Design proper drainage, slope to control erosion and stability of dumps. Flow diversion is usually best achieved by a bank with an upslope channel from which the soil material for the bank has been excavated. The channel upslope of the bank should be on a gradient and of a cross-sectional shape and size that flow velocities do not reach a level which creates scour. *Parabolic or trapezoidal shapes* are preferred over a V shape, whilst channel gradient should not generally exceed 1% unless the channel is lined with a stabiliser such as mesh or stone rip-rap.
6. Design of sediment retention basin.
7. Design of sediment trap sampling and analysis of the finally reclaimed mine spoils for further amendments like pH modification, organic and inorganic fertilisation for creating favourable conditions for seed germination and plant growth.
8. In addition, it suppresses herbaceous vegetation growth and thus eliminates competition between undesirable weeds and trees.

Important guidelines to be followed during the physical reclamation of coalmine overburden dumps are summarize in Box 3.1.

3.4   Different Methods of Dumping

Dumping method planned well in advanced allows easy restoration and designated after use

Box 3.1 Common Guidelines of Physical Reclamation

- Mine plan should be made so that the minimum proportion of overburden (OB) is taken to external OB dump and as early as possible in-pit dumping should start in an opencast mine.
- Topsoil should be recovered and utilized in reclamation of dumps.
- Extensive tree plantation activity should be taken up on vacant land in the leasehold for creation of green belt, avenue plantation and on external and internal OB dumps.
- External OB dumps should have gentle slope merging with the landscape, and in any case, the overall slope of dumps should not exceed 28°.
- Before biological reclamation of dumps, the area should be technically reclaimed by dozing.
- The objective is to bring the mined out land to productive land use as early as possible.
- In EIA, 5-year stage reclamation plans and sections show the progress of land damage mitigation measures.

of land. A waste dump may be formed basically by following methods:

- Area dumping method
- End-tipping method
- Perimeter tipping method

3.4.1 Area Dumping Method

In this method, the entire dump base area is demarcated first, and then the dump is created in the form of small lifts. Once the first lift is completed over the whole area, the second lift is started, and the process is continued until the designed height is achieved. For biological reclamation purpose, as well as achieve stability, a lift of 10 m is preferable, and maximum of three lifts (i.e. ultimate height of 30 m) is advisable. This method of dump formation is possible only when flat or nearly flat lands are available.

The method offers few important advantages, like good compaction, controlled dump slope and simultaneous development of vegetation cover along the dump sides/periphery. However, the formation of such dump does not provide the facility of creation of catchment drain at the bottom horizon of the dump and requires longer travel length for dumpers and more dozer operation time than in case of end-tipping method (Fig. 3.3).

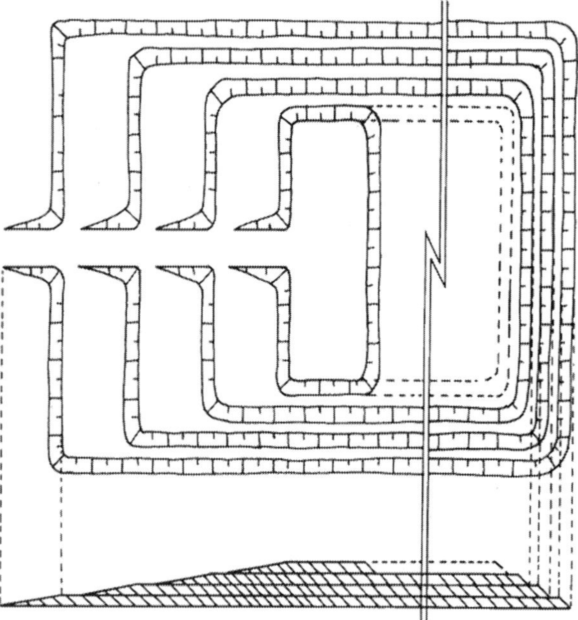

Fig. 3.3 Area dumping method of waste dump formation on horizontal ground

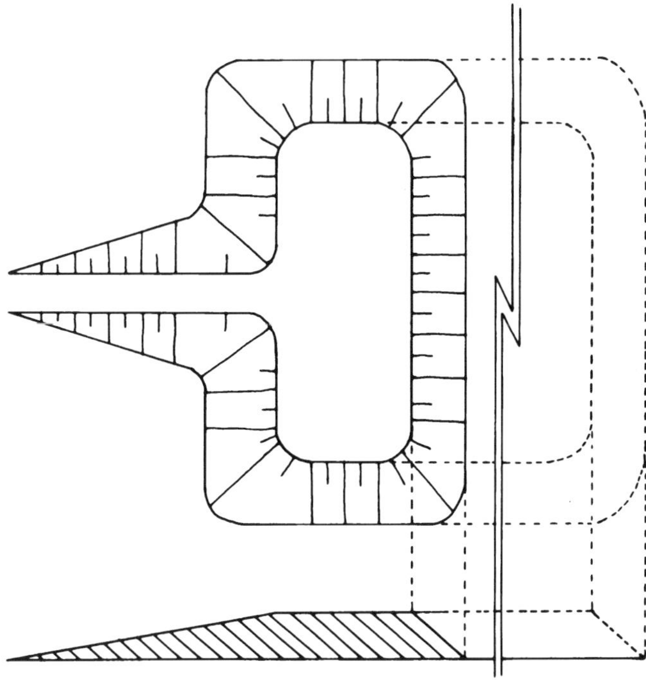

Fig. 3.4 End-tipping method of waste dump formation on horizontal ground

3.4.2   End-Tipping Method

In case of forming a dump on a flat topography by this method, initially an access road up to the designed top level of the dump is gradually made by dumping the waste material following the end-tipping technique. Once the designed level is achieved, the dump edges progresses maintaining the same elevation at the top, and the dumpers unload the materials at the end of the tip of the dump; thus, this method is named as 'end-tipping method'. Generally, the height in single lift is restricted within 30 m. If it becomes necessary and designed so, a second lift or more lifts may be formed following the same method, but a distance of minimum 10–15 m between the crest of one lift and the toe of the consecutive upper lift is required to be maintained always for the safety reasons. This method is applicable in case of all types of foundation topography (flat land, valley fill, hill sides, undulating topography, mine backfill, etc.) and most popular in our country.

The method is popular because it offers the following advantages: facility of formation of French drain at the bottom horizon of the dump (at the bottom horizon of each tier in case of multi-tier dumping), and it involves in comparatively less effort for dumpers and less dozer operation time than in case of area dumping method. However, in this method, the dump slope (slope of each tier in case of multi-tier) is uncontrollable being dependent on natural angle of repose of the dump material, and the material compaction is minimum (though some consolidation in the lower and middle horizon takes place with the time under the self-weight of the dump). Also, it is not possible to vegetate the dump slope surfaces simultaneously with this type of dump formation (Fig. 3.4). In both cases, the dump forming equipment generally consists of dumpers and dozers. Sometimes compactors or wheel dozers (the tyres inflated with brine) are also used. A typical layout of French drain is shown in Fig. 3.5.

3.4.3   Perimeter Tipping Method

In perimeter dumping method, active dumping areas or lift needs to be exposed at any time, and progressive restoration will bring most of the dumping areas into some use as soon as it is completed. This reduces the overall impact and allows the vegetation maximum time for development. In perimeter tipping method, initially outer tip is constructed and then in filling is done for the rest. A schematic diagram showing

Fig. 3.5 Typical layout of
French drain

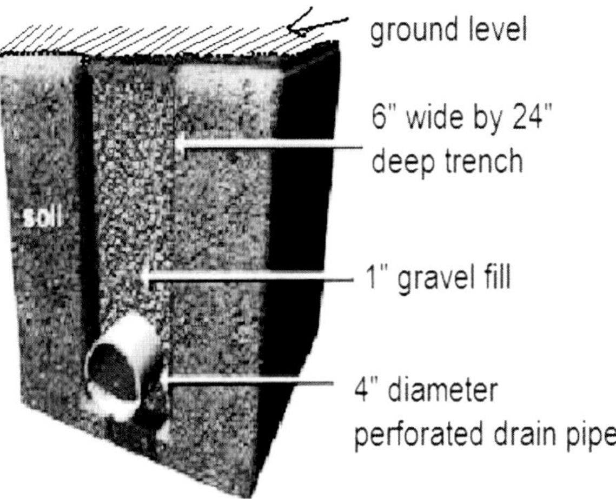

Fig. 3.6 Perimeter
tipping: the outside faces of
the tip are well vegetated
and screen subsequent
tipping. (**a**) Usual method
of tipping on sloping
ground, (**b**) perimeter
tipping on sloping ground,
(**c**) perimeter tipping on
level ground (Coppin and
Bradshaw 1982)

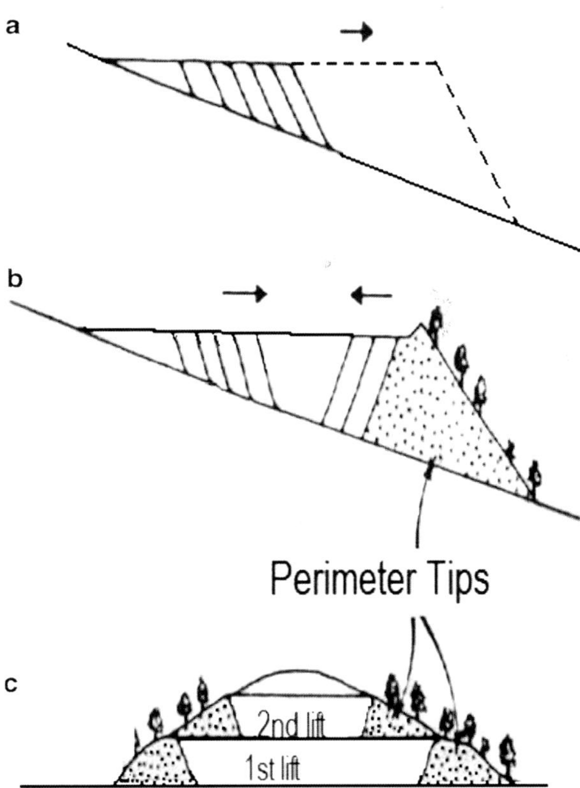

perimeter tipping on sloping ground and level
ground is shown in Fig. 3.6.

3.4.4 Preservation and Optimum Use of Topsoil During Tipping of Mine Spoils

Before tipping commences, topsoil and subsoil
layers should be stripped from the tipping area
for respreading over the tip surface. This method
is appropriate for both regarding of old tips and
construction of new tips, particularly if the material being tipped has poor potential as a soil. Care
should be taken to ensure that the soil being
saved is of better quality than the tipped material.

In case of old spoil dump, in first stage, topsoil
is salvaged from the periphery of the dumps and
heaped. In the second stage, once the spoil is

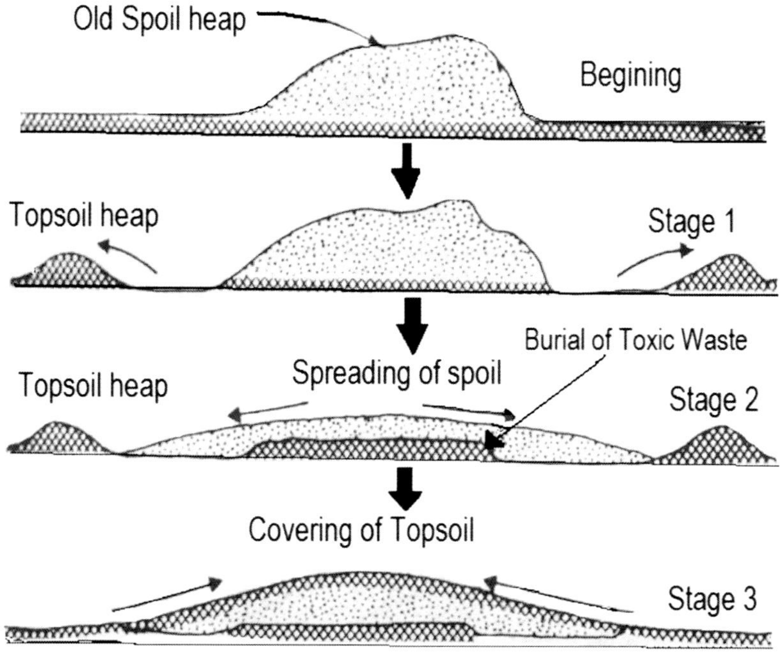

Fig. 3.7 Regrading of old tip, stripping and covering of topsoil in an old spoil dump

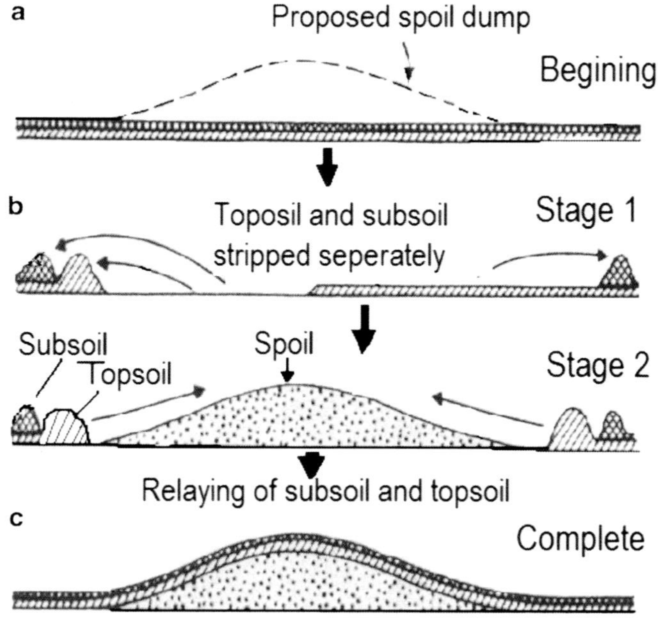

Fig. 3.8 Construction of new spoil dump, stripping and respreading of topsoil and subsoil layer

flatted, the excavated topsoil is spread and covers the spoil heap and vegetated (Fig. 3.7). In case of new spoil dump construction, topsoil and subsoil are removed separately and stored. After spoil dump construction is over, first it is cover with subsoil, then with topsoil in the same sequence as they were salvaged and lastly vegetation cover is develope (Fig. 3.8).

If the properties of spoil materials are known in advance, material that is of particularly poor quality or that is inhibitory to plant growth can be buried within the dump, away from the root zone. This is particularly important with acid-generating pyritic wastes. With contaminated materials, it will be necessary to exclude percolating water as much as possible, to reduce the possibility of toxic leachates. It may be advisable to place a layer of limestone over the toxic material, which will neutralise the acidity or reduce the solubility of toxic metals.

3.5 Drainage and Erosion Control

All the disturbed areas are needed to make some provision of drainage to control run-off on steep slopes to reduce sheet and gully erosion. The greatest source of non-point source pollution at surface mine operations can be attributed to silt leaving the mine site due to erosion. There are several factors that contribute to erosion. These are (a) amount of annual rainfall, (b) soil types, (c) steepness of slopes and (d) lack of vegetation.

Erosion control is very important at any mine site. Topsoil should be preserved and protected to provide the best possible growth medium for vegetation. Correct use of erosion control practices can insure that slopes are graded only once without rills and gullies appearing due to erosion. These practices contribute to successful reclamation in mine-degraded areas. The checklist for technical reclamation is given in Box 3.2.

First consideration for drainage design is to estimate run-off, taking into accounts of rainfall, soil type, total areas and steepness of slope and vegetation cover, etc.

3.6 Estimation of Soil Erosion

An equation termed the Universal Soil Loss Equation (USLE) has been developed that can be used to make a reasonable estimate of the amount of sheet and rill erosion to be expected on a degraded surface mine site. This equation

Box 3.2 Checklist of Activities During Technical Reclamation (Maiti 2010)

1. The advanced analysis of overburden of each major strata on the proposed mine site is essential to determine rock type, texture and thickness along with the chemical and mineral parameters in order to determine what strata are best suited for the surface cover of mine spoil for revegetation purposes.
2. Identify deeper strata, if any available, which is could be used as topsoil substitute, should be stored separately.
3. The depth of application of topsoil should be adequate for root growth and nutrient requirements.
4. Based on analysis, incorporation of organic materials, mulches, pH adjustment, fertilizers and any other soil amendments should be recommended well in advance.
5. If mixing of original topsoil with deeper topsoil substitutes (overburden strata) at final grading is envisaged (which is very common), ensure proper inoculation of soil microbes and the slow releasing fertilizer (manures).
6. Minimize compaction of the final lift of soil by end dumping and placement with dozing equipment to avoid excessive compaction.
7. Use scarification or ripping process on the top surfaces, it is observed that top surface is compacted by heavy equipment.
8. Provide appropriate site drainage to prevent ponding and water erosion on revegetated and other rehabilitated areas.
9. Establish windbreaks to prevent wind erosion on revegetated and other rehabilitated areas.
10. Use annuals/perennials (even climbers) for quick cover, which also provide soil nutrients and conserve moisture for successional vegetation.

can also be used to determine actions to be taken in the process of reclamation. For instance, the need for diversion ditches can be determined by this equation and the length and steepness of slopes that should be created in the regrading process. This equation has been presented in several forms, and the following form appears to be the one most useful for surface mine reclamation:

$$E = R\ K(\mathrm{LS})CP$$

where,
E = estimate of the soil loss rate in tons/ha/year
R = rainfall factor (run-off)
K = soil erodibility factor
LS = slope length factor/slope gradient factor
C = vegetation factor
P = mechanical erosion control factor

3.6.1 Rainfall Factor (*R*)

The R value in the equation takes climatic conditions into consideration. The rainfall factor can vary from year to year, so an average over a number of years is usually used. The rainfall characteristics that have the most effect on erosion are:

Intensity: Intensity of rainfall. It described as amount of rain that falls within a given time. It is expressed as mm/h or (in/h) (1 in. = 25.4 mm). If ½ in. of rain fell in 15 min, the intensity would be 2 in./h. If 2 in. fell in 3 h, the intensity would be 2/3 in./h. The more intense the rainfall (the greater the amount of rain/hour), the more damage the rainfall can do for a given period of time. A less intense rain over a much longer period of time may do more damage, but for any given period of time, the more intense rain will do more damage.

Duration: It denotes length of time that rainstorm occurs. The *duration of high intensity rainfalls* is usually shorter than the duration of low-intensity rainfalls. It is well known that hard, fast rainfalls do not last as long as soft, slow rainfalls.

Intensity of a rainstorm: The *Intensity of a rainstorm* during different parts of the storm is important because it has an effect on the amount of water that soil can absorb. If the storm is intense at the beginning, the large soil pores are likely to be empty and can absorb the rain at a maximum rate. If the greatest intensity occurs at the end of the storm, the soil pores will already be filled with water and infiltration will be slow or completely stopped. Water from this intense rainfall will then flow over the soil surface, causing erosion.

Frequency of rainstorms: The *frequency of rainstorms* is also important because of the effect on water infiltration. When rains occur frequently, there is not enough time for the water from previous rains to move downwards into deeper storage horizons or the water table. When the soil pores are saturated, the surplus water from succeeding rainfalls will flow downhill over the soil surface, causing erosion.

Amount of rainfall during each part of the year: Generally 4 months.

3.6.1.1 Estimation of Run-Off

Run-off is that part of the rainfall that does not infiltrate the soil and flows over the soil surface to a channel and on into a sediment pond or other body of water.

- More easily a rain enters the soil surface → means more storage space (pores) in the soil → the less water available for run-off.
- Sandy soils have large pores and usually → have a higher infiltration rate than silty or clayey soils.

If the sandy topsoil has a compacted subsoil of overburden material, the rainfall will probably be absorbed quickly until the sandy topsoil becomes saturated, at which time run-off will occur. If the overburden beneath the topsoil is not compacted, the rain will move into the overburden after filling the topsoil, and much more rain will be absorbed before run-off can occur.

Any management practice that improves *soil structure* also results in spaces within the soil for rain infiltration. The resulting water storage will decrease the amount of run-off. A decrease in the run-off means less soil erosion and more water stored in the soil for seed germination and plant growth.

- Vegetation can significantly increase infiltration.

Erosion control structures such as diversion ditches are usually designed to carry the greatest amount of run-off that is expected within a specified period, for example, *5, 10, 25, 50 or 100 years*. The most accurate method of predicting run-off from an area is to know the past run-off history for the area. Of course, this is not possible for surface mine reclamation because the watersheds are newly formed in the process of reclamation.

3.6.1.2 Estimation of Run-Off Using Rational Formula

Most of the drainage design calculations performed for reclaimed lands based on inaccurate, antique, but simple 'rational method'. This method is inappropriate for large areas because it assumes continuous rain at uniform intensity. The rational method expresses a design peak run-off in terms of the equation:

$$Q = 0.0028 \, CIA$$

where,

Q = design peak rate of run-off rate in cubic metres per second (m^3/s)

C = run-off constant (Table 3.1)

I = rainfall intensity for the time of concentration (mm/h)

A = area of watershed in hectares (ha); 0.0028, conversion constant to convert fps to SI

Time of concentration (Tc): The time required (time lag) for water to flow from the most remote (in time of flow) point of the area to the outlet once the soil has become saturated and minor depressions filled. For design purposes, the time of concentration is the same as the duration of storm of any given intensity that is likely to occur in any given time. Tc is often calculated from Kirpich formula which expresses Tc as a function of slope angle and slope length:

$$Tc = 0.0195L^{0.77}S^{-0.385}$$

where,

Tc = time of concentration (min)

L = maximum length of flow (m)

S = watershed gradient (m/m), which is calculated as the difference in elevation between source and outlet divided by length (L)

Run-off constant (C): The run-off coefficient is the proportion of rainfall not absorbed by soil. In this equation, the constant value C is used to account for the *soil texture* that affects the ease with which water can enter the soil and the amount of water that can be stored without run-off. This constant also accounts for the topography and type of cover found on the watershed. As the amount of cover on a watershed increases, the amount of run-off decreases.

Table 3.1 Rational run-off coefficient

	Value of C in $Q = CIA$		
	Soil texture		
Topography and vegetation	Sandy loam	Clay and silt loam	Tight clay
Woodland			
Flat (0–5% slope)	0.10	0.30	0.40
Rolling (5–10%)	0.25	0.35	0.50
Hilly (10–30%)	0.30	0.50	0.60
Pasture			
Flat (0–5% slope)	0.10	0.30	0.40
Rolling (5–10%)	0.16	0.36	0.55
Hilly (10–30%)	0.22	0.42	0.60
Cultivated			
Flat (0–5% slope)	0.30	0.50	0.60
Rolling (5–10%)	0.40	0.60	0.70
Hilly (10–30%)	0.52	0.72	0.82

3.6.2 Soil Erodibility Factor (*K*)

The *K* factor represents both susceptibility of soil to erosion and the amount and rate of run-off. Soil texture, organic matter, structure and permeability determine the erodibility of a particular soil. The *K* factor for a particular soil does not change. The influence of soil texture on soil erodibility factor (K) is given in Table 3.2, and combined of influence of soil texture and organic carbon is presented in Table 3.3.

The value of *K* can be determined by applying the nomograph. This would require that the percents of silt, very fine sand, total sand and organic matter be determined as well as the soil structure and permeability. These values can be determined by standard laboratory procedures. Soil organic matter reduces erodibility. Secondly, soil structure affects both susceptibility to detachment and infiltration. Permeability of the soil profile affects *K* because it affects run-off (Table 3.4). The annual distribution of rainfall erosivity directly influences seasonal values of *K*.

Table 3.2 Soil characteristics associated with *K* values

Soil type	Erodibility	*K* value range
Fine-textured; high in clay	Low	0.05–0.15
Course-textured; sandy	Low	0.05–0.20
Medium-textured; loams	Moderate	0.25–0.45
High silt content	High	0.45–0.65

Table 3.3 Soil erodibility actor (*K*) determined by soil texture and organic matter content

Soil texture	Organic matter content (%)		
	0.5	2	4
Fine sand	0.16	0.14	0.10
Loamy sand	0.42	0.36	0.28
Very fine sand	0.12	0.10	0.08
Loamy very fine sand	0.44	0.38	0.30
Sandy loam	0.27	0.24	0.19
Very fine sandy loam	0.47	0.41	0.33
Silt loam	0.48	0.42	0.33
Clay loam	0.28	0.25	0.21
Silt clay loam	0.37	0.32	0.26
Silty clay	0.25	0.23	0.19

Table 3.4 Classification of soil permeability class based on infiltration rate

Permeability class	Infiltration rate (in/h)
Rapid	5–10
Moderate to rapid	2.5–5
Moderate	0.80–2.50
Slow to moderate	0.20–0.80
Slow	0.05–0.20
Very slow	Less than 0.05

Relative soil permeability can be measured by driving a metal cylinder into a water-saturated soil and then maintaining a 0.5-in. head of water over the soil surface within the cylinder. Water is poured onto the soil surface, and the rate at which water enters the soil surface is measured in cm/h. Do not let the water fall directly on the soil surface as this may compact the soil surface and interfere with infiltration.

Permeability can be estimated by the reclamationist after some experience with local soils. Remember that permeability is controlled by soil texture, structure, porosity, surface cracks and organic matter content.

3.6.3 Slope Length and Slope Steepness Factors (LS)

The slope length (*L*) and slope steepness (*S*) factors can be considered together as a single factor (LS). The LS factor represents erodibility due to combinations of slope length and steepness relative to a standard unit plot. Slopes of non-uniform steepness require dividing the slope into segments. The standard factor is calculated based on a standard length of 22 m and a 9% slope. The values of LS for different slope length and degree of slope are shown in Table 3.5.

- If the slope is not uniform or straight, it will be concave or convex. The slope can also be composed of a combination of concave and convex segments. These irregular slopes can be divided into segments of equal length that are reasonably uniform. These values can then be added to determine the LS value for the entire slope.

- To evaluate an irregular slope, the slope is divided into segments that are of approximately equal length with each segment being approximately straight or uniform.

1.0. However, the application of mulch will immediately reduce this value to a more acceptable level. These values are given for slopes of varying percents and lengths (Fig. 3.9).

3.6.4 Vegetative Factor (*C*)

When topsoil has been replaced on a regraded overburden, the topsoil has no protection from the erosive effects of rainfall. Under these conditions, the vegetative factor (*C*) has a value of

3.6.5 Mechanical Erosion Control Factor (*P*)

The mechanical erosion control factor (*P*) is used to evaluate the effectiveness of erosion control measures, such as contour plowing, terracing and diversion ditches. These practices become more important as slopes become steeper and longer. Placing a diversion across the middle of a slope that is 200 ft long effectively reduces the slope to 100 ft in length. This will greatly reduce the amount of soil erosion that could be expected. It is obvious that terraces and diversions must be taken into account when applying the erosion equation because they change the LS factor. If no mechanical erosion control structures are used, the factor (*P*) will have a value of 1.0. In coalmine overburden dump, check dam is constructed on the slope to reduce the velocity of run-off and control of erosion (Photo 3.1).

Table 3.5 Approximate LS soil loss factor determined by the length and steepness of slope (Lyle 1987)

Slope length (m)	Degree of slope (%)	LS—soil loss factor
250	2	0.2
200	4	0.4
150	6	1.5
125	8	2
110	10	2.5
100	6	1.2
90	14	4
60	16	4
50	18	4.5
45	20	5

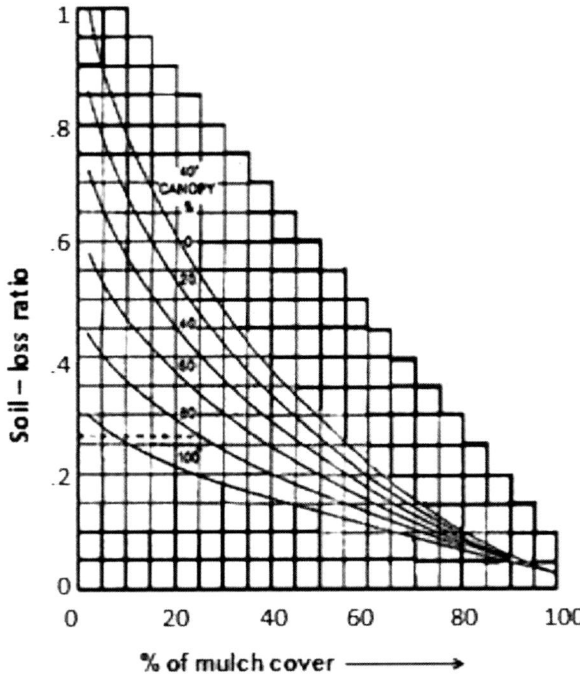

Fig. 3.9 Combined mulch and canopy effect when average fall distance of drops from canopy about 1 m (Lyle 1987)

Photo 3.1 Construction of check dam on the slope of coal mine overburden dump to reduce the velocity of run-off and check erosion

3.7 Diversion Ditch Design

Diversion ditches and terraces are frequently used to control the erosive effects of rainfall on sloping spoils. Diversion ditches have steep-sided channels that are difficult for farm machinery to traverse. These ditches can be as effective as terraces for erosion control. Diversion ditches are usually considered to be single structures used to intercept a surface water flow and divert it away from it. These ditches are usually designed to carry the peak run-off expected in a 10-year period for a 24-h duration rainfall. The three shapes are parabolic, trapezoidal and triangular (Fig. 3.10). They can be constructed by excavating a trench 25–30-cm deep, 60–90-cm wide at the top with a gradient between 45° and 60°, roughly parallel with contour.

Anyone of these shapes can be used to carry the peak run-off at a velocity that is slow enough to prevent serious erosion and yet fast enough to prevent the deposition of sediment in the channel bottom. The channel shape constructed will depend on such factors as steepness of slope, desired velocity, available equipment and mine-soil characteristics. Diversions can be designed in such a way that they will have a cross section and gradient that will move all the water without scouring the channel bottom.

Once run-off loaded with sediment is estimated, the next problem is to design drainage channels that will remove it. The basic relation for estimating the capacity of flow in any channel is:

$$Q = A * V$$

where

Q = quantity of water flowing in the channel (m^3/s)

V = average water velocity (m/s)

A = cross-sectional area of water flow (m^2)

Channels must be designed such a way that V does not exceed the threshold where erosion is triggered. Table 3.6 indicates some typical maximum non-erosive velocities in channels.

3.8 Sediment Load

It is usually not practical for the reclamationist to attempt total erosion control on a surface mine. However, most of the sediment that result from

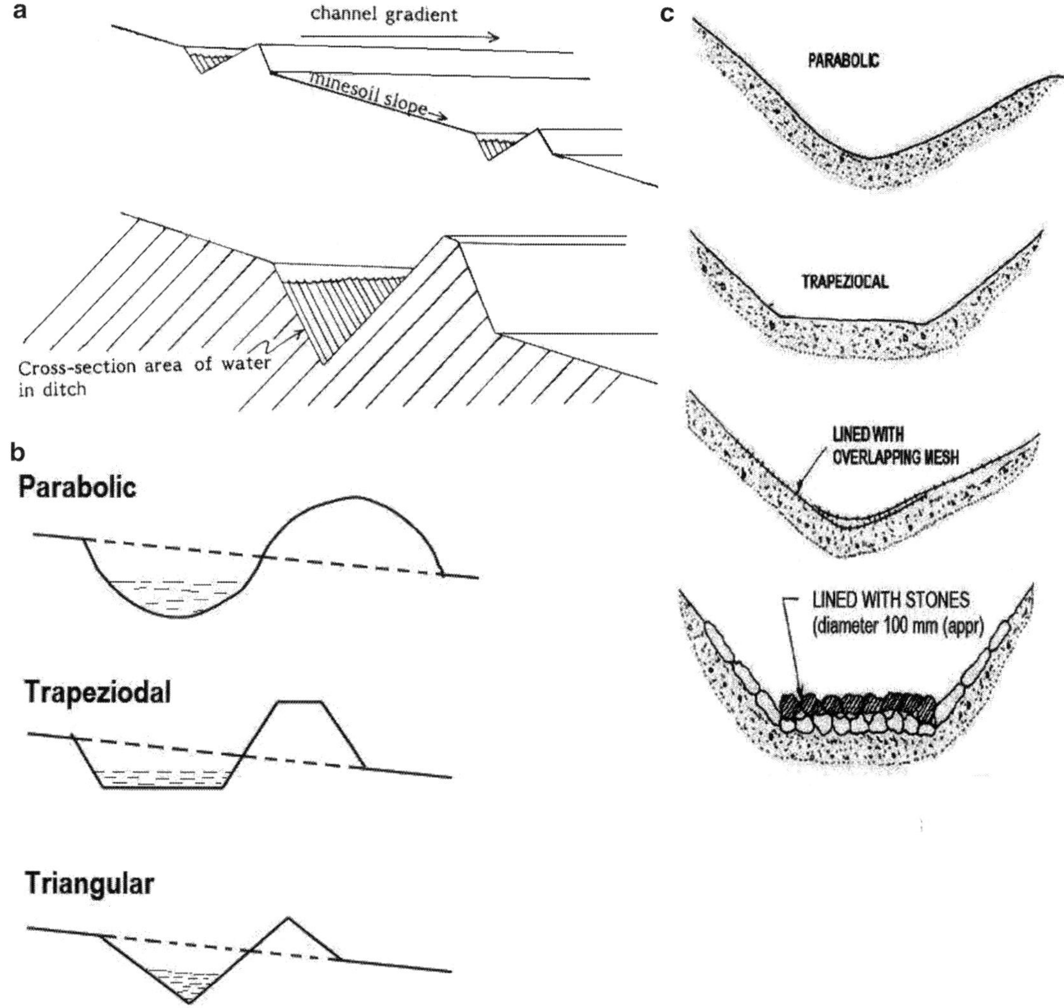

Fig. 3.10 Development of drainage pattern during technical reclamation of overburden dumps. These figures show (**a–c**) different types of drainage channels (Lyle 1987)

Table 3.6 Typical maximum non-erosive velocities in channels carrying clean water and sediment-laden water

Channel bed materials	Maximum velocity, clear water (m/s)	Maximum velocity, sediment-laden water (m/s)
Fine sand	0.46	0.46
Silt loam	0.61	0.61
Fine gravel	0.76	1.07
Coarse gravel	1.22	0.92

erosion from mines should be held on the mine site itself, otherwise, they are going deteriorate the water bodies. There are many methods and techniques for trapping sediment. Smaller amounts of sediment are often trapped by slowing water flow with check dam, stone dams, sandbag dams, ditches and small excavated pits. These are excellent methods of holding sediment at the source. If the sediment is trapped before it joins other streams outside the mine boundary, then sedimentation pond has to design within the mine boundaries.

The sediment being carried along by water run-off can be classified into three different types:

- *Wash load*: The smallest particles of sediment are classified as wash load. These particles are clay and silt size. The smaller clay particles are colloidal in size and they often stay in suspension for long periods of time causing problems. These particles can be removed by coagulation–flocculation process. Field tests will be needed to determine an optimum does and amount coagulant be added to the sediment pond. If not flocculated, much of the colloidal material will pass through the pond.
- *Sediment load*: The next larger sediment sizes are classified as suspended sediment load, and these particles can be fine sand, silt or clay. The water must be moving to keep these particles in suspension.
- *Bedload*: The Largest particles are classified as bedload. These particles are not well suspended in the water. They move in the flowing water by sliding, bounding, or rolling along the bottom of the water channel.

There are two factors that influence sedimentation process:
- Characteristics of the run-off water and types of soil particles being transported. As the velocity or the turbulence increases, there is an increase in the amount of sediment that can be carried by a given volume of water. The characteristics of the soil particles that affect sedimentation are size, shape and density.
- Size is the characteristic that is most often considered and used in calculations. Larger particle settles more quickly small particles such as clay, silt and fine sand. Fine particles are more easily carried by water than the coarser sands. The larger particles also settle more quickly and first to be deposited, when the velocity of the water 'decreases'.

3.8.1 Sediment Ponds

The final structure in the erosion control chain is the sediment pond. The purposes of these ponds are to hold the run-off with suspended sediment long enough to let the sediment settle to the pond bottom and to store the settled sediment. There are two general types of sediment pond:
- *Dry pond* is built in a drainage way that carries water during and shortly after a rainfall.
- *Wet pond* is formed by damming a permanent stream.

The dry pond is preferable because it keeps permanent streams clean, is easier to dredge and is less subject to damage.

Other Influencing Factors
1. The dam is constructed of materials that allow a minimum amount of water seepage from the pond. Vegetation and riprap are used to prevent erosion of the dam.
2. An emergency spillway is provided to allow the passage of excess water that cannot be removed by the riser. This spillway prevents the erosion and loss of the dam during excessive rainfalls. The pond itself holds sediment laden water until part, or all of the sediment is deposited on the pond bottom. The pond also acts as a storage reservoir for sediment.
3. It is difficult to provide guidelines, the location of sediment ponds on a surface mine; however, in most cases, the topography and mining operation will determine the location of pond, preferable in lowest elevation so that maximum amount of sediment-laden run-off is received by gravity.
4. In many cases, the same sediment-trapping operation can be achieved with one large pond or two or more small ponds in series.
5. An advantage of small ponds in series is that they are easier to build and maintain with smaller equipment. Another advantage is that the first pond in a series receives most of the water turbulence and sediment disturbance caused by inrushing storm water. Succeeding ponds receive less disturbance. This means that the effluent from the final pond should be relatively clean, even during peak run-off.
6. A large pond is likely to require more elaborate construction, spillways and outlets.
7. The first of a series of small ponds, where sediment first enters, is likely to hold most of the sediment, and it will be easier to clean with smaller equipment than a large pond.

The succeeding small ponds will contain less and less sediment. Ponds can also be built in parallel. With such a system, one pond can be cleaned, whilst the other pond is in use.

3.8.2 Design of Sediment Pond

The two objectives are

- To trap and hold a specified percentage of the sediment that enters the pond.
- To remove enough of the sediment to meet specified standards for sediment effluent from the pond.

In order to meet either objective, appropriate dimensions for the pond must be determined. The volume of the pond is important for sediment storage purposes, and the area of the water surface is important for sediment settling purposes. Water coming into a pond must travel far enough to the pond outlet to lose velocity and allow the sediment load to settle.

Increasing the area of the pond decreases the velocity of the water flowing out of the pond, which will allow smaller particles to settle in the pond. The distance travelled by the water also provides the detention, time needed to allow particles to settle the required distances. The following equation is used to determine the area of the pond:

$$A = \frac{Q}{V_s}$$

where

A = surface area of water in pond (cm^2)

Q = rate at which cleared water can be produced and, therefore, the rate at which water can be allowed to leave the pond (cm^3/s)

V_s = particle settling velocity (cm/s)

The value for settling velocity (V_s) is obtained by applying Stokes' law:

$$V_s = \frac{g(S - 1)\, D^2}{18\mu}$$

where

V_s = settling velocity of particle (cm/s)

g = acceleration of particle due to gravity (981 cm/s^2)

D = diameter of particle (cm) (from Table 3.7) (sand = 50–2,000 micron; silt = 2–50 micron; clay = <2 micron;)

S = specific gravity of particle (2.65)

μ = kinematic viscosity of water (cm^2/s)

Assumptions.

- It assumes that the particles are smooth and the water is not moving, but both of these assumptions probably will be false, but the values obtained are still useful.
- In order to apply this equation, the minimum size soil particle (D) that is to be removed, water viscosity (μ) and the specific gravity of the particle (S) must be known.
- The particle of smallest diameter that is practical to remove by settling is probably the smallest silt (0.0002 cm). These particles will settle about 17 cm in 24 h in still water at 0°C.
- Most clay particles will take from 30 h to several years to settle this distance.
- The specific gravity (S) for soil particles will be approximately 2.65, and this is the value often used in Stokes' equation.

Example 1. Assume that the smallest particle to be removed from the run-off is the smallest silt fraction (0.0002 cm). The lowest winter

Table 3.7 Size classification of different types of soil particles

Soil particle	Diameter range (cm)
Very coarse sand	0.2–0.1 cm or, 2 mm–1 mm (2,000–1,000 μm)
Coarse sand	0.1–0.05 cm (1,000–500 μm)
Medium sand	0.05–0.025 (500–250 μm)
Fine sand	0.025–0.01 (250–100 μm)
Very fine sand	0.01–0.005 (100–50 μm)
Silt	0.005–0.0002 (50–2 μm)
Clay	<0.0002 (<2 μm)

Table 3.8 Values for coefficient K (run-off coefficient) for soil texture and slope, namely, land use

| | Soil texture and Slope (%) | | | | | | | | |
| | Sandy loam | | | Clay loam and Silt loam | | | Fine clay | | |
Land use type	0–5%	5–10%	10–30%	0–5%	5–10%	10–30%	0–5%	5–10%	10–30%
Forest	0.10	0.20	0.30	0.30	0.40	0.50	0.40	0.50	0.60
Pasture	0.10	0.20	0.30	0.30	0.40	0.50	0.40	0.50	0.60
Bare minesoil	0.30	0.40	0.50	0.50	0.60	0.70	0.60	0.70	0.80

temperature is 0°C (32°F), and the specific gravity of the particles is 2.65. Determine the length of time that the sediment will have to be detained by the settling pond in order to remove *all sand and silt* in the run-off. (Kinematic viscosity of water at 0°C = 0.01794 cm^2/s)

$$Vs = g(S - 1)D^2/18\mu$$

$$Vs = \frac{981(2.65 - 1)(0.0002)^2}{(18)(0.0170)}$$

$$Vs = 0.0002009 \text{ cm/s}$$

$$= 2.009 \times 10^{-4} \text{cm/s}$$

$$= 2.009 \times 10^{-6} \text{m/s}$$

$$= 0.000002 \text{ m/s}$$

Therefore, for settling in 24 h, depth requirement will be calculated as

$$= 24 \text{ h settlingtime} \times Vs$$

$$= (86,400 \text{ s}) \times (0.0002009 \text{ cm/s})$$

$$= 17.3577 \text{ cm. (Thus, particles will settle about}$$

17 *cm in 24 h in still water at* 0°C).

This settling rate (Vs) can now be used to determine the needed surface area (A) of the water in the pond. The equation for this surface area is calculated as

$$A = Q/Vs$$

where
A = surface area of water in pond (m^2)
Q = rate of run-off for a predicted storm (m^3/s)
Vs = critical settling velocity (m/s)

The peak rate of run-off (Q) must be computed in order to use the equation, and the time period for computing the run-off is usually the expected run-off in a 10-year period for a 24-h duration storm. The amount of run-off (R) must

be computed first. The rational formula in a slightly different from that shown in the run-off section can be used:

$$R = K P A$$

where
R = amount of run-off (m^3)
K = run-off coefficient
P = rainfall (m)
A = run-off area (m^2)

Example 2. Assume that the land degraded site is located in the eastern part of India, where the expected rainfall would be 127 mm in a 10-year period of 24-h duration storm. The soil is a silt loam for forest and pasture, and the watershed consists of 10 ha of forest with an average slope of 20%, 5 ha of pastures with an average slope of 6% and 10 ha of minesoil with an average slope of 9% having sandy loam texture (1 acre = 0.4047 ha). Use Table 3.8 for value of K (run-off coefficient) under different soil texture and slope.

1. Calculate the amount of run-off (R) for forest, pasture and minesoil.
2. Calculate total run-off expected in a 24 h period.
3. Determine the rate of run-off.
4. Determine duration of intense rainfall and peak run-off.
5. Determine the surface area (A) of water needed in the sedimentation pond to remove all sediment of 0.0002 cm in diameter or greater. At 25°C, kinematic viscosity (μ) of water = 8.97 $\times$ 10^{-3} cm^2/s, specific gravity of soil particle = 2.65.
6. Determine the minimum depth of the water-holding portion of the pond (in m).
7. Assume safety factor of 1.2 for the final surface area calculation; provide L:W = 2:1,

restrict the depth preferably within 8 m. Assume pond with a length of 150 m or less should have a 0.3-m freeboard.

Solution:

1. Estimate the peak run-off amount (R) for silty loam with different degree of slope, forest (run-off coefficient, $k = 0.5$ for 20% slope), pasture ($k = 0.4$ for 6% slope) and minesoil ($k = 0.6$ for 9% slope).

$$R \text{ for forest} = 0.5 \times 0.127 \text{ m} \times 10 \times 10^4 \text{ m}^2$$
$$= 6,350 \text{ m}^3$$
$$R \text{ for pasture} = 0.4 \times 0.127 \times 5 \times 10^4 \text{ m}^2$$
$$= 2,540 \text{ m}^3$$
$$R \text{ for minesoil} = 0.4 \times 0.127 \times 10 \times 10^4 \text{ m}^2$$
$$= 5,080 \text{ m}^3$$

Total amount of peak run-off from the watershed = (forest + pasture + minesoil)
= **13,970 m^3**.

2. To determine the rate of run-off for the entire period, the total run-off (19,050 m^3) is divided by the seconds of 24 h (86,400 s) to give = **0.16169 m^3/s**.

3. However, as most of the rainfall will take place within a short period of the 24 h, the following equation is used to estimate the period of time when the greatest amount of run-off occurs:

$$T = 6,000(Es - 0.20)$$

where

T = duration of intense rainfall (s);

Es = total rainfall in inches for a 24-h events based on 10-year period
(convert 127 mm rainfall into inches; 127 mm = 12.7 cm = 5 in)

$$T = 6,000(5 \text{ inch} - 0.20) = 28,800 \text{ s}$$

The peak run-off (Q) is estimated by dividing total run-off by the duration of intense rainfall:

$$Q = \frac{R}{T} = \frac{13,970 \text{ m}^3}{28,800 \text{ s}} = 0.48507 \text{ m}^3/\text{s}$$

4. To determine the surface area (A) of water needed in the sedimentation pond, to remove all sediment having 0.0002 cm in diameter (silt particle) or greater.
Calculate Vs for silt particle = 0.0002 cm, at 25°C, kinematic viscosity (μ) of water = 8.97 $\times$ 10^{-3} cm^2/s at 25°C.

$$V_s = \frac{981 \times (2.65 - 1) \times (0.0002)^2}{18 \times 8.97 \times 10^{-3}}$$
$$= \frac{6.4746 \times 10^{-5}}{161.46 \times 10^{-3}}$$
$$= 0.0401 \times 10^{-2} = 4 \times 10^{-4} \text{cm} = 4 \times 10^{-6} \text{ m}$$
$$A = \frac{Q_{\text{peak}}}{V_s} = \frac{0.48507 \text{ m}^3/\text{s}}{4 \times 10^{-6} \text{m/s}}$$
$$= 121,268 \text{ m}^2 (\text{surface area})$$

5. The minimum depth (D) of the water-holding portion of the pond (in m)

$$D = \frac{\text{Peak run - off volume}}{\text{Surace area of water in Pond (A)}}$$
$$= \frac{13,970 \text{ m}^3}{121,268 \text{ m}^2} = 0.1152 \text{ m} = 11.52 \text{ cm}$$

The area is too large for most situations, and depth is too shallow. Also the shallow depth would allow scouring and re-suspension of the sediment that has been deposited on the bottom of the pond. If the smallest diameter particle to be trapped is **very fine sand** **(diameter = 0.005 cm)**, *the size of the pond can be reduced drastically.*

6. Recalculate the surface area and depth of the sedimentation pond by considering the very sand particle having diameter of 0.005 cm. For fine sand,

$$V_s = \frac{981 \times (2.65 - 1) \times (0.005)^2}{18 \times 8.97 \times 10^{-3}} = 0.25 \text{ cm/s}$$
$$= 0.25 \times 10^{-2\sim} \text{m/s}$$
$$A = \frac{Q_{\text{peak}}}{V_s} = \frac{0.48507 \text{ m}^3/\text{s}}{0.25 \times 10^{-2} \text{m/s}}$$
$$= 194 \text{ m}^2 \text{ (Surface area)}$$

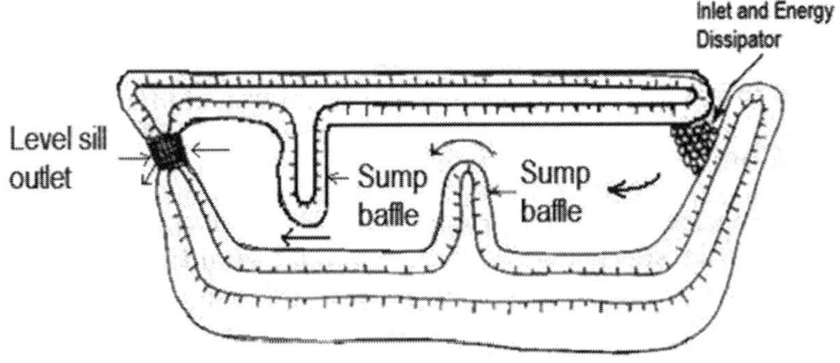

Fig. 3.11 Sedimentation basin with earthen baffles to increase distance of flow

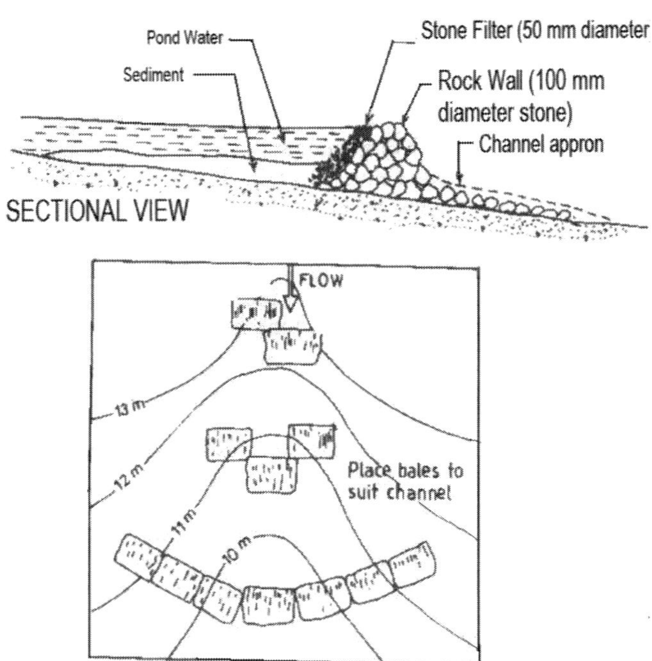

Fig. 3.12 Two simple types of sediment trap hay bales and a stone filter dam

$$D = \frac{\text{Peak run - off volume}}{\text{Surace area of water in Pond (A)}}$$

$$= \frac{13,970 \text{ m}^3}{194 \text{ m}^2} = 72 \text{ m(Depth)}$$

All the calculations up to this point have been based on an ideal settling velocity of particles in still water, with controlled movement of incoming water to the outlet and no dead areas. In order to compensate for disrupting factor, the final surface area should be increased by a factor of **1.2.**

Thus, new surface area (A) $= 194 \times 1.2$

$$= 232.8 \text{ m}^2$$

The depth can now be decreased for the same run-off storage volume:

$$D = \frac{13,970 \text{ m}^3}{232.8 \text{ m}^2} = 60 \text{ m(Depth)}$$

The pond surface area is now very small (15.26 $\times$ 15.26 m), but excessively deep (60 m). Assuming that there is a sufficient space available to build a pond, that is, 30.5-m (100-ft) wide and 61-m (200-ft) long, this would give pond surface area of 1860.5 m^2 (30.5 m $\times$ 61 m).

$$\text{New Pond depth} = \frac{13{,}970 \text{ m}^3}{1860.5 \text{ m}^2} = 7.50 \text{ m}$$

Therefore, dimension of sedimentation pond = 61 m (length) $\times$ 30.5 m (width) $\times$ 7.5 m (deep). Provide 30-cm freeboard all along the pond to get rid-off splashing of water due to wind or storm. This new depth would be more practical, and increase surface area would decrease water turbulence, increase time for sediment settling and remove particles smaller than fine sand.

3.8.3 Sediment Retention

Run-off discharging from a quarry which is sediment laden should be passed through a sediment basin. The sediment pond should be at least three times longer than it is wide. Effective flow length within the pond can be increased by inserting baffles (earthen walls) at intervals to divert flows along a zigzag path (Fig. 3.11).

3.8.4 Sediment Trap

Besides installation of sedimentation basin to retain sediment from major concentration of flow, simple traps can be designed to prevent sediment leaving a disturbed site in overland run-off. These traps include vegetation filter comprising stands of grass of other dense ground cover, filter dams, etc. A view of sedimentation basin with earthen baffles to increase the distance of travels that gives maximum efficiency of removal of sediment is shown in Fig. 3.12.

References

Bochet E, Garcıá-Fayos P (2004) Factors controlling vegetation establishment and water erosion on motorway slopes in Valencia, Spain. Restor Ecol 12(2):166–174

Chaudhuri S (2008) Planning for closure of coal mines. In: Chaudhuri S, Singh G (eds) Environmental management in coal mining areas. EDC programmes, ISM Dhanbad

Coppin NJ, Bradshaw AD (1982) Quarry reclamation. Mining Journal Books, London

Jonathan B et al (2006) Influence of slope and aspect on long-term vegetation change in British chalk grasslands. J Ecol 94(2):355–368

Khandelwal NK, Mozumder BK (1987) Stability of overburden dumps. J MMF (Sp issue):253–260

Lyle ES Jr (1987) Surface mining reclamation manual. Elsevier, New York

Maiti SK (2010) Revegetation planning for the degraded soil and site aggregates in Dump sites. In: Bhattacharya J (ed) Project Environmental Clearance, Wide Publishing, Kolkata (India), pp 189–228

Sidari M, Giuliana R, Giuseppe V, Adele M (2008) Influence of slope aspects on soil chemical and biochemical properties in a *Pinus laricio* forest ecosystem of Aspromonte (Southern Italy). Eur J Soil Biol 44 (4):364–372

Contents

4.1   Introduction

The requirements for the growth of plant are very simple. They require soil for growth of roots and source of nutrients and moisture, while sunlight and carbon dioxide are required for the carbohydrate production (Fig. 4.1). Soil serves the need of the plant by providing support medium for growth (stability), gaseous exchange for roots, water and nutrients which are dissolved in water. However, plant establishment and growth on mine spoils are often limited due to physical factors rather than chemical imbalances (Dollhopf and Postle 1988).

Important soil physical properties that are known to affect plant growth on mine spoils include:

1. Temperature
2. Stoniness and textural classes
3. Bulk density, pore space and compaction
4. Moisture
5. Infiltration rate
6. Rooting depth

S.K. Maiti, *Ecorestoration of the Coalmine Degraded Lands*,
DOI 10.1007/978-81-322-0851-8_4, © Springer India 2013

4.2   Physical Factors

4.2.1   Aspects of Slope and Temperature

The lower and upper temperature considered for plant growth is 5 and 50 °C. Spoil temperature is considerably affected by slope aspect, colour, wetness and plant cover. A south-facing slope attains higher temperature at the excess of 50 °C in summer than north-facing slope. Figure 4.2 shows examples of such measurement in a British colliery waste heap; it may be noted that a 35° south-facing slope exceeded nearly by 12 °C the daily maximum temperature on an adjacent area (Down and Stock 1987).

4.2.2   Stoniness and Textural Class

Stoniness: Higher proportion of stones and rocks dilutes the volume of soil within the root depth which is available for water storage and nutrients. Stones range in size from boulders to fine gravel (down to 2-mm diameter). If the volume of stones is more than about a third of the volume of the soil, then it likely has significant effect on water storage and nutrient availability to the plant. The effect will be particularly severe if stones are closely contact with one another, which reduces the volume of soil available for roots to explore for water and nutrients. Especially they have an important effect on water storage capacity in seasonally dry climates and minesoil. The stoniness fraction in overburden dump materials is generally found between 40 and 60%, sometimes as high as even 80–85% (Maiti 1995, 2007).

The texture is a measure of the size of mineral particles and refers to the relative proportion of particle sizes in a given soil. Natural soils usually posses a fairly wide grade of particle size, but it is very common for mine waste to exhibit an unbalanced distribution. The particle sizes (diameter) of the OB dump materials above 5.6-mm range are put under the textural class 'gravel', which predominates within the top 15 cm of depth, then gradually declines with increasing depth.

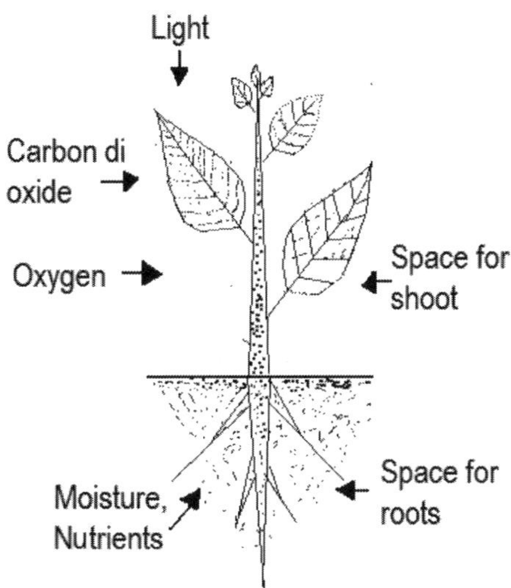

Fig. 4.1   Requirements of plant for growth

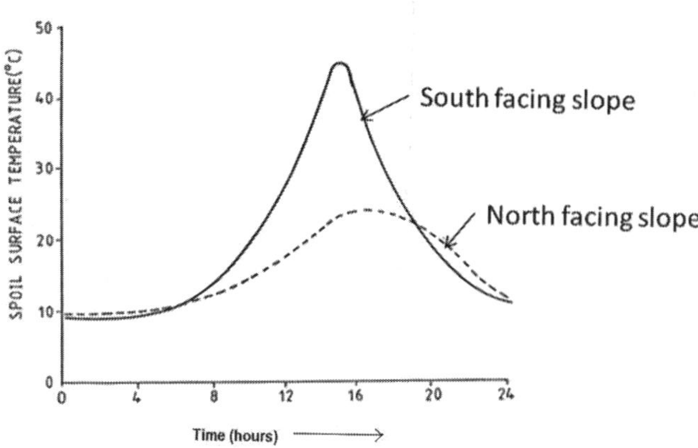

Fig. 4.2   Diurnal temperature fluctuations at the surface of a colliery waste tip showing the difference between north- and south-facing slopes (Down and Stock 1987)

Table 4.1 Particle size distribution (%) in OB dumps (Maiti et al. 2002)

Profile depth (cm)	>5.6 mm	>2 mm	>1 mm	>0.5 mm	>0.212 mm	>0.15 mm	<0.15 mm
0–15 cm	51.23	15.33	3.71	5.77	16.16	5.62	1.98
15–30 cm	45.88	22.55	4.55	6.71	16.43	2.8	6.67
30–50 cm	40.56	17.4	4.2	7.42	23.33	2.4	4.69

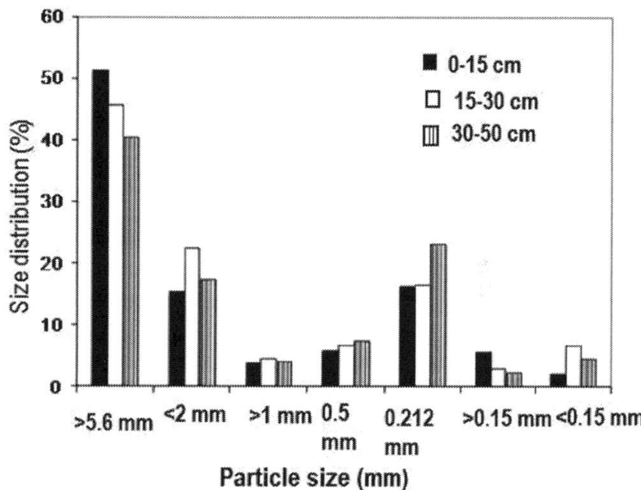

Fig. 4.3 Particle size distribution in overburden dump materials (Maiti et al. 2002)

Table 4.2 Particle size distribution (%) of levelled mined-out derelict sites (Maiti et al. 2002)

Profile depth (cm)	>5.6 mm	>2 mm	>1 mm	>0.5 mm	>0.212 mm	>0.15 mm	<0.15 mm
0–15 cm	17.58	27.32	6.07	8.38	22.3	3.46	14.88
15–30 cm	42.73	18.26	4.32	6.96	18.46	3.87	5.4
30–50 cm	27.77	20.18	5.0	7.58	26.67	3.37	9.45

This phenomenon may be attributed to physical weathering.

In the soil science parlance, gravels are termed 'skeleton material' as they possess little retention capacity of either moisture or nutrients. But surely they enhance the infiltration capacity. The Unified Soil Classification System (USCS) designates inorganic particle size of less than 2 mm as soil. For standard particle size measurement, the soil fraction that passes a 2-mm sieve is considered as soil. Laboratory procedures normally estimate percentage of sand (0.05–2.0 mm), silt (0.002–0.05 mm) and clay (<0.002 mm) fractions in soils. Soil size fraction of dump materials, particles having diameter range between 0.212 and 0.5 mm dominate (Maiti et al. 2002). Particles in this size range exert major influence on both moisture holding and nutrient retention capacity. Variation of particle size distribution versus depth in one of the overburden dump and mined-out area of Jharia coalfield is given in Table 4.1 and Fig. 4.3. The non-soil fraction (diameter >2 mm) in the top 15-cm layer has been observed to be 56–65%. The fraction constitutes 68.43% at 15–30-cm depth and 58% at 30–50-cm depth. Reclamationists, including Hu et al. (1992), rated soil with more than 50% stoniness as poor quality. In contrast, levelled mined-out sites non-soil portion was found to be less, that is, —45% (0–15 cm), 61% (15–30 cm) and 48% (30–50 cm) (Table 4.2 and Fig. 4.4). The finer-size fraction was found highest at the top 15 cm due to weathering. This type of spoil texture definitely encourages growth of grass and herbs.

Fig. 4.4 Particle size distribution in levelled mined-out derelict sites (Maiti et al. 2002)

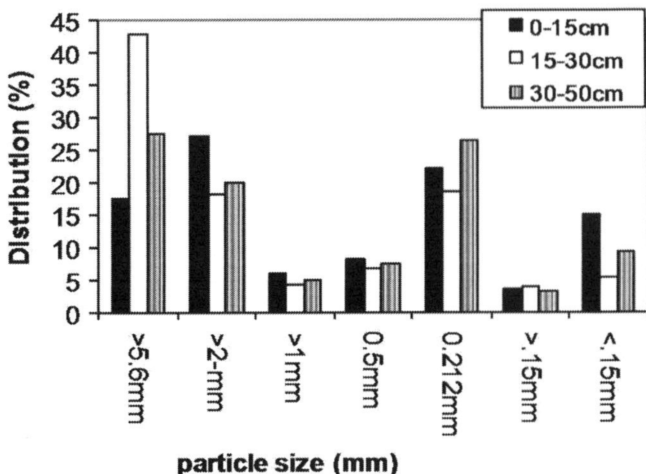

Table 4.3 Texture analysis of overburden profile, Rajapur opencast project, BCCL (Maiti 1995)

Sl. no	Profile depth (cm)	Stoniness (%)	Soil (%)	Sand (%)	Silt (%)	Clay (%)
1.	0–15	76.2	23.8	63.5	22.5	11.4
2.	15–30	80.0	20.0	58.0	31.0	11.0
3.	30–45	61.1	38.9	60.5	30.3	9.2
4.	45–60	66.7	33.3	66.0	24.5	9.5
5.	60–75	56.8	43.2	76.0	13.9	10.1
6.	75–90	56.3	43.8	61.0	24.4	13.6
7.	90–105	50.0	50.0	64.0	22.7	11.3
Mean		63.9 ± 11	36.1 ± 11	64.4	24.3	11.3

Mine spoil contains higher percentage of gravels followed by coarser sand and lower silt and clay content. The average percentage of gravel, sand, silt and clay reported for some of the Indian coal mine spoils was as follows: 12–20%, 25–30% and 52–60% (Parej OCP, CCL); 25, 31 and 44% (Bina, NCL). The above percentages depend on the geology of the overburden strata and mining conditions. However, in Rajapur OCP of Jharia coalfield, gravel contents exceed more than 60%. In KD Haslong OB dump (North Karanpura, CCL), the gravel contents were reported 60%, silt 25% and clay only 6% (Maiti 2006a, b). Maiti (1995) studied the reclaimed OB dumps on profile basis and found average stoniness about 64% (50–80%) (Table 4.3). However, several researchers reported lower clay content, lack of structure, high bulk densities, low water-holding capacity and poor physical conditions of mine spoils even after several years of revegetation.

4.2.3 Bulk Density, Pore Space and Compaction

The knowledge of soil bulk density and porosity is essential to the reclamationist for the understanding of root growth in the ecorestored site. Soil pores contain air for plant root metabolism and provide a reservoir for water storage. While bulk density values are used for conversion of soil mass and soil volume, it is also used to calculate the total pore volume in a soil as well as the weight of soil to move during an excavation. High bulk density values may indicate the presence of compact layers that could restrict root growth and water penetration.

There are two densities commonly used in soil analysis:
- Bulk density
- Particle density

Bulk density is defined as the dry mass of soil per unit volume, which includes both soil solids and pore space. It is the density of soil which exists naturally and includes air space, organic matter and soil solids. The bulk density (B_d) is calculated as

$$B_d = \frac{Oven\ dried\ soil\ mass\ (80^\circ C),\ g}{Soil\ volume,\ cm^3}$$

As discussed earlier, B_d affects the root growth due to compactness of soil and reduces water availability. Natural root growth is restricted when B_d becomes greater than 1.5 g/cm^3 in fine-texture (clay and silt) soil and 1.7 g/cm^3 course-textured soil (sandy) (Lyle 1978). B_d less than or equal to 1.3 g/cm^3 is good, between 1.3 and 1.55 g/cm^3 is fair and greater than 1.8 g/cm^3 is considered as extremely bad for the growth of plants. Bulk density values are affected by soil texture and modified by structure. In sandy soil, large-size particles are fairly packed in a defined volume, which contain few micropores but many macropores. The small aggregated soil particles in a fine-textured soil tend to be separated by many micropores. However, soils with well-developed structure, large macropores may exist between the peds.

Particle density is the density of solid soil particles only, that is, measurement does not include pore space (air space). The standard value of particle density is 2.65 g/cm^3. It is measured as the weight of a known volume of solids without voids. It is determined by adding a known weight (say about 100 g) of oven-dried unsieved soil to a partly filled measuring cylinder (250 mL) and noting the displacement volume of water (after stirring to remove air bubbles).

Porosity: The pore space of soil is that portion of the soil volume occupied by the air and water (Brady 2000). Soil pores are cracks or tubular holes in the soil filled with water or air. They can store water or circulate air to roots, and larger pores drain excess water. Thus, porosity has a substantial effect on infiltration and water-holding capacity. There are two categories of soil pores: (1) *micropores*—pores size less than 0.06 mm in diameter that are important for water retention and (2) *macropores*—pores greater

Table 4.4 Bulk density of mine spoils (Maiti 2003)

Location of dumps	Bulk density (g/cm^3)
Parej project (CCL)	1.93–2.0
KD Haslong (CCL)	1.7–1.74
Bina project (NCL)	1.51
Kusunda project (BCCL)	1.89–2.16.
Other areas	
Forest soil	1.13–1.20
Grass land	1.20–1.28
Cultivated land	1.35–1.48.

than 0.06 mm in diameter that are important for aeration and saturated water flow. It is calculated as follows:

$$Pore\ space\ (\%) = 100 - \left(\frac{Bulk\ density}{Particle\ density} \times 100 \right)$$

Examples Calculation of bulk density with the given data

Cylinder height = 5.0 cm
Cylinder inside diameter = 24.4 cm
Weight of oven-dried soil = 87.6 g
Volume of the cylinder = 3.14 $\times$ (12.2)^2 = 76 cm^3
Bulk density = 87.6 g/76 cm^3 = 1.15 g/cm^3

Vogel (1987) stated that high bulk density is the most troublesome feature of reconstructed soil on surface-mined land. Surface bulk densities of minesoil are usually greater than those of undisturbed soils because of higher compaction, lack of structure and higher coarse fragment content. Generally, the B_d of coal mine dumps was found higher in the order of 1.93–2.0 g/cm^3 (Table 4.4). Maiti (2007) analysed the bulk density for different types of minesoil and found that it is significantly higher than topsoil (Table 4.5).

Why B_d Happens to Be Higher in Ecorestored Site?

During overburden handling operations, i.e, regarding and topsoil replacement by heavy equipment under wrong conditions (e.g. high moisture contents) causes surface compaction.

Table 4.5 Comparative analysis of soil fraction, bulk density and field moisture content in reclaimed, unreclaimed OB dumps and topsoil in a coalmine area (KD Heslong project, CCL) (Maiti 2007)

Sampling locations	Soil fraction (%)	Bulk density (Mg m^{-3})	Field moisture (%)
Reclaimed OB dumps	44.4[c]	1.76[ab]	6.08[a]
Unreclaimed OB	46.9[bc]	2.04[a]	3.24[b]
Topsoil	56.2[ab]	1.45[b]	6.88[a]
Fresh OB	64.8[a]	*Not analysed*	*Not analysed*
5% LSD	11.7	0.39	1.23

Number of samples analysed = 10
Superscript letter represents analysis of variance (ANOVA)
Values with different letter indicate significant difference at $p < 0.05$

In compacted spoil, the particles are packed closely together and have a few macropores. This will cause less water storage space and less space for air movements into the spoil. Finally, if compacted spoil surface is formed, it will be harder for the roots to penetrate than a loose soil.

If B_d Is Not Corrected?

High bulk density favours the growth of weeds because most of weeds have superficial root system and restricted to 10–15-cm depth. Secondly, it reduces soil infiltration and enhances surface run-off thus increasing soil erosion. Undisturbed soils tend to be more porous than minesoils, with a developed system of cracks and fissure; however, pore in minesoils are typically larger (macropores). Wet mine spoils compact more easily than dry spoil because water acts as lubricant. Only waste with a high clay content is liable to become over-compacted even with light machinery.

4.2.4 Moisture Content

Moisture content in mine spoil plays 'key role' for the plant growth. It acts as reservoir of water required for plant growth as well as source of nutrients which is dissolved in it. Most of the soil reactions like weathering, cation exchange, mineralisation and fertilisation are taking place through soil solution. Soil temporarily stores water, making it available for root uptake, plant growth and habitat for soil organisms. Thus, proper assessment must be made and correction measures be taken accordingly. Most of the reclamationist says moisture content 'is the only limiting factors for plant growth and establishment on dumps'.

Soil moisture retention capacity of minesoil depends on:
- Texture of soil—the higher the clay and silt, the higher the capacity
- Amount of soil particles
- Higher organic matter
- Litter accumulation
- Vegetation cover
- Topography and slope

Water potential: Water is held in soil by strong cumulative forces of the H bond. The adhesive forces (between water and minerals) are very strong near the mineral surface, and cohesive forces (between water molecules) occur throughout the water films. Because the water held to soil particles has less freedom than free water (water potential = 0), it is measured in negative bars. The strength at which water is held in the soil is called *water potential*:
- Gravitational water = $>-1/3$ bars
- Plant-available water = $-1/3$ bars to $-1/15$ bars
- Unavailable water = -15 bars
- Field capacity = amount of water held at $-1/3$ bars
- Wilting point = -15 bars

In majority of cases, mine spoils have very low moisture contents (3.4–4.4%), and in dry summer, it comes down to less than 3% (Maiti 2003). In a 5-year-old overburden dump of Jingurda mine of NCL, Jha and Singh (1992) reported moisture contents in the range of 2.6–5.4%. In case of Bina mine OB dump (NCL), it was found only 4.4%. In OB dumps of BCCL area (Rajapur dumps), it was found only 2.3–6% during lean seasons (Maiti 1995). In other land use, like forest, grassland and cultivated land, this value was found 12–15%, 13–16% and 10–12%, respectively. The moisture contents in different Indian coal mining OB dumps are presented in Table 4.6.

In the USA, vegetation potentiality of mine spoils is categorised on the basis of available water capacity (AWC) (Schafer 1979): AWC > 10%, good, 5–10% fair and <5% is poor. If moisture content is low, amendments like organic manures, mulching, hay materials (paddy straw and rice husk) and saw dust have to be practised. There are other methods also available under the 'in situ moisture conservation practices'.

Barnhisel (1979) reported that water availability to plants in mine spoils is less than in unmined or otherwise disturbed soil (Table 4.7). Maiti (2008) also studied moisture content in the reclaimed OB dumps and other areas vicinity to the mine and found that with increase depth, moisture content also found to be increased 2–3% more from the surface (Table 4.8).

4.2.5 Infiltration

Infiltration is defined as the vertical movement of water into the soil, and velocity at which water is entered is called infiltration rate and typically expressed in cm/h. An infiltration rate of 2 cm/h means that a water layer of 2 cm on the soil surface will take 1 h to infiltrate. In dry soil, water infiltrates rapidly. As more water replaces the air from the soil pores, the water from the soil surface infiltrates more slowly and eventually reaches a steady state. The infiltration rate controls water content of mine spoils and the amount of surface run-off. The rate at which water can enter into the spoil primarily depends on porosity. The other factors that control the rate of water movement into the spoils are (Donahue et al. 1990):

1. The percentage of sand, silt and clay—coarse sands permit rapid infiltration.
2. Soil structure.
3. Organic matter—the greater the organic matter, the higher the infiltration.
4. The depth of soil to hardpan, bedrock or other impervious layers.
5. Amount of water in the soil.
6. Vegetation cover.
7. Temperature—warm soil takes more water.
8. Compaction, which usually reduces pore space, slows infiltration.

In mine spoil, vegetation cover can considerably increase the infiltration rate, by reducing velocity of surface run-off and increasing the pore space. Long-term solutions for maintaining or improving infiltration include practices that increases soil organic matter, enhances soil aggregation, reduces soil disturbance and compaction. The rate of infiltration has been classified as follows (Donahue et al. 1990):

Category of infiltration	Rate of infiltration (cm/h)
1. Very slow	<0.25
2. Low	0.25–1.25
3. Medium	1.25–2.5
4. High	>2.5

Table 4.6 Field moisture (%) content of coal mining spoil materials

Mine spoils	Field moisture (%)
Jharia coalfields	2.3–6
Bina, NCL	4.4
Jingurda mines, NCL	2.6–5.4
KD Heslong project, CCL	3.24–6.08

Table 4.7 Relative availability of water in mine spoils, disturbed soil and undisturbed soil in the Kentucky area, USA (Barnhisel 1979)

Sample	Field capacity (%)	Wilting point (%)	Available water (%)
Mine spoils ($n = 238$)	19.8	8.8	11
Disturbed soil (86)	33.7	13.5	20.2
Undisturbed soil ($n = 48$)	36.9	13.8	23.1

n = nos. of observation

Mine spoils usually have low infiltration rate (IR) due to high bulk density and low porosity (Maiti 1995). In Parej dumps (CCL), low IR rate in the range of 1.65–4.5 cm/h has been reported. In other OB dumps, the infiltration rate was also reported very low: 1.60 cm/h (Bina, NCL), 0.9–1.2 cm/h (KD Heslong, CCL), etc. As per Hu et al.'s (1992) classification, infiltration rate in coal mine overburden dump could be categorised under 'slow category'. Maiti (2008) conducted infiltration test for reclaimed dumps, agriculture land, wasteland and natural sal (*Shorea robusta*) forest of IB valley area (MCL)

and found that due to compaction in reclaimed dumps, IR was found to be lowest (0.275 cm/h), while in agriculture land and sal forest, IR value was found to be 1 cm/h and 2.3–3.4 cm/h, respectively (Table 4.9). Maiti (2008) opined that infiltration rate in the reclaimed coal mine overburden dumps can be categorised as 'very low'. A typical infiltration test reading along with infiltration curve is shown in Table 4.10 and Fig. 4.5.

4.2.6 Soil Depth (Rooting Depth)

Soil depth can be defined as 'the distance from the soil surface to anything that prevents a plant root from growing and absorbing water and nutrients'. A layer of rock, toxic chemicals, tight clays or water table can be the factor that prevents roots from penetrating deeper into a soil. Soil depth are classified under the following suggested heading (Lyle 1978). Rooting depth always includes topsoil, which is usually 15–25 cm and distinguished from subsoil by structure, darker colour and lesser root density.

Table 4.8 Moisture content under different land use (Maiti 2008)

| | Profile depth (cm) | | |
Samples	0–20	20–40	40–60
Reclaimed OB dumps	4.5	4.65	5.24
Agriculture soil	5.70	6.70	7.01
Waste land	3.89	6.01	6.56
Sal forest[a]	7.22	7.08	7.84
Sal forest[b]	7.47	6.47	8.36
Forest land[c]	4.97	8.24	9.52

[a]Ananta OC Expansion, MCL
[b]Bhubaneswari OCP, MCL
[c]Kaju and sal forest, Talabira project, MCL

Table 4.9 Infiltration rate and cumulative infiltration (Maiti 2008)

Sampling location	Initial moisture (%)	IR (cm/h)	Cumulative IR (cm/h)
Ananta project, IB valley, MCL			
(a) Reclaimed OB dumps (*C. siamea, A. auriculiformis, A. arabica*, etc.; 2/ 3-year-old plantation)	3.6	0.283	14.863
(b) Topsoil (mine face)—wasteland (core zone)	4.2	0.991	19.858
(c) Sal forest	4.8	2.760	92.382
(d) Sal forest	4.3	2.972	100.97
Bhubaneswari project, IB valley, MCL			
(a) Sal forest (buffer zone)	4.5	2.265	110.757
(b) Agriculture land (core zone), Jhilinda village	5.2	0.991	25.265
(c) Wasteland (core zone), Naraharipur village	4.9	0.991	28.358
Talabira project, IB valley MCL			
(a) Wasteland (Khinda village)	9.08	1.019	20.132
(b) Sal forest (near Budia palli village)—core zone	5.8	3.397	116.277
(c) Agriculture land (Dumra Munda village)	5.79	1.019	28.033
(d) Wasteland (dominated by kaju and sal tree)	4.9	1.840	87.473
Reclaimed OB dumps, Rajapur, BCCL	3.7	0.27	14.21

Table 4.10 Typical data set for the infiltration test conducted in a reclaimed OB dumps (*C. siamea, A. auriculiformis, A. arabica*, etc.; 2/3-year-old plantation) (Maiti 2008)

Elapsed time (min)	Time interval(min)	Time interval (h)	Amount of water added (mL)	Infiltration amount (cm)	IR (cm/h)	Cumulative IR (cm/h)
5	5	0.083	540	0.764	9.172	9.172
15	10	0.167	300	0.425	2.548	11.720
40	25	0.417	290	0.410	0.985	12.705
70	30	0.500	210	0.297	0.594	13.299
100	30	0.500	175	0.248	0.495	13.795
120	20	0.333	115	0.163	0.488	14.283
150	30	0.500	105	0.149	0.297	14.580
180	30	0.500	100	0.142	0.283	**14.863**

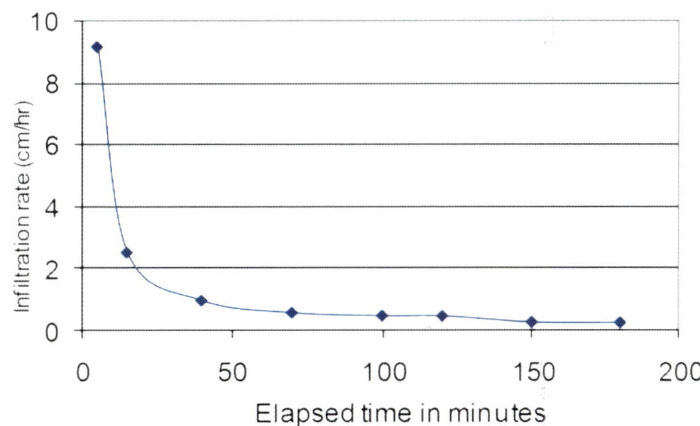

Fig 4.5 Infiltration rate in a reclaimed coal mine OB dump (Location: Ananta project, MCL)

Category of Rooting Depth (Lyle 1978):

1. Very shallow— <15 cm
2. Shallow—15–30 cm
3. Moderately deep—30–52.5 cm
4. Deep—52.5–90 cm
5. Very deep— >90 cm

Hu et al. (1992) have categorised rooting media depth (RMD) for minesoil as follows:

- RDM ≤ 5.0 cm = very poor
- RDM 5.0–40 cm = good
- RDM ≥ 40.0 cm = extremely good

4.2.6.1 Root Development in Mine Spoils

Root development is restricted in spoils due to changes in physical and chemical properties during soil handlings and compactions of overburden dump materials. Dollhopf and Postle (1988) suggested that when clay content in mine spoils was greater than 40%, it caused low permeability and infiltration rate, structural and compaction problems. When sand content is greater than 70%, mine spoils retain insufficient water for plant establishment and growth.

4.2.6.2 Root Activity (Rooting Depth)

Soil temperature, compaction and water potential affect root activity. Maiti et al. (2002) studied the rooting depth in the reclaimed dumps under natural vegetation. Root density was found nearly twice in old minesoil and natural soil than newly reclaimed spoil. The roots in reclaimed minesoils are restricted within 40 cm for grasses and herbaceous vegetation. Soil moisture increases with depth, which may be the reason for growth of roots in deeper depth (Tables 4.11 and 4.12).

Table 4.11 Rooting depth (cm) in the mined-out area covered with natural grass vegetation (*Cynodon*) (Maiti et al. 2002)

Profile depth (cm)	Moisture content (%)	EC_e (1:1; w/v) (dS/m)	Rooting depth in each pit (cm)
0–15	4.13	0.29	25, 30, 34, 25, 32, 28, 34
15–30	4.62	0.25	
30–50	6.17	0.25	

Number of pits was 7 and dug up to 50 cm
Average root length in each pit: 203.3 cm; weight: 2.42 g

Table 4.12 Rooting depth (cm) in a mined-out area under natural herbaceous vegetation (*Hyptis suaveolens*)

Profile depth (cm)	Moisture content (%)	EC_e (1:1; w/v) (dS/m)	Rooting depth in each pit (cm)
0–15	2.3	0.39	5, 12, 13. 5, 10, 14, 13
15–30	5.55	0.37	
30–50	13.74	0.35	

Hyptis suaveolens: tap root length 16 cm and branch root 19 cm
Sida cordifolia: tap root length 35 cm and branch root 22 cm

4.3   Chemical Properties and Plant Growth

Chemical tests are described in detail by Black (1965), Jackson (1973), Ghosh et al. (1983), Page et al. (1982), Allen (1989) and Maiti (2003). The analysis of mineral concentration in oven-dry soil is usually expressed in units of parts per million (ppm) (weight/weight basis), which is equivalent to mg/kg. Higher concentrations are expressed as percentage, 1% = 10,000 mg/kg. Concentration in soil can be converted to kg/ha for a specified depth, usually 15 cm, using the bulk density according to the formula:

$$kg/ha \ in \ a \ depth \ of \ D \ cm =$$
$$\frac{BD(g/cm^3) \times concentration \ in \ soil(mg/kg) \times Depth}{10}$$

As for example, if an average soil BD = 1.5 g/cm³, concentration = 5 mg/kg, depth = 15 cm, then concentration will be 11.25 kg/ha. This expression may be convenient when comparing mineral concentrations in soil for fertiliser additions.

Important factor which effects plant growth on mine spoil are

1. pH—acidity or alkalinity
2. Electrical conductivity or salinity
3. Organic matter
4. Essential macro- and micronutrients
5. Properties affecting supply of nutrients
6. Presence of any chemicals in toxic quantities

4.3.1   pH

The alkalinity or acidity of soil is measured by using a scale called pH. This scale ranges from 0 to 14. Most natural soils range in pH from 5.0 to 8.5. In general, the optimum pH for plant growth is 5.5–7.5. Soil acidity is common in the regions where precipitation is high enough to leach appreciable quantities of exchangeable base-forming cations (Ca^{2+}, Mg^{2+}, K^+ and Na^+) from the surface of soils, while alkalinity occurs when there is a comparatively high degree of saturation with base-forming cations (Brady 2000). The pH may influence nutrient absorption and plant growth in two ways, (1) through direct effects of hydrogen ions and (2) indirectly through its influence on nutrient availability and the presence of toxic ions (Brady 2000). The effect of pH on spoil is mostly indirect by changing solubility and availability of metals and other ions, including deficiencies or toxicities (Brady 2000). Maximum availability of the primary nutrients, nitrogen, phosphorus and potassium, as well as secondary nutrients, sulphur, calcium and magnesium, is at a pH range of 6.5–7.5. The availability of minor elements iron, manganese, boron, copper, chlorine and zinc is more in the acidic range than in the neutral or alkaline range (Lyle 1978).

The pH is normally measured in soil: water slurry of 1:2.5 (weight/volume) ratio, stirred thoroughly for 30 min (some authors suggested 15 min), after which the supernatant is collected by centrifugation for pH measurement in a pH metre. The paste pH value (1:1) is considered to be close to the true pH value of field conditions where plant roots are exposed. The pH value usually increases with increase in dilution ratio, due to increase in H^+ concentration in the solution. Hence, it is always advisable that during reporting of pH value, dilution ratio has to be mentioned. Other than water, 1 M KCl and 0.1 M $CaCl_2$ are also used as dilution solutions to measure pH in mine spoil. Measurement of pH in 1 M KCl solution in 1:1 or 1:2 ratio is sometimes referred as the *buffered pH*, which will provide better information concerning the chemical properties of system (Maiti 2003). To overcome the influence of soluble salts, measurement of pH in 0.1 M $CaCl_2$ using a 1:1 or 1:2 ratio gives the better results and used for determination of *lime potential*. Hence, during reporting of the pH value, along with the dilution ratio, mention the dilution solution (i.e. distilled water, KCl or $CaCl_2$) (Table 4.13). For reclamation point of view, buffer pH and lime potential give better indication of mine spoil properties. (Maiti 2003).

In majority of cases, pH of mine spoil is found acidic to neutral range. However, in some spoil dumps, slightly alkaline pH was found, e.g., 7.8–8.4 (Block -II, BCCL, Maiti 1995) and 8.1–8.5 (Ghanashyam OCP, ECL). The acidic pH of mine spoils is due to the geological characteristics of rock (presence of pyrites), and majority of the cases found acidic (<5) and sometimes very acidic (<4) due to presence of thick pyrite band in the overburden strata (Table 4.14). As long as paste pH value is greater than 5, it will not pose problems for plant growth (Maiti 1995). During profile analysis of mine

Table 4.13 The measurement pH with different types of dilution solution in minesoils collected from the reclaimed overburden dumps JCF ($n = 7$, mean $\pm$ SD) (Mukhopadhyay and Maiti 2010)

Rhizospheric samples	pH			
	1:1 (H_2O)	1:2.5 (H_2O)	1:2 (KCl)	1:2 ($CaCl_2$)
1	4.21 ± 0.02	4.46 ± 0.03	3.66 ± 0.02	3.96 ± 0.04
2	4.75 ± 0.03	4.99 ± 0.03	3.95 ± 0.04	4.3 ± 0.07
3	4.95 ± 0.09	5.22 ± 0.10	4.17 ± 0.10	4.48 ± 0.09
4	4.91 ± 0.09	5.14 ± 0.09	4.18 ± 0.10	4.44 ± 0.10
5	5.22 ± 0.09	5.48 ± 0.10	4.42 ± 0.09	4.80 ± 0.15
6	4.83 ± 0.15	5.08 ± 0.19	4.06 ± 0.22	4.37 ± 0.23
7	4.75 ± 0.05	5.0 ± 0.05	3.91 ± 0.09	4.29 ± 0.13

Table 4.14 pH of minesoils in reclaimed and unreclaimed overburden dumps

Name of mining project	pH_{water} (1:1; weight/volume)	Reference
Kusunda OCP, JCF	4.61 (4.21–4.92) (unreclaimed OB dump)	Mukhopadhyay and Maiti (2011)
	4.80 (4.18–5.35) (reclaimed OB dump)	
KDH project, CCL	4.5–5.23 (reclaimed dump)	Maiti (2006a, b)
Dipka project, SECL	4.45 (unreclaimed)	
	4.70 (reclaimed with *Grevillea robusta*)	
Manikpur, SECL	3.35–3.59 (unreclaimed)	
	3.90–3.97 (reclaimed with *Grevillea robusta*)	
Kusmunds, SECL	4.40 (barren OB)	
	4.57–6.94 (Reclaimed with *D. sissoo, C. fistula*)	
Gevra project, SECL	4.85–6.50 (reclaimed with *C. siamea, T. arjuna*)	
	5.68–6.36 (unreclaimed)	

Table 4.15 Effect of profile depth on pH values (1:2.5; w/v) (Maiti 1995)

Profile depth (cm)	Unreclaimed mine spoils exposed to natural succession					Subsided area	Gardensoil
	Up to 2 year	2–5 year	5–8 year	9–12 year	>12 year		
0–10	4.81	5.08	5.19	5.68	5.96	4.58	6.51
10–20	4.61	4.91	4.80	5.10	5.12	4.69	6.11
20–30	4.48	4.77	4.61	4.95	4.80	4.84	6.27
30–40	4.34	4.71	4.60	4.78	4.80	4.89	6.08
40–50	–	–	–	–	–	4.87	6.42
50–60	–	–	–	–	–		6.36
60–70	–	–	–	–	–		6.33
70–80	–	–	–	–	–		6.15

spoil, pH was found to be more acidic as depth increases. Maiti (1995) studies the role of natural succession process on the improvement of pH and found that pH was improved due to leaching of H^+ ions and accumulation of organic matter (Table 4.15). The acidity of mine spoil is neutralised by adding lime, in the form of ground limestone (calcite, $CaCO_3$) and dolomite ($CaCO_3$. $MgCO_3$). The magnesium limestone ($MgCO_3$) is avoided on sites where spoil salinity is high as the magnesium contributes to the salinity. The quantity of lime can be over 100 t/ha if the spoil is highly pyritic, so an accurate assessment of lime requirement is - economically important. The method of lime requirement is determined by SMP single-buffer method (Page et al. 1982).

4.3.2   Electrical Conductivity (EC)

Electrical conductivity is the measure of solubility in the soil. Most soluble salts are composed of the cations Na, Mg and Ca and the anions Cl, SO_4 and HCO_3. Usually smaller quantities of K, NH_4, NO_3 and CO_3 also occur. Soils that are saline can be recognised by their high EC and may cause growth problems because of osmotic effect. High salt concentration increases the potential forces that hold water in soil and makes it more difficult for plant roots to extract the moisture. During dry period, salt concentration in the soil solution may be increased which may kills the plant by pulling water from roots by exo-osmosis process (i.e., loosing of water from the root

cells). Salinity effects are modified to a great extent by water content.

In coal mine spoil dumps, higher variations of EC have been observed. In Dhanpuri mines, EC of 0.14 (vegetated dump) to 0.27 dS/m (bare OB dump) has been observed by Ramprasad and Awasthi (1992). In Jharia coalfield, Maiti and Banerjee (1992) observed slightly higher value of 0.3–0.5 dS/m (acidic spoil) and 0.8–0.9 dS/m (alkaline spoil). In Ramagundam project (SCCL), the EC was found 0.243–0.6 (Maiti and Reddy 2003). Thus, it is expected that EC of these dumps will not pose problems for plant growth (Maiti 2006a). The EC value of reclaimed and barren coal mine overburden dumps is given in Table 4.16.

4.3.3   Essential Elements for Plant Growth

Sixteen minerals/elements have been recognised as essential for plant growth. These 16 essential nutrients are categorised into:

- Basic body building elements: carbon, hydrogen and oxygen
- Macronutrients: nitrogen, phosphorus, potassium, calcium, magnesium and sulphur
- Micronutrients: boron (B), chlorine (Cl), copper (Cu), iron (Fe), manganese (Mn), molybdenum (Mo) and zinc (Zn)

First categories of elements (C, H, O) are used in relatively high concentration. The source of carbon, hydrogen and oxygen is mainly air and water, while the source of the other six

Table 4.16 Electrical conductivity of minesoils in reclaimed and unreclaimed overburden dumps

Name of mining project	ECe (1:1; soil/water), dS/m	Reference
Kusunda OCP, JCF	0.66 (0.45–0.78) (unreclaimed OB dump)	Mukhopadhyay and Maiti (2011)
	0.37 (0.26–0.47) (reclaimed OB dump)	
KDH project, CCL	0.17 (0.08–0.25) (fresh OB)	Maiti (2006a)
	0.22 (0.19–0.26) (under *D. sissoo* plantation)	
Dipka project, SECL	0.059 (unreclaimed)	
	0.024 (reclaimed with *Grevillea robusta*)	
Manikpur, SECL	0.386–0.675 (unreclaimed)	
	0.156 (reclaimed with *Grevillea robusta*)	
Kusmunda, SECL	0.061 (barren OB)	
	0.137–0.565 (reclaimed with *D. sissoo, C. fistula*)	
Gevra project, SECL	0.039–0.081 (reclaimed with *C. siamea, T. arjuna*)	
	0.022–0.045 (unreclaimed)	

Table 4.17 Concentration of basic elements and macronutrients in plant biomass (Donahue et al. 1990)

Elements	Concentration (%)	Sources/remarks
(a) Basic body building		
(i) Carbon (C)	43.6	Air CO_2
(ii) Hydrogen (H)	6.2	H_2O
(iii) Oxygen (O)	44.4	H_2O and O_2
(b) Macronutrients		
(i) Nitrogen (N)	1.5	Deficient, applied in terms of fertiliser
(ii) Potassium (K)	0.9	Sufficient, weathering of rock
(iii) Phosphorous (P)	0.2	Deficient
(iv) Magnesium (Mg)	0.2%	Not needed for spoil
(v) Calcium (Ca)	0.2%	Sometimes contains
(vi) Sulphur (S)	0.2%	Excess amount

macronutrients are soil solids. The remaining essential mineral nutrients are referred to as 'micronutrients' because they are essential only in very low concentration. Macronutrients are generally constituents of organic compounds found in plants, such as protein and nucleic acid, while micronutrients are predominantly constituents of enzyme molecules (Table 4.17).

4.3.4 Nutrient Poverty of Mine Spoil

Ecosystem disruption results in increase in nutrient export from the system and causes depletion of 'soil carbon pool'. Mining activities cause loss of litter layer, which is an integral storage and exchange site of nutrients. Due to loss of litter layers and organic matter, nutrient-holding capacity of mine spoil is drastically reduced, which is a major disadvantage of surface mine sites and commonly termed as "inherently low fertile soil". Therefore, mine spoils are nutritionally and microbiologically impoverished in nature (Visser et al. 1979).

Although a large number of elements is required for healthy plant growth, only three elements—organic matter, nitrogen and phosphorous—are usually lacking in spoil to such an extent that corrective measures are required. The major sources of macronutrients (NPK) in the OB dumps are decomposition of organic matter, while major sources of organic matter are the leaf litter accumulation and their decomposition. According to Lyle (1978), low availability of N, P

Table 4.18 Nutritional characteristics of reclaimed and unreclaimed OB soils, fresh overburden and topsoil (Maiti 2007)

Sampling locations (n = 10)	OC (%)	Av-N (ppm)	Av-P (ppm)	Na [cmol (+) kg^{-1}]	K [cmol (+) kg^{-1}]	CEC [cmol (+) kg^{-1}]
Reclaimed OB	1.82[a]	30.1[b]	1.20[b]	0.186[a]	0.094[b]	2.04[b]
Unreclaimed OB	1.53[ab]	22.4[b]	0.76[b]	0.166[a]	0.099[b]	1.67[b]
Topsoil	1.16[b]	48.5[a]	3.64[a]	0.153[a]	0.134[a]	4.07[a]
Fresh overburden	0.56[c]	na	1.20[b]	0.270[a]	0.056[c]	1.99[b]
5% LSD	0.39	14.0	1.43	0.120	0.030	1.39

n = number of samples
Superscript small letters represent analysis of variance (ANOVA)
Values with different letters indicate significant difference at $p < 0.05$

and K is also the cause of poor plant growth on mine spoil. Others observed that phosphorus is a major limiting nutrient during colonisation and early succession process on surface-mined land (Maiti 1995).

In coal mining areas, plant-available N is found higher in topsoil (49 ppm) than subsoil, reclaimed OB dumps and other mined-out areas (25–30 ppm) (Maiti 1995). The lower value of available N in OB dumps may be attributed due to lack of microbial activity and mineralisation of organic matter. Available P is also found lower in comparison to topsoil, which may be due to the fixation of phosphorus by coal shell (Coppin and Bradshaw 1982), and in acidic pH, it forms insoluble iron aluminium phosphate (Brady 2000). Exchangeable potassium is found lower in reclaimed minesoils than topsoil, while cation exchange capacity (CEC) for the topsoil is found significantly higher than minesoils (Table 4.18).

4.3.5 Organic Matter (OM)

Organic matter consists of decomposed plant and animal residues. Most cultivated soils contain up to 1–5% OM within upper 25 cm of soil. Schafer et al. (1980) found that OM was present only in the upper few centimetres of 53-year-old spoils. OM in soils is the source of (1) nearly 90–95% nitrogen, (2) 5–60% phosphorous, (3) 80% sulphur and large portion of boron and molybdenum (Donahue et al. 1990). The average organic carbon (OC) level in reclaimed minesoil is found higher than topsoil due to accumulation and decomposition of leaf litter.

This higher level of OC in reclaimed and unreclaimed minesoils compared to topsoil may be due to the deposition of coal dust and presence of carbonaceous shell and coal fraction in the overburden materials (Maiti 2003). High level of OM in mine spoil improves aggregation and infiltration capacity and increases the availability of nutrients. Status of organic carbon in most of the coal mine spoil was found low (Table 4.19).

Accumulation of OC in OB dumps depends on the nature of soil, existing vegetation and climate. Down (1974) found 0.79, 1.52 and 1.81% OM on sites of age 0, 5 and 12 years, respectively, in mine spoils at Somerset Coalfields, whereas Rimmer (1982) reported OM accumulation at the rate of 0.8% per annum in northern England in 1–8-year-old spoils. However, after 10 years of plantation in Dhanpuri mines (NCL), OC level has been found to be increased from 0.049 to 0.285%, which seems to be very low (Ramprasad and Awasthi 1992). In another study of SCCL mines, Maiti and Reddy (2003) found that % of OC accumulation has increased from 0.01 to 0.8% after 8 years of reclamation. It can clearly be seen that accumulation of OC is not uniform and it depends on nature of spoil and existing vegetation. Due to lower accumulation of OM, reduced nutrient availability was reported on reclaimed site compared to soils of natural sites (Maiti 2007). The organic matter content in the mine spoils could be enhanced by the following practices:

- Addition of organic amendments like green manure, farmyard manures (FYM), poultry manure and compost (municipal compost, vermin-compost etc.)

Table 4.19 Status of organic carbon in reclaimed and unreclaimed OB dumps

Location	Organic carbon (%)	References
Kusunda, BCCL	1.82 (0.84–2.62)	Mukhopadhyay and Maiti (2011)
Dipka, SECL	0.68 (unreclaimed)	Maiti (2006a); Maiti and Singh (2006)
	1.35 (reclaimed with *Grevillea robusta*)	
Manikpur, SECL	0.30 (unreclaimed)	
	0.38–0.40 (reclaimed with *Grevillea robusta*)	
Kusmunds, SECL	0.51 (barren OB)	
	0.61–0.82 (reclaimed with *D. sissoo, C. fistula*)	
Gevra, SECL	0.61–0.68 (reclaimed dumps with *C. siamea, T. arjuna*)	
	0.38 (unreclaimed)	
Chirimiri, SECL	1.1 (0.82–1.43) – 0–15 cm (7 years old)	
	0.73 (0.4–1.1) – 15–30 cm	

- Addition of mulch materials – like hay and straw, spreading of herbs and shrubs cutting, addition of leaf litters, saw dusts etc.
- Addition of sewage sludge (is used in the USA, which contains high organic matter and nutrient (NPK) and all essential ingredients for successful vegetation establishments).
- Plantation of high yielding biomass species, like Bamboo (*Dendrocalamus*).
- Grass–legume mixtures in form of seed cake (with cow dung) can enhance OM.
- Green mulch: daincha (*Sesbania grandiflora*), popularly known as 'green mulch', could be used, but these plant have to be buried/mixed with spoil cover mechanically (common practice in agriculture).

4.3.6   Nitrogen (N)

Nitrogen, after water availability, is act most important limiting factor for plant growth in arid and semiarid ecosystems. In most natural ecosystems, N inputs are minimal and N retention and efficient cycling are critical for maintenance of ecosystem productivity. In addition, unlike undisturbed soil, in reclaimed soil, NH_4^+ and NO_3 concentrations are not correlated with soil organic matter (SOM). The higher inorganic N concentration found in the reclaimed plots, as well as its spatial characteristics and lack of significant correlation with SOM, suggests that the N cycle of this system is less efficient, or less tightly coupled, than in the undisturbed ecosystem (Smith 1993).

Spatial analysis of biotic and abiotic characteristic suggests that root exploration is likely less in the reclaimed ecosystem, which may result in carbon limitations on microbial biomass and activity in areas away from plant bases (Mummy et al. 2002). Subsequently, plant and microbial N uptake are likely less in these areas, and mineralised N is not cycled as rapidly back into the organic pool. This may be important to long-term ecological stability because inorganic N is potentially subject to greater losses from leaching, volatilisation and conversion to gaseous forms than organic forms of N. Additionally, high levels of ammonium are known to increase mineralisation of the indigenous soil N, potentially resulting in N limitations in subsequent years. The underlying overburden of the reclaimed sites is at a shallow depth and very permeable. N leached into this medium may be carried below the root system of grasses. Because deeper roots from forbs and shrubs are lacking, this N may not be cycled back into the system. Further research is required that measures organic N pool sizes and N cycling rates, inputs and outputs before the long-term sustainability of this ecosystem can be ascertained. The nitrogen poverty in OB materials and other mined-out areas could be overcome by

- Fertiliser application
- Planting legumes
- Addition of organic matter

Approximately organic matter contains 5% of nitrogen, and it is unfortunate that most new spoil contains very little organic matter, and microbial decomposition of OM is the major source of nitrogen. Disturbance drastically alters the flow of nitrogen through the soil–plant–microbial ecosystem. To achieve successful revegetation and maintain a self-sustaining plant community, it is necessary to establish an active biological nitrogen cycle (Skeffington and Bradshaw 1981). Nitrogen deficiency can limit revegetation and long-term stability of land disturbed by surface mining. According to Inouye et al. (1987), nitrogen is a major soil-limiting nutrient which influences plant productivity remarkably.

Studies on natural and artificial colonisation of sand waste from Kaolin mining (UK) by Marrs et al. (1981) reported that at first, stages of natural colonisation rely on nitrogen-fixing species until a capital of approximately 1,000 kg/ha has accumulated in the whole system or 700 Kg N/ha in the soil. This figure of 700 kg/ha of soil N is another important implication for reclamation, which depends on the rate of mineralisation of the N. Although soil OM contains 5% wt. of N, only about 1–3% of the total amount is released yearly by decomposition. The rate of mineralisation or nitrogen turnover can increase the productivity of the soil. This can be achieved in two ways:

1. Mineralisation and Nitrification

   Nitrogen mineralisation and nitrification on mine spoils were found to be very low, and deficiency is often due to the lack of adequate amounts of mineralisable organic nitrogen and lower mineralisation rates rather than an actual shortage of total nitrogen. Reeder and Berg (1977) reported that due to lack of ammonification, indigenous as well as fertiliser nitrogen supplied to spoils is much less available to plants as compared to similar amounts of nitrogen present in or added to the undistubed soils.

   The N deficiency in mine spoils is often due to

   - Lack of adequate amount of mineralisable organic nitrogen
   - Lower mineralisation rate
   - Lower nitrification
   - Lack of microbial activity

2. Improvement of Nitrogen Poverty

   Generally initial seeding with grass and legume mixture is the common practice for improvement of nitrogen poverty in mine spoil. Studies conducted in mine spoils of Dhanpuri mines (SECL, MP) showed that nitrogen content in leguminous plot was more (222 kg/ha) than that of nonleguminous plot (102–169 kg/ha). Even simple vegetation can enhance the mineralisable nitrogen in the spoil from 24.53 kg/ha (bare dump) to 165 kg/ha (Ramprasad and Awasthi 1992).

The productivity of the mine spoil could be increased by enhancement on nitrogen mineralisation rate:

(a) *Reducing C/N ratio*: Only when C/N ratio is 25:1 or less, there will be a significant net release of nitrogen from microbial activities. According to Lanning and Williams (1979), when C/N ratio is greater than this, there will be a demand for available nitrogen by microbial population and addition of nitrogen will improve the C/N ratio. Generally, mine spoils have wider C/N ratio than natural soil, because a large fraction of total carbon and N is present in the form of plant litter.

(b) *Improving spoil condition*: Mixing a fine silt residue with coarse sand waste will improve nitrogen mineralisation rates (Reeder 1985). It also increases water availability and cation exchange capacity and reduces leaching and ammonia volatilisation.

4.3.7 Phosphorous (P)

4.3.7.1 Source and Fixation

Phosphorous is the 'second key plant nutrient' and one of most troublesome nutrient for the reclamationist. In spoil, P comes from the breakdown of rocks (apatite) and mineralisation of organic matter. The soluble H_2PO_4 rapidly reacts in clay minerals to form insoluble phosphate, called 'phosphate fixation'.

In acidic soil, phosphorous forms insoluble Fe and Al phosphate, and in alkaline pH, it forms Ca phosphate (Brady 2000). Phosphorous is most available at pH 6.5 for mineral soil and pH 5.5 for organic soils. Although critical concentration for P is not well established, it is believed that when carbon/organic P ratio is about 200:1 or narrower, P is readily released mineralised. If ratio is 300:1 or wider, the organisms use most P by immobilising it into their cells instead of releasing it for plant use. Plant recovers about 10–30% of P that is added to a soil as fertiliser. Very little P is lost by leaching and most of the P (70–90%) is not absorbed by the plants; they formed insoluble Fe, Al or Ca compound (Lyle 1978).

For neutral to alkaline soil pH, the standard method is extraction of the soil with 0.5 N sodium bicarbonate solution at pH 8.5 (Olsen method) using 1:20 soil to solution ratio, shaken for 30 min. On acidic soils, an extraction with a solution of 0.03 N HCL plus 0.025 N ammonium fluoride gives a better correlation with plant growth (Bray and Kurtz 1945).

4.3.7.2 P Status in Coal Mine Spoil

The concentration of extractable P was found to be low in OB material ranging from 0.01 to 0.05 ppm or one-tenth of P concentration observed in topsoil. The concentration of extractable P was reported in coal mine spoil in the order of 8.6 kg/ha (Dhanpuri OB), 10.8–15.3 kg/ha (KD Heslong, CCL) and 19 Kg/ha (Bina OB). The average concentration of P for optimal plant growth should be in the order of 20 ppm (45 kg/ha).

4.3.7.3 P Fixation by Shale

The P fixation by colliery shale will depend largely on two factors: One is surface area available for fixation and per unit weight of shale. The more shale is broken down by physical weathering, the larger is its surface area per unit weight; as phosphate absorption is a surface phenomena, phosphate requirement is therefore influenced by the degree of weathering (Doubleday and Jones 1977). Although the total phosphorus content of some mine spoils may equal or exceed those of the unmined soils, the plant-available phosphorus in almost all the spoils is invariably in the deficiency range (Yamamoto 1975). Consequently most of the spoils exhibit tremendous responses to phosphorus fertilisation under conditions of adequate water availability:

P requirement (kg/ha) $= 11.5 - 320A + 13.11 B$

   (Coppin and Bradshaw 1982),

where A = bicarbonate-extractable P (mg/kg) and B = P-sorption capacity.

4.3.8 Cations

The most important cations are K, Ca, Mg and Na. These cations are determined in the laboratory by extraction them with standard neutral (i.e. pH 7) 1N ammonium acetate solution, at an extracting ratio of 1:5 (w/v; soil to solution) and shaken for 20–30 min. This will remove the exchangeable cations held on the cation exchange complex and also the soluble cations (the exchangeable plus water-soluble cations are the available form). The forms of soil K in the order of their availability to plants and microbes are solution > exchangeable > fixed (non-exchangeable) > mineral (Fig. 4.6).

The third key nutrient element 'K' is an enigma. The amount of total K found in most soil is sufficient for many decades, but it not easily available to the plant, because 98–99% of K is always in fixed form. Most K used by plant in a given season comes from exchangeable K and water-soluble K. In most soil, exchangeable K is the major source of K to plant. In mine spoil, exchangeable K concentration was found in low to moderate range (Table 4.20). The elements Ca and Mg were creating any problems for plant growth in mine spoils. In Bina OB, NCL Ca and Mg concentration was found 0.051 and 0.01%, respectively, which is sufficient for plant growth. In alkaline spoils, Ca and Mg deficiencies in plant can occur in the absence as well as presence of sufficient amount of $CaCO_3$.

4.3.9 Storage and Supply of Nutrients (CEC and BS)

There are certain factors that can be measured that help in the prediction of how the soil will behave as a plant growth medium.

4.3.9.1 Cation Exchange Capacity (CEC)

CEC is the total capacity of the soil to adsorb cations onto electrically charged sites, from where they are available to plants but resistant to leaching. Certain soil minerals, such as clay, particularly in combination with organic matter,

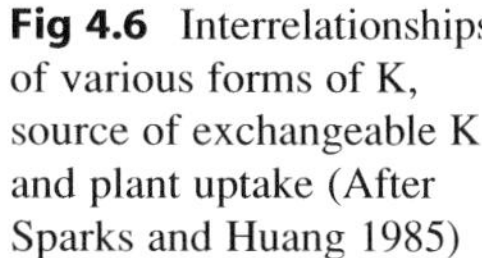

Fig 4.6 Interrelationships of various forms of K, source of exchangeable K and plant uptake (After Sparks and Huang 1985)

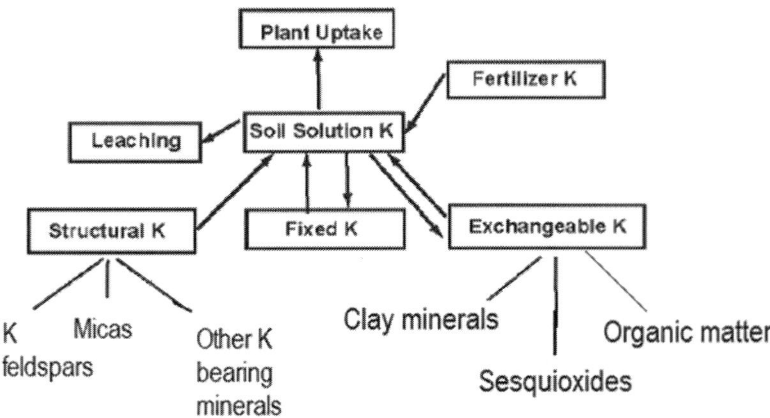

Table 4.20 Exchangeable K level in different types of soil and OB soil

Locations and soil type	Exchangeable K (ppm)	References
Black soil (Rajasthan)	156	Biswas and Mukherjee (1994)
Alluvial soil (West Bengal)	273	
Red soil (Jharkhand)	274	
Lateritic soil (Orissa)	39	
Alkali soil (Delhi)	234	
Forest soil (Tamil Nadu)	117	
Reclaimed OB	36.7	KD Heslong project, CCL Maiti (2007)
Unreclaimed OB	38.61	
Topsoil	52.26	
Fresh OB	21.84	
Natural sal forest (0–15 cm)	126	Maiti (2006a)
OB dumps of 6–8 year old		Maiti (1995)
0–10 cm	48.6	
10–20 cm	34.5	
20–30 cm	33.6	
30–40 cm	21.3	
K fertility rating for agricultural soil		Ghosh et al. (*IARI manual* 1983)
High fertility	≥ 125	
Medium fertility	$= 53.6–125$	
Low fertility	≤ 53.6	

possess a number of electrically charged sites which can attract and hold oppositely charged ions. The negatively charged sites make up the CEC and have the ability to hold H^+, Ca^{+2}, Mg^{+2}, Na^+, NH_4^+, etc., and positively charged sites hold OH^-, $SO_4^=$, NO_3^-, PO_4^-, etc., that make up the anion exchange capacity. Ions held at these sites can be exchanged and act as an important index of nutrient status because exchangeable cations are most important source of readily available plant nutrients (Tisdale et al. 1985). The CEC is expressed in terms of milliequivalents per 100 g (meq/100 g) of soil. The higher CEC value indicates more cation-holding capacity of soil. A clay soil will have a larger CEC than a sandy soil. Increasing the OM content of any spoil will help to increase the CEC value because it holds cations like clay particle. CEC also depends on soil texture. In sandy soil, CEC is found minimum, and in humus, it is maximum. In mine spoil, CEC is reported very low (7–8 meq/100 g) due to low organic matter.

4.3.9.2 Percent Base Saturation

Base saturation refers to the proportion of the cation exchange sites in the soil that are occupied by the various cations (H^+, Ca^+, Mg^+, K^+). The surfaces of soil minerals and organic matter have negative charges that attract and hold the positively charged cations. Cations with one positive charge (H^+, K^+, Na^+) will occupy one negatively charged site, while cations with two positive charges (Ca^+, Mg^+) will occupy two sites. Percent base saturation tells what percent of the exchange sites are occupied by the basic cations. If calcium has a base saturation value of 50% and magnesium has a base saturation value of 20%, then calcium occupies half of the total exchange sites (CEC) and magnesium occupies one-fifth of the total exchange sites (CEC). For example, where the soil has a CEC of 5 meq/100 g, 2.5 meq/100 g of the CEC is occupied by calcium and 1 meq/100 g of the CEC is occupied by magnesium. If all the exchangeable bases (Ca, Mg, K and Na) total 100%, then there is no exchangeable acidity.

4.4   Microbiological Properties

For establishment of self-sustaining soil–plant system in OB dumps, the development of functional soil–microbial community is essential (Wilson 1965). The importances of microbes are

- Responsible for decomposition of plant litter.
- Mineralisation (release of essential plant nutrients, i.e. organic to inorganic form).
- Nutrient cycling (cycling of mineral element from soil–plant–litter falls back to soil).
- Accumulation of organic matter.
- Formation of humus.
- Decomposition of organic matter produces slimes and gums, which aid in the formation of soil structures. Decomposed organic matter—humus—absorbs and holds water and nutrients for plant use.

Microbial processes are so important to ecosystem recovery that the activity of microorganisms may be used as an index of the progress of soil genesis in mine spoils. There are three groups of microbes generally present:

- *Bacteria*
- *Fungi*
- *Actinomycetes*

Their main function is decomposition (i.e. release of organically bound nutrients to inorganic plant-available form). They are all heterotroph (i. e. carbon source is organically bound carbon like dead plant and animal residues) and not like green plants. Green plants are autotroph, that is, they take carbon as CO_2 (inorganic) from atmosphere. Thus, if the pre-mining organic layer (O horizon) is destroyed, the only C source for microbial utilisation is the plant biomass that is expected to accumulate over several growing seasons on the site. Until such accumulation occurs, microbial activity remains at a very low level, with little improvement of adverse soil physical and nutrient conditions:

Remedy—initial requirement of organic carbon in the mine spoil should be supplied by

- Application of organic mulches
- Organic manure and compost
- Garbage mulch/sewage sludge, etc.

Assessment of microbial activity for the progress of reclamation/success of reclamation:

1. By studying *soil community respiration* (in situ CO_2 evolution methods). As microbes are heterotroph, they release CO_2 during respiration, and it is absorbed in dilute alkaline solution (0.1N KOH) and titrated against dilute acid.
2. *Microbial decomposition* can be used as an indicator of the degree of soil ecosystem recovery. Since, microbial activity largely controls nutrient cycling in derelict sites, rate of microbial decomposition activity can be measured by using litter-bag or nylon-bag methods.

There are other methods also being popularly used to study the rate of ecosystem recovery:

1. *Enzyme analysis*—The principle is that the microbes decompose by the secretion of enzyme: *cellulase, amylase, invertase*, etc., for carbon/carbohydrate and *protease, urease*, etc., for protein. This method is very first, and it can be done in laboratory.

2. *Soil microbes study*—by standard plate count methods.

Other costly but very sophisticated and accurate estimation and very less time-consuming techniques are ATP measurement and nitrogen content.

Mummy et al. (2002) states that after 20 years of reclamation, total plant cover recovers to pre-disturbance levels, but soil total microbial, that is, biomass, bacterial and fungal, biomarkers averaged only 20, 16 and 28%, respectively, of amounts found in undisturbed soil. Similarly, in reclaimed soil, microbial biomass carbon (MBC) was estimated to average only 44% of amounts found for undisturbed soil. A number of studies have shown that disturbance associated with surface mining is highly detrimental to arbuscular mycorrhizal (AM) fungal populations. Soil storage is known to be especially detrimental to AM populations, in some cases to such an extent that the soil has little or no capacity to infect plants with fungal symbionts. That fungal biomass in undisturbed soil is dominated by AM.

References

Allen SE (1989) Chemical analysis of ecological materials, 2nd edn. Oxford-Blackwell, London

Barnhisel RI (1979) Characteristics of soil properties of reconstructed prime ana non-prime land in Western Kentucky. In: Sym on surface mining, hydrology, sedimentology and reclamation. University of Kentucky, Lexington, pp 119–122

Biswas TD, Mukherjee SK (1994) Text book of soil science, 2nd edn. TMH, New Delhi

Black CA (ed) (1965) Method of soil analysis (Part 2, Chemical and microbiological properties). American Society of Agronomy, Madison

Brady NC (2000) The nature and properties of soils, 10th edn. PHI, New Delhi

Bray RH, Kurtz LT (1945) Determination of total, organic and available forms of phosphorus in soils. Soil Sci 59:39–45

Coppin NJ, Bradshaw AD (1982) Quarry reclamation. Mining Journal Books, London

Dollhopf DJ, Postle RC (1988) Physical parameters that influence successful mine spoil reclamation. In:- Hossner LR (ed) Reclamation of surface mined land, vol 1. CRC Press, New York

Donahue RL, Miller RW, Shickluna JC (1990) Soils – an introduction to soils and plant growth, 5th edn. PHI, New Delhi

Doubleday GP, Jones MA (1977) Soils of reclamation, Chapter 6. In: Hackett B (ed) Landscape reclamation practices. IPC Science and Tech Press/IPC House, Guildford

Down CG (1974) The relationship between colliery-waste particles size and plant growth. Environ Conserv 1 (4):281–284

Down CG, Stock J (1987) Reclamation, Chapter 10. In: Environmental impact of mining. Allied Sciences Publisher Ltd., London

Ghosh AB, Bajaj JC, Hassan R, Singh D (1983) Soil and water testing methods – a laboratory manual. IARI, New Delhi

Hu Z, Caudle RD, Chong SK (1992) Evaluation of firm land reclamation effectiveness based on reclaimed mine soil properties. Int J Surface Mining and Reclamation 6:129–135

Inouye RS et al (1987) Oil field succession on a Minnesota sand plain. Ecology 68:12–26

Jackson ML (1973) Soil chemical analysis. PHI, New Delhi

Jha AK, Singh JS (1992) Influence of microsites on redevelopment of vegetation on coal mine spoils in a dry tropical environment. J Environ Manage 36:96–116

Lanning S, Williams ST (1979) Nitrogen in revegetated clay and sand waste- III: the use of clover in revegetation. Environ Pollut 21:89–95

Lyle ES (Jr) (1978) Surface mining reclamation manual. Elsevier, New York

Maiti SK (1995) Some experimental studies on Ecological aspects of reclamation in Jharia coalfield. Ph.D. dissertation, Indian School of Mines, Dhanbad

Maiti SK (2003) Handbook of methods in environmental studies, vol 2. ABD Pub, Jaipur

Maiti SK (2006a) MoEF report on an assessment of overburden dump rehabilitation technologies adopted in CCL, NCL, MCL and SECL mines (No. J-15012/38/98-IA II (M). MOEF, New Delhi

Maiti SK (2006b) Ecorestoration of coalmine OB dumps – with special emphasis on tree species and improvements of dump physico-chemical, nutritional and biological characteristics. MGMI Trans (India) 102 (1–2):21–36

Maiti SK (2007) Bioreclamation of coalmine overburden dumps—with special emphasis on micronutrients and heavy metals accumulation in tree species. Environ Monit Assess 125:111–122

Maiti SK (2008) Reports on primary baseline data on soil quality of Ananta OC Expn, Bhubanesweri OC Expn and Talabira Project of MCL, July 2008

Maiti SK, Banerjee SP (1992) Reclamation and natural succession on spoil dumps – a case study from Jharia coalfield. In: Mozumder BK (ed) Proceedings of the 4th national seminar on surface mining. ISM, Dhanbad

Maiti, SK, Reddy MS (2003) Nutrient accumulation in reclaimed overburden dumps of Ramagundam OCP-1, SCCL, In: Srivastava BK et al (ed) Proceedings of the environmental management in mines. Mining Engineering Department, BHU, Varanasi, India, pp 249–256

Maiti SK, Singh S (2006) Ecorestoration status of coalmine OB dumps of Korba, Gevra and Kusmunda area of SECL, India. In: Shringarputale et al (ed) Proceedings of the international symposium on environmental issues in mineral industries, VNIT/CSM, Nagpur/Glossop, pp 217–224

Maiti SK, Karmakar NC, Sinha IN (2002) Studies into some physical parameters aiding biological reclamation of mine spoil dump – a case study from Jharia coalfield. IME J 41(6):20–23

Marrs RH, Roberts RD, Skeffington RA, Bradshaw AD (1981) Ecosystem development on naturally colonized china clay wastes. II. Nutrient compartmentation. J App Ecol 69:163–169

Mukhopadhyay S, Maiti SK (2010) Ecorestoration of coalmine overburden dumps- with emphasis on minesoil properties, natural VAM colonization, litter accumulation and tree growth. Minetech 31(2):16–26

Mukhopadhyay S, Maiti SK (2011) Trace metal accumulation and natural mycorrhizal colonisation in an afforested coalmine overburden dump: a case study from India. Int J Min Reclam Environ 25(2):187–207

Mummy D, Stahl PD, Buyer JS (2002) Soil microbiological and physicochemical properties 20 years after surface mine reclamation: comparative spatial analysis of reclaimed and undisturbed ecosystems. Soil Biol Biochem 34:1717–1725

Page AL et al (eds) (1982) Methods of soil analysis, Part 2, 2nd ed. Agronomy monograph, vol 9. ASA and SSSA, Madison

Ramprasad B, Awasthi BK (1992) Nutritional status of afforested OB dumps of Dhanpuri Mines in MP. J Trop For 8(2):116–118

Reeder JD (1985) Fate of nitrogen-15 –labelled fertilizer nitrogen in revegetated cretaceous coal spoils. J Environ Qual 14(1):126–131

Reeder JD, Berg WA (1977) Nitrogen relation and nitrification in cretaceous shale and coalmine spoils. Soil Sci Soc Am J 41(5):922–927

Rimmer DL (1982) Soil physical conditions on reclaimed colliery spoils heap. Soil Sci 33:567–579

Schafer MW (1979) Cover-soil management in Western surface mine reclamation. Symposium on surface mining, hydrology, sedimentology and reclamation. University of Kentucky, Lexington, pp 305–310

Schafer MW, Nielsen GA, Nettleton WD (1980) Minesoil genesis and morphology in a spoil chronosequence in Montana. Soil Sci Soc Am J 44(4):802–807

Skeffington RA, Bradshaw AD (1981) Nitrogen accumulation in Kaoline waste in Cornwell-IV: sward quality and development on nitrogen cycle. Plant Soil 62:439–451

Smith JL (1993) Cycling of nitrogen through microbial activity. In: Hatfield J (ed) Advances in Soil Science, 18. Springer, New York, pp 91–120

Sparks DL, Huang PM (1985) Physical chemistry of soil potassium. In: Munson RD (ed) Potassium in agriculture. American Society of Agronomy, Madison, pp 201–276

Tisdale SL, Nelson WL, Beaton JD, Havlin JL (1985) Soil fertility and fertilizer, 5th edn. Macmillan Publishing Co., New York

Visser S, Zak J, Perkinson D (1979) Effect on surface mining on soil microbial communities and process. In: Wali MK (ed) Ecology and coal resource development, vol 2. Pergamon Press, New York, pp 643–651

Vogel NG (1987) A manual for training reclamation inspector in the fundamental of soils and revegetation. Soil and Water Conservation Society America (September), p. 178

Wilson HA (1965) The microbiology of strip mine spoil. West Va Agric Exp Bull 506T:5–44

Yamamoto T (1975) Coal mine spoils as a growing medium: AMAX Bellee AYR South mine, Gilleette, Wyoming. Third symposium on surface mining reclamation, vol 1, Kentucky, pp 49–61

Topsoil Management

5

Contents

5.1   Introduction

The term 'topsoil' generally refers to the 'A-soil horizon' which is usually darker than the underlying soil because of the accumulation of organic matter Brady (2000). Topsoil primarily provides a suitable growth medium for vegetation. It is the major zone of root development and biological activity. All the microbial activities that enhance plant growth are present in this layer, including insects, earthworms, bacteria, fungi, nematodes, and actinomycetes. They help in decaying of organic matter to form humus by mineralisation process and release plant nutrients in inorganic form. In addition, topsoil enhances soil structure and improves texture, infiltration characteristics and overall fertility status of mine spoils. Topsoiling is essential for rejuvenation of microbial life in the bare mine spoils. Topsoil, being a precious natural resource, is also required under legislation to remove, store and reuse appropriately in the ecorestoration work at the mine degraded lands. It may require establishing vegetation on shallow soils, soils containing potentially toxic materials, stony soils and soils of critically low pH (high acidity).

In ecorestoration purposes, topsoil not only consists of only upper 15-cm layer of the surface horizon but may also consist of the A, B and C-soil horizons or any combination thereof and which has been determined through soil surveys, laboratory analyses and field trials found to be suitable as a plant growth medium for the post-mining land use.

S.K. Maiti, *Ecorestoration of the Coalmine Degraded Lands*,
DOI 10.1007/978-81-322-0851-8_5, © Springer India 2013

Soil is unconsolidated mineral material in the immediate surface of the earth that serves as a natural medium for the growth of plants and differs from material from which it was derived in many physical, chemical, biological and morphological properties and characteristics. *Subsoil* means any subsurface earthen materials, excluding any material within the topsoil layer, which is capable of supporting plant life (Wyoming ELQD 1994).

The topsoil *lower limit is set at 30-cm depth or at a root growth-inhibiting layer whichever is shallower.* This layer can be hard rock, a pedogenetically indurated layer, a chemically unfavourable layer or a strongly contrasting layer. Litter layer, if exists, occurs above the topsoil. Topsoil can usually be differentiated from subsoil by texture as well as colour. Clay content usually increases in the subsoil. Where subsoil is often high in clay, in such case, the topsoil layer may be significantly coarser in texture. The depth of topsoil may be quite variable. On severely eroded sites, it may be gone entirely. Figure 5.1 illustrates the ideal pedon units showing topsoil, subsoil and weathered overburden layers above the mineral deposit.

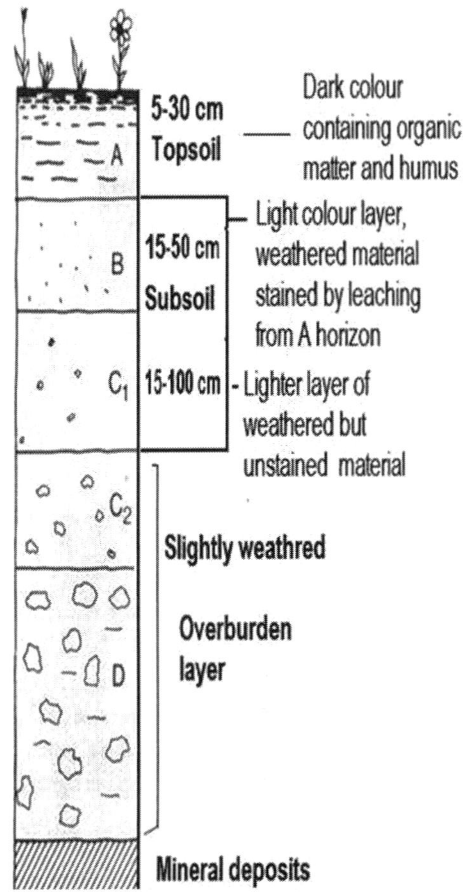

Fig. 5.1 Depicted the ideal horizon of a pedon unit

5.2 Factors Influencing Topsoil Properties

The characteristic of topsoil is strongly influenced by soil-forming factors, both externally and internally. The characterisation and subsequent stratification of topsoils, therefore, has to take into account all these factors which are interdependent; they are related to each other and influence one another. The factors influence the characteristics of topsoil are climate, vegetation and organic matter, topography and physiography, mineralogical constituents, surface processes, biological activity and human activity (FAO 1998):

1. *Climate*: Climate is the dominating factor that influences, directly or indirectly, the topsoil properties. It has a profound influence not only on the topsoil but also on topsoil-forming factors such as vegetation, topography and human activity. Climatic parameters which are important for the topsoil are temperature, moisture, radiation and wind.

2. *Vegetation*: Vegetation contributes in several ways to the formation of topsoil characteristics, such as
   - Penetrating roots loosen the soil and improve porosity and aeration.
   - Litter, decaying branches and stems are transformed into organic matter (OM).
   - Enhances aggregation.
   - Increases structural stability.
   - Increases water-holding capacity.
   - Contributes to the nutrient-holding capacity.
   - Provides the soil with N, P and S and other nutrients which were stored in the above ground biomass.

3. *Biological activity* comprises soil faunal and soil microbial activity. It enhances
   - Physical mixing of organic matter within the soil profile.
   - Inoculation of the plant litter with decomposer populations.
   - Adjustment of soil physical properties to a level more conducive for OM decomposition.
   - Physical disintegration of organic matter.
   - Stimulation of decomposer populations.

   Soil microbial activity is responsible for the biochemical breakdown of plant tissue, involving fungi, yeasts, bacteria, etc., thus liberating plant nutrients and synthesising relatively stable organic compounds which are added to the soil organic matter fraction.

4. *Organic matter (OM)*: A change in OM invariably affects physical and chemical properties of the topsoil:
   - One of the important functions of OM is to bind the soil particles. If the amount of OM is reduced, this binding will become less; therefore, structural stability is decreased.
   - Second important function of OM is to increase the water-holding capacity of soil; hence, loss of OM decreases water-holding capacity of soil. It has been reported that soil OM can hold up to 20 times more water than a similar amount of mineral soil particles, so this effect can be fairly drastic (FAO 1998).
   - Third most important physico-chemical property influenced by OM is the CEC. It has been reported that 1% of OM gives the soil about 3–4 cmol(+) CEC kg^{-1} soil. This drop can be very important especially in soils that have a low mineral CEC, like many soils in tropical regions.

5.3 Components of Topsoil Management

Topsoil management includes-inventory and characterisation, salvaging, storing, redistribution and quality assurance during the entire process. Detailed steps of topsoil management are shown Fig. 5.2.

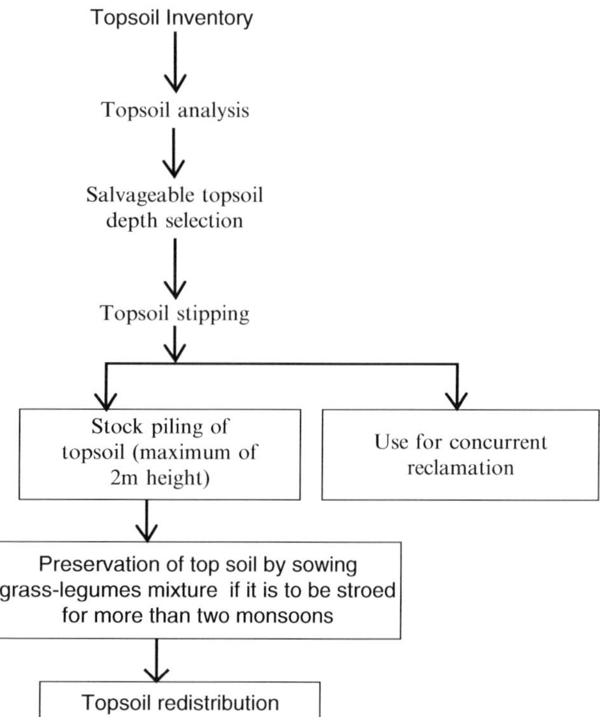

Fig. 5.2 Important steps of topsoil management (After Maiti 2010)

The important planning issues of topsoil management are

- Inventory of topsoil resource
- Topsoil removal (salvaging operation)
- Topsoil storage (stockpile)
- Topsoil redistribution
- Topsoil monitoring and quality

5.3.1 Inventory of Topsoil Resources

The following important activities are to be followed for proper inventory of topsoil resources for a project (Wyoming ELQD 1994):

1. *Soil survey*: In-depth soil survey will assist the identification of suitable topsoil material for salvage. Thus, site-specific characteristics of topsoil may influence soil stripping and stockpiling, and redistribution should be specifically noted.
2. *Preparation of topsoil map*: Prepare a topsoil map indicating areas to be affected and not to be affected during mining operations. Affected areas should be intensively mapped covering details of physico-chemical characteristics and depth of topsoil and subsoil (stripping depth), in the scale of 1:4,000. The soil map should include mapping unit numbers which will be based on stripping depth, characteristics and topography. The soil should show the boundary of all mapping units.
3. *Description on mapping unit*: similarities or dissimilarities of topsoil in terms of characteristics, depth of suitable topsoil, etc.
4. *Soil sampling*: Sampling locations should be clearly marked on the soil map. Numbers of sampling locations are site specific and depend on nature and characteristics of soil. Profile description should be taken for each soil being sampled. The major horizons (A, B, C) should be separately described, sampled and analysed. Generally, soil profile should be sampled up to 1.5 m or bedrock. Even below the 1.5-m depth, suspected that soil is suitable for reclamation, sampling should be continued beyond the depth of 1.5 m, if topsoil is limited in the area.
5. *Soil analysis*: Soil sample should be placed in a clean polythene bag, labelled and transported to the laboratory immediately. The sample is air dried at 30–35 °C (room temperature) as soon as possible for 4–5 days (or 1 week). After drying, grind the sample and sieve through 2-mm size. Greater than 2-mm size retains in the sieve, weights and expresses as coarse fraction. Less than 2-mm size soil particle is used for further analysis.
6. *Data presentation*: All analytical results should be presented in a tabular form with respect to depth (profile basis) and same to be the reference with mapping unit, sampling site location and soil horizon. The analysis report should provide the name of the laboratory conducting the analyses, sampling dates, dates of analyses, analytical methods adopted and their references.
7. *Topsoil stripping*: It is recommended that as a part of mining plan, the guideline should outline how the operator proposes to salvage all suitable topsoil. The plan should include:
   - Delineating procedure for the removal of entire topsoil material.
   - Training of equipment operators in proper topsoil salvage procedures.
   - Monitoring of topsoil salvage activities in the field by qualified personnel.
8. *Volumetric presentation*: Using data derived from soil survey (salvaging depth) and analysis, volume of topsoil that will be salvaged and available for reclamation is estimated.

Identification of the quality and depth of topsoil is essential for any salvaging operation. There are five factors which could be taken as preliminary guidelines, depicted in Table 5.1.

5.3.1.1 Removal and preservation of flora

It is well understood that, for the successes of any ecorestoration project, it is essential that native flora should be preserve by developing a 'floral bank' (if it is not use immediately) or it may be transplanted and use for the new ecorestoration of sites. It is always advisable that, during topsoil inventory, floral inventory also be carried out. Generally, it should include classification of tree species (number, height, DBH and crown cover), types of shrubs, climbers, herbs, grasses, ferns, mosses (bryophytes), legumes etc. All

Table 5.1 Identification of topsoil quality in the field (Modified after Ferris 1996)

Factors	Description
Location and depth	Deep topsoil is usually located in valley floors, while ridge tops generally have very shallow topsoil. Shallow ridge topsoil usually covers subsoil or unweathered overburden that may not be favourable for reclamation
Colour	Brownish earth-tone surface colour indicates topsoil and when colour changes to light brown usually means topsoil has ended. Soil moisture darkens the soil so recent rainfall saturation should not be mistaken for an actual change of colour
Structure	Structure is the best indication of topsoil, and well-developed topsoil that lies below the immediate surface can be identified by its blocky hexagonal shape
Texture	It is a very important property of topsoil and has an effect on plant growth; silty loam soil is the best for stripping. Heavy clay soils have limited water infiltration capacity, while very sandy soils have poor water retention capacity and often very low in nutrients
Density of root growth	Roots can be indicators of topsoil. Presence of dense mass of roots indicates the depth of stripping. However, isolated roots, especially shrub roots, can penetrate well beyond topsoil. Therefore, only root growth cannot be used as only factor for determining topsoil depths

these species must be documented in a scientific manner (local name, botanical name and family), category of species (rare or common, uniqueness if any), any red-data book species, economic and medicinal value etc. This information should be well documented along with the photographs of the species, so that, they are easily recognizable/ identified by the non-botanist, which is an essential for the development of floral bank/during transplantation work. Before transplantation of plants, operator should ensure that soil is properly moistened, otherwise roots will get damaged (especially feeder roots). Therefore, all the transplantation work must be carried out during monsoon, when soil pore is saturated with water.

5.3.2 Topsoil Removal (Stripping)

It is advisable that limit of disturbance should always be kept minimum. Strip topsoil only from those areas that will be disturbed by excavation, filling, road building or compaction by equipment. A 10–15 cm (102–152 mm) stripping depth is common, but depth varies depending on the site. For example, in Photo 5.1 shows a vertical section of topsoil and subsoil layer in KD Heslong project, where thickness of topsoil layer is very thin and in such cases, both topsoil and subsoil layer (up to the depth of bedrock) may be stripped. While, in another Photo 5.2, a thick topsoil cover is shown, where well-developed

topsoil and subsoil layer of 10-m depth can be stripped and used for ecorestoration work. Before stripping operation, determine the depth of topsoil by taking soil cores at several locations within each area to be excavated. Topsoil depth generally varies along a gradient from hilltop to toe of slope. Before starting topsoil stripping operation, sediment basins, diversions and other erosion control structure should be constructed.

Determine depth of topsoil on 10-m spacing. The depth of topsoil material should be at least 7 cm. Soil factors such as rock fragments, slope and layer thickness affect the ease of excavation of topsoil. Generally, the upper part of the soil that is richest in organic matter is most valuable. Keep topsoil separate from overburden, and store layers separately to ensure that material is restored in the same order that it was removed.

- *Removal of vegetation*: Topsoil should be removed after the removal of vegetation cover otherwise it will interfere the activities.
- *Timing of topsoil stripping*: Should not strip when it is too dry or too wet, as this can lead to compaction, loss of soil structure, loss of viability of seed and importantly loss of microbes and mycorrhiza fungi.
- *Double stripping*: The topsoil should be stripped at an interval of 5–10 cm, next 10–30 cm.
- All topsoil shall be removed in a separate layer from the areas to be disturbed. '*A*'

Photo 5.1 View of topsoil layer in KD Heslong project, CCL (*Photo Maiti, 2009*)

Photo 5.2 Thick topsoil (*dark brown colour*) and subsoil (*light brown colour*) found in eastern part of Raniganj Coalfield (*Photo: Maiti*)

horizon, that is, rich in organic matter and other soil nutrients. If A-layer is very thin, then the quality of B-horizon is to be evaluated (in between A-horizon and bedrock).

- *Subsoil segregation*: The '*B*' horizon and portion of '*C*' horizon and other underlying layer that has potential capacity for root development shall be segregated and replaced as subsoil. When there is no topsoil or it is very thin, in such cases, topsoil substitute and supplement is needed.
- The quality of topsoil materials should be ascertained by physico-chemical analysis like pH, conductivity, organic matter, CEC, nutrient content (NPK) and texture class that are generally monitored.
- *Limits on topsoil removal area*: The removal of vegetation material, topsoil or other material may result in erosion, which may cause air/water pollution.

5.3.2.1 Precautions

- The size of the area from which topsoil is removed at one time shall be limited.
- The surface soil layer shall be disturbed at a time when physical and chemical properties of topsoil can be protected and erosion can be minimised.

Many workers stated that, in most area, the '*A horizon*' of natural soil is vastly superior to any underlying state. Even it is only 7–10-cm thick (3- or 4-in. thick), careful handling and return of this horizon to the surface is required for most successful reclamation. The soil survey indicates that the properties of '*A horizon*' that are important for reclamation are:

- *Texture*—loam, sandy loam and silt loam are best; sandy clay loam, silty clay loam, clay loam, and loamy sand are fair. Do not use heavy clay and organic soils such as peat or muck as topsoil.
- *Structure*
- *Organic matter*—generally, the upper part of the soil, which is richest in organic matter, is most desirable; however, material excavated from deeper layers may be worth storing if it meets the other criteria.
- pH

Plass (1978) stated that proper topsoiling may involve the removal and storage of A, B, C horizons.

5.3.2.2 Topsoil Stripping Equipment

There are three general types of equipment suitable for stripping and transportation of topsoil (Hanks 2003); these are

(a) Scrapers
(b) Loaders, trucks and dozers
(c) Shovels and trucks

(a) *Scrapers*

Scrapers are best for striping the topsoil and universally used. They are capable of removing very thin layer of materials, which added advantage for separation of topsoil and subsoil. Scraper cuts should not be more than 50% of topsoil depth for a 15 cm thick topsoil layer (6″) or more. Stripping should proceed from higher to lower topographical areas, that is, shallower to deeper depth of topsoil; thus, scrapers are always being pushed downhill side, which will give better productivity. It is important that after stripping operation is completed, the topsoil should be dragged out of the cut and onto stripped ground. Otherwise, loose topsoil inadvertently dragged onto previously stripped areas which will be difficult to salvage and quality likely to be lost.

(b) *Loaders, Truck and Dozers*

Loaders work well in 60 cm (2 ft) and thicker topsoil on flat and gently rolling topography with 60 cm or more of a subsoil which is suitable as substitute of topsoil. Thus, if topsoil and subsoil is dozed into piles or rows and loader cuts suitable subsoil under the topsoil while loading, the topsoil quality will not be impacted. While in irregular topography with shallow depth topsoil on overburden, the loader and dozer fleet will significantly cut and load overburden and deteriorate the topsoil resource.

(c) *Shovels and Trucks*

Shovels and trucks are very cost-effective combination in salvaging and moving large topsoil stockpiles. If the shovel and truck operation is trying to remove all the topsoil stockpiles, the economics of the operation will be reduced by 50% for the volume that is represented by the stockpile edges and floor.

5.3.3 Topsoil Storage (Stockpiles)

It is advisable that, in Indian conditions, stockpile the topsoil if it is absolutely essential. Topsoil stripped from an area prior to mining has to be stockpiled for reclamation purpose in later stage. It has to be stabilised from wind/water erosion and used wherever practical for establishing permanent vegetation. The topsoil stockpile should be surrounded by a silt fence. The principles of stockpiling of topsoil and the rationale behind it are as follows:

- Topsoil from the roads to and from stockpiles has to be removed prior to use. Location boundary of stockpiles should be marked before stockpiling is begun.
- Topsoil and other materials removed should be stockpiled only when it is impractical to promptly redistribute such materials on regraded areas.
- Height of the stockpiles should be kept as low as possible with a large surface area, preferably within *2* m or less. (Higher height leads to formation of anaerobic zone, and all soil properties will be lost because of anaerobicity.)
- Stockpiled materials should be selectively placed on a stable area (not disturbed) and protected from wind and water erosion and unnecessary compaction.
- The establishment of a quick-growing vegetation cover on the topsoil stockpiles is advantageous to reduce erosion and losses and may also be required by regulation. Proper construction of slopes as well as a ditch/berm (1.5 ft of higher) around the stockpile will also aid in erosion control and topsoil conservation.
- Revegetated topsoil dumps with legumes (*Stylosanthes*) and grasses (*Pennisetum pedicellatum*), which not only protect the stockpiles from wind and water erosion but also maintain active soil microbe population and help restore the nutrient cycling. This can be done either by temporary seeding or development of permanent vegetation cover:

  (i) *Temporary seeding*—protect topsoil stockpiles by temporarily seeding as soon as possible, within 30 days after the formation of the stockpile.

  (ii) *Permanent vegetation*—if stockpiles will not be used within 12 months, they should be stabilised with permanent vegetation to control erosion and weeds.

- Provide adequate access to and fro to the storage area.
- Ascertain about the quality and quantity of the material to be stored.

Note:Mycorrhiza fungi are obligate symbiont; therefore, quickly establish the vegetation cover, otherwise these important fungi will die. In Indian condition it is advisable to always identify a land from where topsoil could directly borrow and use, so that all the microbes will remain active.

5.3.3.1 Geometry of Topsoil Dump

During the storage of topsoil, care should be maintained to protect maximum level of biological activity; therefore, they should be constructed as follows:

- To provide the maximum surface area
- To have slopes capable of avoiding erosion and gully formation

In opencast mining project, geometry of dumps is controlled by space constraints, particularly, if it is stored for longer duration of time. If possible, the maximum height should be 5 m, with a slope of 1 in 3 (i.e. 18.5° to the horizontal) (Fig. 5.3). A stockpile of heavier soils should be as shallow as possible, ideally less than 1 m in height. As a rule of thumb, the following stock geometry should be maintained as far as possible to preserve the topsoil:

(a) Height of topsoil dump:

1. 5.0 m (max.) for sandy soil
2. 2.0–3.0 m for loamy soil
3. 1 m for heavy clayey soil
4. 0.5–1.0 m (max.) for intermediate soil texture

(b) Slope: less than 20–25°

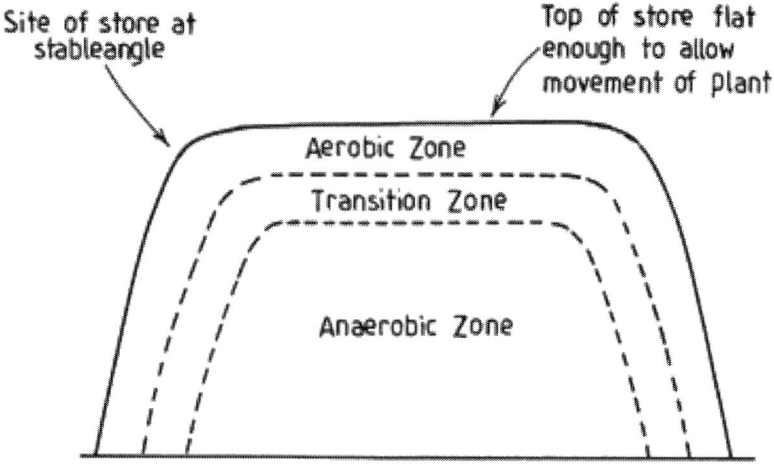

Fig. 5.3 Schematic diagram of topsoil dump (Ghose 2004)

5.3.3.2 Process of Topsoil Preservation

It is advisable to avoid topsoil storage, especially long term. However, if storage is unavoidable, upon completion of the surface of the heap, the following steps are to be followed to keep the soil in good condition:

(a) The surface should be thoroughly ripped with suitable subsoiling machinery for the purpose of (1) relieving surface compaction caused by the passage of scrapers and other machines, (2) aeration of the soil and (3) encouragement of deep-rooting plants by introduction of vegetation.

(b) Following ripping, the heap should immediately be cultivated with suitable low-maintenance species, such as dwarf grasses, to prevent erosion and gully formation.

(c) The surface vegetation should be actively maintained with seeding and weed control operations.

The important guidelines for topsoil preservation are given in Box 5.1.

5.3.3.3 Importance of Soil Microorganisms During Topsoil Preservation

Symbiotic Microorganisms

Plants form beneficial symbiotic associations with a number of soil microorganisms, including fungi, bacteria and actinomycetes (single-celled

---

Box 5.1 Important Guidelines for Topsoil Preservation

Location of stockpiling: Avoid slopes, natural drainage ways and traffic routes during the selection of stockpile location. On large sites, respreading is easier and more economical when topsoil is stockpiled in small piles located near areas where they will be used.

Sediment barriers: Use sediment fences or other barriers where necessary to retain sediment.

Temporary seeding: Protect topsoil stockpiles by temporarily seeding as soon as possible, no more than 30 days after the formation of the stockpile. Use grass–legume mixture (Dinanath grass–Stylosanthes legumes) as fast-growing cover species as well as to retain the quality of topsoil.

Permanent vegetation: If stockpiles will not be used within 12 months, they must be stabilised with permanent vegetation to control erosion and weed growth. Sowing of grass–legume mixture along with seeds of small shrubs at the periphery of the stock piles should be done.

Table 5.2 Considerations and practice in topsoil conservation (After Cooke and Jhonson 2002)

Soil characteristics	Determine depth and horizon of each profile, which will be needed to handle topsoil, subsoil and overburden separately
	Texture class, plastic limit, structural status and bulk density in relation to loss of porosity and other physical and biological changes during soil moving
Soil movement	Carefully lift, transport, store and reinstatement the topsoil to avoid compaction, killing of soil fauna and the release of dormancy of buried seed bank
	Avoid spreading over dissimilar underlying material to prevent hydraulic discontinuity and slope instability
Changes during storage	*Physical*: loss of organic matter and the alteration of binding of soil particles, loss of aggregate stability, soil compaction
	Chemical: At the centre of stockpile, anaerobic conditions develop, which leads to increase ammonium nitrogen, redox potential and pH
	Biological: initial increase in bacterial populations in response to dead fungal biomass, soil fauna and plant roots. Sharp declines in soil invertebrates especially earthworms

plants usually found in soil). Mycorrhizae are a natural component of the ecosystem in most Indian soils. They are very important in India, as they are necessary to ensure the establishment of some plant species. The majority of native plant species used in restoration probably form associations with vesicular-arbuscular mycorrhizae (VAM) and ectomycorrhizal fungi. These fungi have been shown to be effective in increasing the uptake of phosphorus by plants growing in phosphorus-deficient soils. Some species of orchids only become established in the presence of particular mycorrhizal fungi.

The ability of VAM fungi to associate with plants is rapidly depleted by topsoil disturbance and stockpiling. This often results in low levels of infection in the early years of restoration (Miller and Cameron 1976). Similarly, only limited numbers of ectomycorrhizal fungi species have been seen in recent restoration. As a result, some species may not recolonise restored areas until specific mycorrhizae have recolonised. To conserve mycorrhizal inocula, topsoil should be (a) directly returned wherever possible, and (b) when stockpiling is unavoidable, the piles should be low and revegetated as soon as possible.

Nitrogen fixation by legumes relies on a symbiotic association between the plant and the bacteria known generally as rhizobia. Rhizobia appear to be more tolerant to disturbance and stockpiling than mycorrhizal fungi. Cooke and Johnson (2002) suggested important aspects of topsoil conservation during movement and storage (Table 5.2).

5.3.4   Spreading of Topsoil

Before spreading of topsoil on derelict site, site preparation is essential for effective utilisation of topsoil. Thus, before spreading topsoil, erosion and sedimentation control devices must be established, which are consisting of diversions, berms, dikes, waterways and sediment basins. Other important measures, which are to be taken, are as follows:

- *Grading*: Maintain grades on the areas to be topsoiled according to the approved plan. Adjust grades and elevations for the spreading of topsoil.
- *Roughening*: Immediately prior to spreading of topsoil, loosen the surface by disking or scarifying to a depth of at least 10 cm (4 in.), to ensure bonding of the topsoil and subsoil, so that water can flow one layer to another. If no amendments have been incorporated, loosen the soil to a depth of at least 15 cm (6 in.) before spreading topsoil. This can be accompanied by applying 5–8 cm topsoil, tilling them into subsoil by using rotary tiller, and then remaining topsoil, that is, additional 10 cm, should be applied, so that a minimum depth of 15 cm of topsoil is applied above the subsoil.
- Uniformly distribute the topsoil to its premining thickness. If sufficient topsoil is available, a minimum compacted depth of a 50 cm on 3:1 slopes and 1 m on flatter slopes is suggested. To determine the volume of topsoil required to various depths, use the Table 5.3. Do not spread topsoil while it is muddy condition.

Table 5.3 Volumes of topsoil required for various depths (NRCS 2001)

Depth (mm)	Volume of topsoil (m^3/100 m^2)	Volume of topsoil (m^3/ha)
2.5	5.0	7.5
10	12.5	15.0
25	50	75
100	125	150
250	500	750
1,000	1,250	1,500

- *Compaction*: Compact the topsoil enough to ensure good contact with the underlying soil, but avoid excessive compaction, as it increases run-off and inhibits seed germination. Light packing with a roller is recommended where turf is to be established. In an area, where topsoil to be spread is compacted, in such case, up to 50 cm (20 in.) of surface can often be treated with ripping, before the application of topsoil or plantation.
- Strength of typical topsoil after compaction may be around 6,000 kPa, whereas limit of root growth is restricted at 3,000 kPa. Therefore, care should be taken that compact should not exceed 1,400 kPa (according to cone penetrometer guidelines) (Hanks 2003).
- Ensure that soil horizons are replaced in the same order that they were removed.
- All the tilling operation should be parallel to the contour.

Mulching: If possible, after the application of topsoil, a layer of mulch (2–3″ or 5–8 cm thick) may apply to cover topsoil and provides organic matter initially. The mulch layer will prevent soil to dry out and crusted, reduce evaporation, ameliorate extreme temperature, prevent erosion and create congenial microhabitat for soil biological community.

On slopes and areas that will not be mowed, the surface may be left rough after spreading topsoil. A disc may be used to promote bonding at the interface between the topsoil and subsoil.

After topsoil application, follow procedures for temporary or permanent seeding, taking care to avoid excessive mixing of topsoil into the subsoil.

5.3.4.1 Sources of Topsoil

Before commencement of any ecorestoration work, sources of topsoil have to be identified. There are generally two sources, either from (a) previously stored topsoil dump or (b) identify nearby land to borrow topsoil material. In Indian conditions, during pit plantation of tree species, topsoil is put in the plantation pit only along with weathered overburden in the ration of 1:4. However, it is recommended that, for ecorestoration purposes, if sufficient topsoil is available, it should be spread in the entire area. The use of topsoil during ecorestoration practices is shown in Table 5.4.

5.3.4.2 Live Topsoiling

Live topsoiling is extracting topsoil from its place of origin and placing it directly onto an area that has already been mined, backfilled and graded for reclamation. This is the most desirable topsoil management option, as the topsoil is handled only once and does not compact during storage within stockpiles.

Topsoil and other materials shall be redistributed in a manner that

- Achieve an approximate uniform, stable thickness consistent with proper contouring, surface water drainage system
- Prevent excess compaction of soil
- Protect from water/wind erosion

5.4 Alternatives to Topsoil

If topsoil is unavailable, the cost of transportation is prohibitive, or topsoil is very inferior quality, which is unsuitable for restoration, then

Table 5.4 Issues related to the use of topsoil during ecorestoration process (Maiti 2010)

Spreading in entire area	*Spreading of topsoil (by scraper)*—at least 10 cm, sometimes 15 cm or even 60 cm, thick topsoil cover is spread, which is based on type of vegetation to be raised (fruit orchards or good timber-yielding plants). Uniformly distribute topsoil to a minimum compacted depth of 2″ (51 mm) on 3:1 slopes and 4″ (102 mm) on flatter slopes
Pouring in plantation pit	Stockpile the topsoil in the plantation area itself, and put about one basket (20–30 kg) in each plantation pit; quantity depends on types of plants and quality of topsoil. However, in overburden materials, topsoil in the ratio of 4:1 is recommended
Non availability of good quality topsoil	In such situation, find out alternative to topsoil. There could be various reasons for unavailability of topsoil in plantation site, like–the cost of transportation is prohibitive or topsoil is of very poor quality or topsoil was not preserved properly. Such case improves the quality of waste materials by use of organic matter, amendments, farm yard manures (FYM), mulches, etc.
Drainage—important activity when topsoil is spread	*Drainage*—minimise percolation/infiltration by providing about 1 m deep compaction layer above which topsoil to be spread
	Provision for catch drain/sediment retention trap/garland drain with weir netting/ boulder mounting on the side wall
	Provision for installation of diversion structure

subsoil, overburden, waste rock or similar materials must be used as a substitute for revegetation (Riley 1978). These materials will generally require techniques to increase their organic matter and nutrient content. Their physical characteristics may require amelioration, and their pH may need to be adjusted. The physical and chemical properties of the proposed substrates should be thoroughly investigated prior to their use in restoration.

The following techniques which can improve the ability to support plant growth in the long term are

- Application of organic matter such as animal manures, sewage sludge or other organic wastes.
- Chemical amendments such as
  - Gypsum to improve the structure and reduce the pH of highly alkaline substrates
  - Lime to raise the pH of acid substrates
  - Inorganic fertilisers
- *Soil conditioners*: Many proprietary soil conditioners, such as polyvinyl alcohol polymers, are available which may be useful in certain situations. However, field trial should be conducted to assess the ecorestoration purposes before it is applied in larger area.
- Growing green manure crops which can be incorporated into the substrate.

- Establishing nitrogen-fixing species such as legumes to increase the organic matter and nitrogen content of the substrate.
- Applying mulch.
- Seeding rates will probably have to be increased compared to those for topsoil in order to establish a satisfactory cover of plants on these alternative substrates.

5.4.1   Evaluation of Present and Potential Productivity of Existing Overburden Material

As discussed earlier, if suitable quality of topsoil is not available easily, selected overburden material can be used as supplement of topsoil. In such cases, detailed analysis of overburden materials and laboratory trials of vegetation growth along with the various doses and combinations of amendments could be performed well in advance.

The planning process must address the following:

- Nutrient contents of overburden materials and weathering characteristics.
- Types of species tested and type and quantity of soil amendments/fertiliser used.

- Clearly mention the quantity of soil amendments/fertiliser that shall be required along with cost against topsoil redistribution.

The availability of mineral elements essential for plant growth varies considerably in strata of overburden. So, if strata of overburden contain good mineral nutrients, they can be used after testing and stored separately. Soil survey and overburden investigation before mining are used to determine the suitability as topsoil for plant growth media.

5.5 Monitoring of Post-mining Soil Development on Ecorestored Site

It is very essential to monitor and check whether the purpose for which topsoil has been redistributed is showing satisfactory progress or not in terms of following parameters, like:

- Monitor soil genesis process, in terms of accumulation organic matter, leaf litter, root density, enrichment of microbes, formation of soil structure and development of soil horizon.
- Minesoils have wider C:N ratio than natural soil, so monitor the improvement of C:N ratio.
- Monitor the improvement of bulk density of minesoils: Bulk density is closely related with machinery used. Compact zones with bulk density of 1.7–1.9 g/cm^3 all are found within 20 from the surface.
- Monitor the improvement of water in relation to (1) infiltration rates because ground cover enhances water infiltration, (2) increase in moisture content and (3) recuperation of water table.
- Microbial activity: Regularly monitor the enhancement of microbial activity in terms of CO_2- flux (soil respiration), root density, mycorrhiza colonisation, soil enzyme activity, ATP measurement, etc.

5.6 Advantages and Disadvantages of Use of Topsoil

Advantages of topsoil include its high organic matter content and friable consistence (soil aggregates can be crushed with only moderate pressure) and its available water-holding capacity and nutrient content. Most often it is superior to subsoil in these characteristics. The texture and friability of topsoil are usually much more conducive to seedling emergency and root growth. In addition to being a better growth medium, topsoil is often less erodible than subsoils, and the coarser texture of topsoil increases infiltration capacity and reduces run-off.

Disadvantages: Although topsoil may provide an improved growth medium, there may be some disadvantages. Stripping, stockpiling, hauling and spreading topsoil (importing topsoil) may not be cost-effective. Handling may be difficult if large amounts of branches or rocks are present or if the terrain is too rough. Most topsoil contains weed seeds, which compete with desirable species.

5.7 Limitations of Application of Topsoil

Do not apply topsoil to slopes steeper than 2:1 (to avoid slippage) or to a subsoil of highly contrasting texture. For example, sandy topsoil over clay subsoil is a particularly poor combination especially on steep slopes. Water may creep along the junction between the soil layers and cause the topsoil to slough. In addition, in some cases, handling costs may be too high to make topsoiling beneficial. In site planning, compare the option of topsoiling with that of existing subsoil.

References

Brady NC (2000) The nature and properties of soils, 10th edn. PHI, New Delhi

Cooke JA, Johnson MS (2002) Ecological restoration of land with particular reference to the mining of metals and industrial minerals: a review of theory and practice. Environ Rev 10:41–71

FAO (1998) Introduction, Chapter 1. In: Topsoil characterization for sustainable land management. Land and Water Development Division, Soil Resources, Management and Conservation Service, Rome. ftp://ftp.fao.org/agl/agll/docs/topsoil.pdf

Ferris KF (1996) Topsoil, Chapter 4. In: Vicklund L (section ed) Handbook of western reclamation techniques, 2nd edn. www.techtransfer.osmre.gov/NTTMainSite/Library/.../topsoil.pdf. Publish date 31 Dec 1996

Ghose MK (2004) Restoration and revegetation strategies for degraded mine land for sustainable mine closure. Land Contam Reclam 12(4):363–378

Hanks (2003) Protecting urban soil quality: examples for landscape codes and specifications. USDA Natural Resources Conservation Service, December 2003. www.soils.usda.gov/sqi/management/files/protect_urban_sq.pdf

Maiti SK (2010) Revegetation planning for the degraded soil and site aggregates in Dump sites. In: Bhattacharya J (ed) Project environmental clearance. Wide Pub, Kolkata, pp 189–228

Miller RM, Cameron RE (1976) Some effects of topsoil storage during surface mining on the soil microbiota. Fourth symposium on surface mining and reclamation, NCA /BCR coal conference and Expo III, Louisville, Kentucky, pp 131–139

NRCS (2001) Planning and design manual – topsoiling NRCS, National Resource Conservation Service. ftp://ftp-fc.sc.egov.usda.gov/AZ/ewp/erp/topsoiling.pdf

Plass WT (1978) Reclamation of coal- mined land in Appalachia. J Soil Water Conserv 33(2):56–61

Riley CV (1978) Chapter-Chemical alternations of strip mine spoil by furrow grading – revegetation success. In: Hatnik RJ and Davis G (eds.) Ecology and Reclamation of Devastated Lands, vol 2, Gordon and Breach, London, pp 315–331

Wyoming ELQD (1994) Wyoming Department of Environmental Land Quality Division: GUIDELINE NO.1-Topsoil and Overburden; Unknown/11-84; Rules Update/8-94. www.deq.state.wy.us/lqd/

Contents

S.K. Maiti, *Ecorestoration of the Coalmine Degraded Lands,*
DOI 10.1007/978-81-322-0851-8_6, © Springer India 2013

6.1   Introduction

Revegetation of mined-out areas and overburden (OB) dumps is essential for effective and successful reclamation. During revegetation planning, knowledge of the pre-existing vegetation and a vision for the development of desired ecosystem is essential. The plan should include specific strategies that are used to

- Stabilise surface materials and prevent erosion.
- Protect topsoil and accelerate pedogenesis process.
- Enhance natural vegetation growth and establish self-sustainable vegetation cover using native species.
- Ultimately support future land use identified in reclamation plan.

While determining appropriate revegetation programme for a site, the following aspects should be considered:

1. Future land use or land cover intend to be developed on derelict sites (as per closure plan report).
2. Climatic conditions including mean daily temperature, the growing season, the duration of critical moisture deficits and precipitation.
3. Size of the revegetation area in order to assess material requirements (e.g. planting stock, seed and soil amendment, geotextile).
4. Contouring of the area to mimic local topography and blend into surrounding landscape consistent with future land use or land cover.
5. Creating water bodies in the low-lying area (if possible) or other special considerations to be given for post-mining land use.
6. Availability of stockpiled materials for revegetation.
7. Success of natural revegetation and species present.
8. Contouring to ensure proper drainage or re-establish previous drainage (if possible).
9. Identify erosion-prone areas and the necessity for erosion control work, including the use of bioengineering techniques (geotextile, coir mat).
10. Minesoil characteristics including texture, pH, moisture regime, soluble salts and content of nutrients and organic matter and required amendments that may affect revegetation success.
11. Use of original or native species present on the site.
12. Reuse of soils on the site that were shifted during mining activities.
13. Timing of seeding to coincide with optimal germination times (depends on local climatic conditions).

While attempting to restore a native ecosystem, the initial revegetation effort is unlikely to produce a vegetation identical to the original (Donahue et al. 1990). The final canopy species cannot be established in the first attempt only, because other species may dominate the vegetation in the early stages of ecorestoration. The initial revegetation effort must establish the building blocks for a future self-sustaining system development so that natural successional processes lead to the desired vegetation complex. The best time to establish a vegetation cover is determined by the seasonal distribution and reliability of rainfall. All the preparatory works must be completed before the time (May–June).

Successful tree seedling establishment on drastically disturbed lands is contingent on seven major variables: (1) selection of proper native species, (2) purchase of the best-quality planting stock, (3) correct handling of planting stock, (4) correct planting techniques, (5) effective control of competing vegetation, (6) proper soil conditions and preparation and (7) weather (Miller 1999).

Ashby and Vogel in their excellent reference *Tree Planting on Mined Lands in the Midwest: A Handbook* (1993) argue that there are seven major environmental factors and physical characteristics in tree planting success: (1) climate (macro- and microclimate), (2) soil physical factors (texture, organic matter, coarse fragments, surface roughness, compaction and drainage), (3) soil chemical factors (reaction or pH, toxic elements and soil infertility, particularly of nitrogen and phosphorus), (4) competition with herbaceous ground covers, (5) lack of soil organisms, (6) mammals and birds and (7) fire.

Vogel (1987) opined that the reclamation specialist *can control* six factors that influence

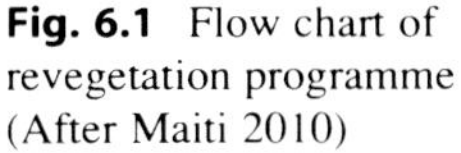

Fig. 6.1 Flow chart of revegetation programme (After Maiti 2010)

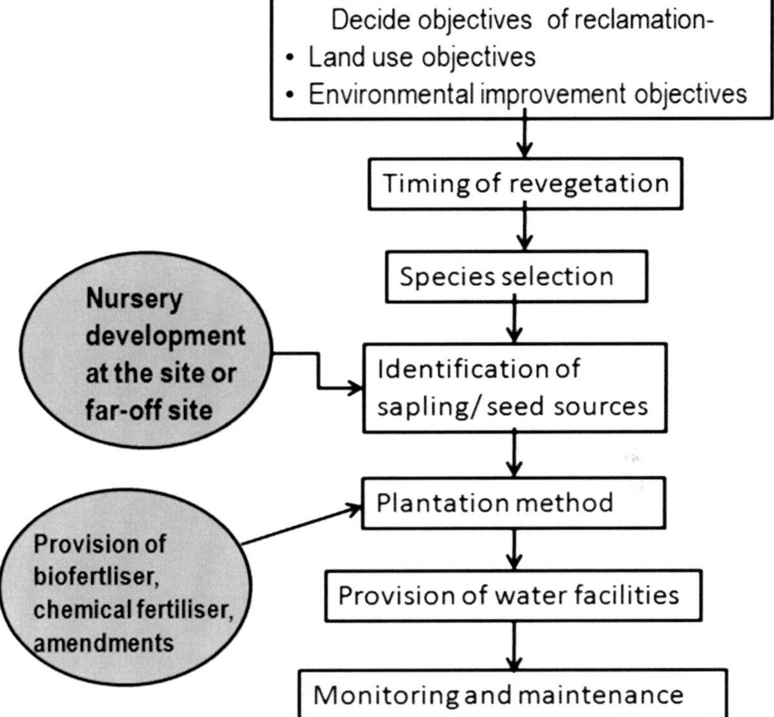

revegetation success: (1) quality of planting stock, (2) care of planting stock, (3) method of planting, (4) time of planting, (5) competition from herbaceous vegetation and (6) soil compaction.

There are several important activities which are to be planned before the onset of revegetation programme (Maiti 2002, 2010). The details of activities are shown in Fig. 6.1

1. *Fencing of the Area*: For protection of the revegetation area, there are *three types* of fencing could be used based on situations, such as (a) cattle-proof, (b) stone wall and (c) barbed-wire-cum-vegetative fence (Fig. 6.2). The seeds of local thorny fencing species like *Prosopis, Acacia nilotica, Zizyphus* and *Pithecellobium dulce* are sown before the monsoon. Along with thorny species, *Agave* and *Phoenix sylvestris* also could be planted. After the demarcation of the area by fencing, is over, the total area could be again sub-divided by placing boulder/rock (one layer only) to initiate various reclamation measured based on the suitability of the area. Once subarea demarcation (at least on soil cover basis) is over, amendments should be added or special measures should be initiated for at least to those the problematic areas. If the area is fire prone, trench fencing is recommended. For plantation in narrow patches, barbed-wire fencing is recommended.

2. Timing of revegetation: Timing—best time is 1st week of July–3rd week of July. All preparatory work must be completed.

3. Species selection: Depends on the end use of area.

4. Seed sources: Identify seed sources (collect either from mother plants or reputed seed suppliers) and create provisions for storage facilities. If seeds are to be collected directly from plants, such cases, ensure seed processing facilities.

5. Nursery development and provision of biofertiliser: Temporary nursery or permanent nursery, identification and demarcation of sites, timing, resources (manures, fertiliser, seeds, container, irrigation facilities), tools, shading arrangement and biofertiliser.

6. Plantation method: Pit plantation for tree saplings and broad-casting for grass-legumes mixture seeds.

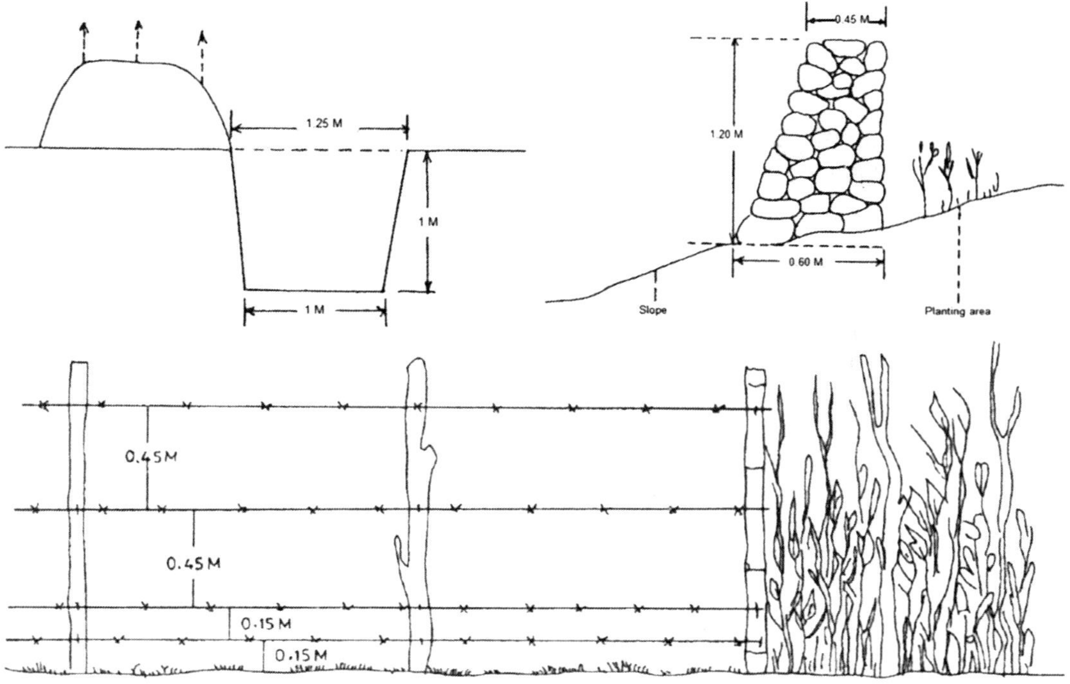

Fig. 6.2 (**a**) Cattle-proof trench, (**b**) stone wall fencing and (**c**) barbed-wire-cum-vegetative fencing

7. Provision of watering facilities: At least for the 1st year, construction of water tank at the plantation sites/identification of water sources, mode of transportation and application.

8. Application of Geojute/geotextile: Especially for steep slope and area prone to erosion, identify manufacturer, specifications (mesh size), method of laying, cost, etc.

9. Soil amendments/fertiliser: A detailed chemical analysis of the soil is essential for the planning of a revegetation programme. Analysis is needed to determine the presence or absence of essential elements for plant growth and determine those soluble elements that may be toxic to plants. After the analysis of overburden materials, the requirement of soil amendment can be decided, accordingly, provision of green manures, compost or farm yard manure (FYM), mixed fertiliser and slow-releasing fertiliser (for P: rock phosphate) could be made.

6.2 Selection of Plant Species: A Vital Component for Ecorestoration Success

The choice of plant species depends on the characteristics of spoil, climatic conditions and ultimate/desire land use. Grasses are more tolerant to adverse soil pH and moisture stress than legumes therefore, grasses are easier to establish. Pasture grasses and legume could be used for stabilisation of spoil dumps. Trees and shrubs are used where a windbreak is desired or where visual shield is required. They are also suitable to steep or rocky terrain and coarse waste. The successful vegetation establishment of waste dumps also depends on micro- and macroclimate, disease and insect resistance, competition, growth pattern and propagation ability of the species.

The availability of moisture in the spoil, especially during warm dry summer is important. If the moisture level is not high enough, irrigation may be necessary to provide moisture. As the vegetation matures, the need for the irrigation diminishes since grasses get adapted to dry summer conditions. If the objective is to restore the native vegetation and fauna, then the choice of species are predetermined.

Some indigenous species may not thrive in areas where soil conditions are substantially different after mining. If this is the case and the objective is to re-establish vegetation which fulfils the functions of the original native vegetation, then some species from outside the mining area will have to be introduced. Species that have similar growth forms to the original vegetation, and thrive in areas with comparable soil types, drainage status, aspect and climate to the restored area, are the most appropriate. One of the major approaches, at least in terms of establishment of the native ecosystem, should be to search the local area for natural analogues of the post-mining landscape and minesoils and use these as models for the proposed post-mining ecosystem. Care must be taken to avoid introducing a species, which could become an unacceptable in future, invade surrounding areas of native vegetation, or become a weed for the local agricultural industry.

Plant species can be established on restored areas from the following:

- Propagates (seeds, lignotubers, corms, bulbs, rhizomes and roots) stored in the topsoil, useful for the establishment of climbers (like Asparagus, Hemisdesmus), ferns, etc.
- Sowing seeds
- Spreading harvested plants with bradysporous seed (seed retained on the plant in persistent woody capsules) on to areas being restored
- Planting nursery-raised seedlings
- Transplants of individuals from natural areas
- *Habitat transfer*—the transfer of substantial amounts of relatively undisturbed soil with its vegetation intact from natural areas (around 1 m^2 or more in area and 200–300 mm depth)
- Invasion from surrounding areas through vectors including birds, animals and wind

6.3   Exotic Plantations and Wildlife Habitat: An Issue to Be Dealt Judiciously?

An exotic species of plant or animal is one that was '*introduced into an area where it did not occur previously*' (SER 2004). Since ecological restoration of natural ecosystems attempts to recover as much historical authenticity as can be reasonably accommodated, the reduction or elimination of exotic species at restoration project sites is highly desirable. Current reclamation and rehabilitation efforts often make use of exotic species. These practices include afforestation or plantation programmes that include large-scale introduction of fast-growing exotic species such as Subabul (*Leucaena leucocephala*), Australian acacia (*Acacia auriculiformis*), *Gliricidia sepium*, Silver oak (*Grevillea robusta*) and *Eucalyptus* species and species of commercial importance such as Teak (*Tectona grandis*) and Common Iron wood or Jungli-saru (*Casuarina equisetifolia*).

The use of non-native or exotic may be justifiable in some areas, but areas which are categorised as fragile ecosystems and biologically species rich need to be treated differently. Usually these are the unique areas that support wide variety of flora and fauna and provide ecological services. Compensatory afforestation programmes, though prescribed under various rules, do not really compensate the losses incurred due to land use conversion—in this case, forest to opencast mines. The problems of introducing exotic plants are the following:

- Alter community composition and structure by influencing energy or nutrients proportion to their biomass
- As exotics become established, reduces the share of available resources to the native or local species
- Results in loss or decrease in diversity
- Loss of food species to other small mammals and makes habitat unfit for utilisation and reduces species-rich diversity and density of bird species

6.4   Direct Seeding of Tree Species

Direct seeding is a generic term for sowing native tree, shrub and grasses seeds into prepared ground as a method of direct plant establishment. Direct seeding is now recognized as a practical and cost-effective way to establish large numbers of plants in derelict site. Direct seeding techniques can be used to revegetate large areas of cleared land for as little as one-tenth the cost of planting seedlings (Grewer 1988). The successes of direct seeding depends on a number of factors, namely, ground preparation, depth of sowing and seed size, weed control, and sowing time. These factors were considered likely causes of seedling establishment and survival problems. Reports suggested that sowing time is the most important factor influencing germination of species. Availability of soil moisture after sowing is a critical factor that determines the success of seeding coal mine degraded sites.

The growth performance of direct seeding of plant species were carried out by Jha et al. (1999) in the 12-year-old flat surface coal mine spoil located in Jayant project, Singrauli area, NCL. The seed mixture consists of leguminous trees, nonleguminous trees, leguminous forbs, grasses and crops was sown without any topsoil or fertiliser. Out of 30 plant species, 24 species showed satisfactory growth after 1 year of seeding. Seedling emergence for tree species ranged between 20 and 85%. Maximum germination occurred in *Acacia nilotica* and minimum in *Acacia tortilis*. Among five leguminous forbs, seedling emergence was greatest in *Clitoria ternatea* (80%) and lowest in *Desmodium tortuosum* (20%). Among grasses, seedling emergence ranged from 25 to 30%, and the values were not significantly different among various species. The seeding of leguminous forbs, *Stylosanthes hamata*, and grass *Pennisetum pedicellatum* and *Heteropogon contortus* in the experimental plots of flat and sloppy areas enhanced the colonisation of a large number of plant species. Jha et al. (1999) also reported that dense canopy of *Gmelina arborea* completely eliminated the seeded grass *Pennisetum pedicellatum*. However, under the dense canopy of *Gmelina arborea*, *Heteropogon contortus* grows well. The dense canopy of *Dendrocalamus strictus*, *Pongamia pinnata* and *Leucaena leucocephala* completely eliminated the seeded grass *H. contortus* and *P. pedicellatum*.

6.5   Hydroseeding

Hydroseeding, also known as '*hydromulching*', is a fast, efficient and economical process of applying seed to either good- or poor-quality areas. For difficult terrain, broadcasting by hydroseeding is commonly used. It is frequently used for mine land reclamation and for seeding inaccessible, low fertility soil and slopes. While grasses are most commonly seeded this way, other plants and most notably wildflowers are also hydroseeded. The purpose of hydroseeding is twofold:

- To improve seed germination and plant establishment
- *To slow or reduce erosion*

The basic process involves mixing mulch, water, seed and fertiliser into the tank of a hydromulch machine (Fig. 6.3). The mixed slurry is pumped from the tank and sprayed evenly onto the ground. The 'mulch' is typically a commercial fibrous product made from recycled wood or paper. The intent of the process is to provide a microenvironment beneficial to seed germination and plant establishment and to keep erosion in check during the process.

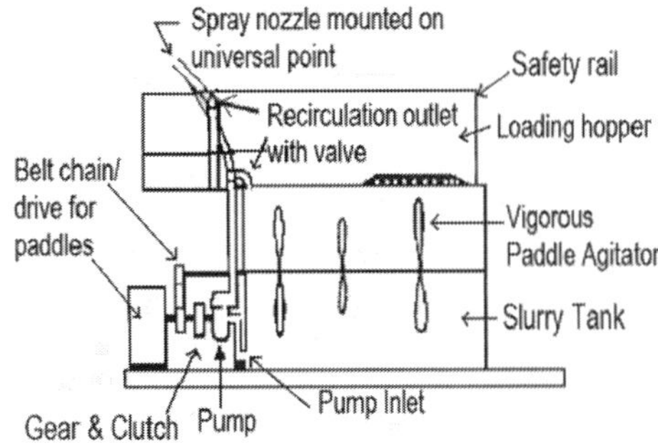

Fig. 6.3 The essential features (longitudinal section) of a typical hydroseeder. *This unit could be mounted on a trailer or on the back of a truc*

The hydroseeding is a single-step process for spraying slurry of seed, fertiliser, fertile soil, adhesive (cow dung) and water, usually a layer of 0.5–2 mm thick is sprayed. In stony and nutrient-poor area, soil-enriching media should be used in higher amount. This process should not be use during heavy rain or strong wind. Mulch applied to the ground provides significant moisture for seed germination and reduces moisture evaporation from the soil. Poor germination is almost always due to inadequate moisture during critical times. It is important to achieve the right fibre texture in your mulch mix, which will resist erosion but can still be easily pumped through the hydromulching machine.

The process begins by mixing mulch, seed, fertiliser and water in the tank of a hydromulching machine. The mixed material is then pumped from the tank and sprayed onto the ground. The material is often referred to as slurry, much like a soupy batch of green papier-mâché. Once applied to the soil, the material enhances initial growth by providing a microenvironment beneficial to seed germination.

Typical hydroseeding mixture specification, for example, comprise 100 kg/ha of seed, 50 kg/ka fertiliser and 3,000 kg/ha of mulch which is a total of 3,600 kg/ha would require a slurry suspension of 10–15%, which is too high for most machines. It would be necessary, therefore, to split the application, and apply in two stages. (Coppin and Bradshaw 1982).

6.6 Vegetation Cover Development Techniques

Herbaceous Vegetation Cover Development
- Good quality of topsoil (top 5 cm) contains seeds of grasses, rhizomes, tuber of herbs and grasses and roots, which could also be a good source of herbaceous vegetation cover.
- *Habitat transfer*—the transfer of substantial amount (around 30 cm^2 or more in area and 10–15 cm depth) of relatively undisturbed soil with its vegetation intact from nearby forest.
- Grass mulches.
- Sowing of seeds (grass–legumes mixtures).
Tree Cover

- Nursery-raised seedling or transplant of individual plants.

Different vegetation cover development techniques for restoration of coal mine degraded sites are enumerated as below (Mine Rehabilitation 2006).

6.6.1 Planting of Seedlings

The use of hand-planted seedlings has advantages and disadvantages over direct seeding. Advantages include less wastage of seed, more accurate planting densities, better survival rates (in some but not all cases) and usually better survival where weed competition is a problem. Where rapid growth is important (for example when forestry is one of the long-term rehabilitation objectives), planting seedlings may be more appropriate than direct seeding.

Disadvantages include the higher costs associated with establishing a nursery (or buying plants from a commercial nursery), and the labour costs of hand planting. However, combination of seeding and planting, could be used depending on the species being established.

Sources of local plants, the age and size of the seedling when planted, site preparation, the planting method, and the time of planting in relation to climatic conditions are all important for successful rehabilitation establishment using seedlings. For better survivability consideration should be given to:
- plant water availability.
- whether to provide water to the plants by physically watering or establishing a trickle reticulation system.
- providing the correct amount and type of fertiliser.
- providing protection from domestic stock, feral herbivores and native mammals.
- inoculating with symbiotic microbes (biofertliser).

6.6.2 Transplanting

Transplanting whole plants or clumps of plants can be an effective means of establishing certain species in some circumstances. Transplantation could be very worthy for those trees, which are very slow grower, difficult to establish through saplings, poor survival rate, may take many decades to reach maturity. The whole tree could be transplanted by using excavators and trucks to transplant in the new site.

6.6.3 Habitat Transfer

Although generally expensive and only used in specialized circumstances, habitat transfer is another option for establishing botanical diversity when other methods fail. It involves the collection and transplanting of whole clumps of plants in patches using, for example, a front-end loader. This can prove useful on a small scale where establishment of particular recalcitrant species or combinations of species is a high priority.

6.6.4 Natural Recolonisation

Natural recolonisation can, over time, result in many native plant species establishing through seed brought into a site by wind, water or fauna (such as seed in bird droppings). It is essential to understand which species will quickly recolonise in acceptable numbers and which will take much longer. However, where natural recolonisation takes a very long time, seeding or planting may be needed to establish some key species in order to meet rehabilitation objectives and stakeholders' expectations. Protection of native vegetation communities adjacent to a mine during mining operations is essential for providing a source of seed and, thereby, facilitating natural recolonisation.

6.7   Criteria for Selection of Species

The selection of plant species for a particular area depends on the quality of spoil, climatic conditions, location and needs of the surrounding. The following aspects are important to consider for any biological reclamation project (Maiti 2002; Maiti et al. 2007). Both *short-term and long-term* goals of reclamation are to be fulfilled. In present day context, the species should have:

- Economic importance
- Aesthetic value
- Ecologically sound or all three
  Along with above values, the species should be:
- Able to survive in high pollution load (dust)
- Suitable to the local climate
- Able to attenuate dust and noise pollution, which is depends on shape and size of tree, canopy density, leaf characteristics (size, shape, nature, and texture) etc.

To achieve the *long-term reclamation goal*, the nature of species should preferably be:

1. Multipurpose use
2. Faster growth
3. N-fixing capacity
4. Easy to establish
5. Tolerance to adverse climatic and spoil conditions
6. Deep root system
7. Compatible with undergrowth (synergistic effect with grass–legume mixtures)
8. Should improve spoil fertility through organic matter and nutrient cycling
9. Should have aesthetic value
10. Should be sound economically and ecologically

Primary consideration in selection of plant species also be considered based on the nature of mine spoil as well as future land use plan of the ecorestored site (Table 6.1).

Other Considerations

- Insect resistance and disease resistance
- Landscape planting (trees with rapid growth can be effective in visual screen plantation for tailing ponds, waste heaps and mine building)
- *Growth habit* – easily propagated, quick to establish, deep root system and perennial
- *Competition* – species should be chosen such that it grows favourably with other components of the mixture
- *Availability* – species should be easily available

Table 6.1 Selection criteria of plant species based in spoil characteristics and future land use pattern (Maiti 2006)

Primary consideration	Plant species selected
Nature of spoil	
Toxic metals at high concentration	Metal-tolerant plants
Toxic metals moving into herbage	Unpalatable species Fencing with spiny shrubs around site perimeter
Extreme acidity/alkalinity	Natural invaders of acidic or alkaline conditions
High level of salts	Salt-tolerant species Natural invaders of salty area
Drought conditions	Drought-tolerant species Certain-metal-tolerant cultivars
Poor nutrient status	Legumes or other nitrogen fixer Species that grow in nutrient-poor areas
Based on land use	
For wildlife propagation	Variety of native and naturalised species that provide seeds, fruits, palatable herbage, nesting site, etc.
For aboriginal or tribal use	Native species Timber, medicinal or food crops Species that regenerate after practices such as burning of forest
For amenity and recreation	Low productivity

6.8   List of Common Tree Species Used for Revegetation Programme

In India, mine-degraded areas are commonly revegetated planting nursery growing tree species by pit plantation method. For example, in the reclaimed overburden dumps of Singareni coalfields (Maiti and Reddy 2003), tree species are comprised of: *Melia azedarach, Prosopis juliflora, Gmelina arborea, Cassia siamea, Acacia auriculiformis, Dendrocalamus strictus, Ailanthus excelsa, Albizia lebbeck* and *Eucalyptus* hybrid were planted, and satisfactory growth was found for *P. juliflora, C. siamea, D. sissoo, A. auriculiformis, Dendrocalamus strictus* and *Eucalyptus hybrid.*

- *D. sissoo* and *Eucalyptus* were found suitable for places having soil cover more than 0.9 m.
- *Dendrocalamus strictus* is suitable for soil cover more than 0.6 m.
- *Acacia auriculiformis* and bamboos in alternative lines are suitable for areas having soil depth less than 0.6 m.
- In case of poor and gravelly soil, having soil depth less than 0.6 m, *A. auriculiformis* was found to be more suitable. With soil less than

0.3 m or less and eroded sites, khair (*A. catachu*) and *Acacia auriculiformis* are planted in an alternative line and are reported to be more suitable. In clay soil, only *A. catachu* gives better results.

- Shrubs: *Vitex negundo* (nirgundi), *Ipomoea carnea* (besharam), *Cajanus cajan* (arhar) and *Ricinus communis* are preferable.
- Grasses: *Pennisetum pedicellatum* (dinanath grass), *Cenchrus ciliaris, P. purpureum*, etc., also be preferred.

The survey of vegetation in the reclaimed areas of Jharia coalfield by Maiti (1995, 2003, 2006) showed that a large number of species were found to be growing well on the reclaimed overburden dumps and other mined-out areas. Those are *Melia azedarach, Gmelina arborea, Prosopis juliflora, Cassia siamea, Dalbergia sissoo, Ailanthus excelsa, A. auriculiformis, Pongamia pinnata, Eucalyptus, Albizia lebbeck, Leucaena leucocephala, Dendrocalamus*, etc. Out of these, growth of *Leucaena, Melia, Prosopis, C. siamea, C. fistula, D. sissoo, Acacia auriculiformis, Alstonia* and *Eucalyptus* were found satisfactory. These tree species are relatively easier to establish and more useful for revegetation of mine spoil and are given in Table 6.2.

Table 6.2 Type of plant species commonly used for OB dump reclamation

Trees	Characteristics
1. *Acacia auriculiformis*	Hardy, easy growing, nitrogen fixer, useful for dry and arid areas
2. *Albizia lebbeck* (siris)	Leguminous, nitrogen fixer, soil binder, dry or alkaline soil, fodder, small timber
3. *Alstonia scholaris* (chatin)	Evergreen, good dust catchers
4. *Azadirachta indica* (neem tree)	Slow growing, multipurpose
5. *Cassia siamea* (chakundi)	Easy to grow, fast growing, evergreen, nitrogen fixer
6. *Dalbergia sissoo* (shisham)	Leguminous, nitrogen fixer, hardy, dust resistant, timber
7. *Dendrocalamus sp.*(bamboo)	Good soil binder and has multipurpose use
8. *Eucalyptus spp.*	Easy growing, dense underground root system, used for land reclamation, timber, pulp and paper
9. *Gmelina arborea* (gamar)	Increases soil nitrogen, timber, pulp
10. *Grevillea pteridifolia* (silver oak)	Evergreen, high letter production, hardy, grow well in coal mine degraded land
11. *Leucaena leucocephala* (Subabool)	Nitrogen fixer, used as fodder, fast growing, restores water shades and grassland
12. *Melia azedarach* (bakain)	Medium-size tree, deciduous, fast growing
13. *Prosopis juliflora*	Xerophytes, fodder, hardy, grows in very poor areas
14. *Pongamia pinnata* (karanj)	Slow growing, evergreen
15. *Syzygium cumini* (jamun)	Fruit tree, evergreen
16. *Terminalia arjuna* (arjun)	Slow growing, medicinal
Shrubs	
1.*Vitex negundo* (nirgundi)	
2. *Cajanus cajan* (arhar)	
3. *Sesbania* sp. (dhaincha)	
Grasses	
Pennisetum pedicellatum	
Cenchrus ciliaris	
Pennisetum purpureum	

The survey of vegetation in reclaimed area of KD Heslong project of CCL showed that a large number of species were found to be growing well on the reclaimed overburden dumps and other mined-out areas. Those are *Melia, Prosopis juliflora, Gmelina, Cassia siamea, A. auriculiformis, Dendrocalamus, Ailanthus, Albizia lebbeck, Eucalyptus, Dalbergia sissoo, Pongamia pinnata* and *Leucaena leucocephala*. Out of these, *Leucaena, Melia, Prosopis, Cassia siamea, C. fistula, D. sissoo, Acacia auriculiformis, Alstonia* and *Eucalyptus* were found satisfactory. These following tree species were proved to be relatively easier to establish and more useful for revegetation of mine spoil.

6.9 Choice of Tree Plantation in OB Dumps

The compositions of trees depend on the location of the area to be revegetated. The composition of species also depends upon the local needs (that can be by the socio-economic survey). For example:

6.9.1 OB Dump: Near to the Community

The emphasis should be given on fruit trees, MPTs and aesthetically beautiful plants.

- *Fruit Trees*: Guava, jackfruit, mango, ber, citrus, jamun, bel, karand, etc. The spacing of plantation is preferably by 2.5 × 2.5 m.

- *Multipurpose Trees*

Timber: Siris (*A. lebbeck*), gamhar, Eucalyptus and sissoo

Fuel Wood: Bakin, chakundi, *A. auriculiformis, C. fistula, Lagerstroemia, Peltophorum*

Oil Seeds: Karanj (*Pongamia pinnata*), Jatropha (*Jatrohpa curcas*), Margosa tree (*Azadiracchta indica*)

Fodder: Subabul, Siris, Tamarinds and *Prosopis juliflora*

Religious Tree: Peepal (*Ficus religiosa*), Banyan (*Ficus benghalensis*), Wood apple (*Aegel mermelos*).

6.9.2 In OB Dumps

To make plantation as a barrier between office complex, colony and near potential dust sources, composition of tree species will be: Inner side of the green belt/inside of green belt *Fruit tree*—mango, jamun, etc. Outer side of green belt (towards dust source) Leafy and flowing trees, like *Alstonia* (near the edge or near to dust source) and *Cassia siamea*. Other trees that could be planted are:

- *D. sissoo* (grows well in dumps)
- *Albizia lebbek*
- *Peltophorum ferrugineum* (grows well in dumps)
- *Acacia catechu*
- *Tectona grandis*

It has been found that 8-m wide green belts between road and building can reduce the dust fall by 2–3 times.

6.9.3 Extreme Environmental Conditions

In the hazardous dumps like problems of fire, unavailability of topsoil or practically no soil cover, the following hardy species should be planted: *Acacia nilotica*, *A. auriculiformis* and *A. tortilis*. Species of *Prosopis* like *P. chilensis*, *P. cineraria* and *P. juliflora* are planted to get the benefits of fodder for cattle and supply of firewood. The *Prosopis* sp. has synergistic effect on grass. In the inter-spacing of *Prosopis*, grasses like *Cenchrus ciliaris*, *Dichanthium*, *Pennisetum* and *Stylosanthes* could be sown.

6.10 Suitability of Direct Seeding of Species in the OB Dumps

The seeding rates and germination percentage for direct-seeded grasses, legumes, crops and forest plant species on 5-year and 10-year-old sites are shown in Table 6.3 (Jha et al. 1999).

The studies conducted in Jhingurda coal mine spoil, NCL during 1985–1987, Jha and Singh (1992) suggested the following species for revegetation of old coal mine spoil dumps:

Trees	Herbaceous legumes	Grasses
Pongamia pinnata	*Stylosanthes humilis*	*Chysopogon fulvus*
Acacia tortilis	*Sesbania aegyptiaca*	*Bothriochloa pertusa*
Acacia nilotica	*Clitoria ternatea*	*Dendrocalamus strictus*
Albizia lebbeck	*Stylosanthes hamata*	*Bothriochloa intermedia*
Albizia procera	*Desmanthus virgatus*	*Cenchrus setigerus*
Acacia auriculiformis	*Dichrostachys cinerea*	

6.11 Tree Species Suggested by CPCB for Minesoil Reclamation

Suitability of tree species as suggested by Central Pollution Control Board (CPCB), New Delhi, (2000) for revegetation of mine spoil and degraded area are depicted in Table 6.4.

6.12 Pollution-Tolerant and Pollution-Sensitive Tree Species

Based on intensive vegetational surveys, supplemented with the transplant and laboratory experiments, a list of 50 species of pollution-tolerant flowering plants has been prepared. Fourteen sensitive plant species have also been identified which can be used as bioindicators of air pollutants. Plant species listed under tolerant or sensitive heads are mainly relevant to pollutants emanating from thermal power plants and coal-fired industries (chiefly SO_2 and particulates) and have been studies in the agro-climatic conditions of north Indian plains. Most common tolerant and sensitive species are mentioned below (Table 6.5).

Table 6.3 Seeding rate and seed germination (%) of trees, leguminous crop and grass–legume mixture sown in Jhingurda coal mine spoil, NCL (After Jha et al. 1999)

Species	Seeds/m^2	Germination percentage	
		5-year-old site (1986)	12-year-old site (1987)
Trees			
Acacia nilotica	100	52	50
Pongamia pinnata	25	20	60
Prosopis juliflora	100	16	–
Albizia procera	100	7	10
Albizia lebbeck	100	–	60
Acacia auriculiformis	100	2	40
Acacia tortilis	100	–	20
Acacia catechu	100	3	–
Leucaena leucocephala	250	2	–
Leguminous crops			
Phaseolus aureus	100	32	65
Phaseolus mungo	100	24	60
Cajanus cajan	50	10	60
Pennisetum typhoides	100	–	20
Legumes/grasses			
Clitoria ternatea	100	61	80
Desmanthus virgatus	100	16	55
Stylosanthes humilis	100	8	–
Stylosanthes hamata	100	4	45
Dichrostachys cinerea	100	10	50
Cenchrus setigerus	200	–	50
Sesbania aegyptiaca	50	–	20
Desmodium tortuosum	50	–	25
Chysopogon fulvus	200	–	20
Bothriochloa pertusa	200	–	30
Bothriochloa intermedia	200	–	25
Dendrocalamus strictus		–	

Table 6.4 Suggested multipurpose tree (MPTs) for revegetation of mine spoil (CPCB 2000)

Acacia catechu	*Acacia nilotica*	*Acacia tortilis*
Albizia procera	*Albizia lebbeck*	*Azadirachta indica*
Casuarina equisetifolia	*Dalbergia sissoo*	*Dendrocalamus strictus*
Gmelina arborea	*Holarrhena antidysenterica*	*Holoptelea integrifolia*
Leucaena leucocephala	*Madhuca indica*	*Melia azedarach*
Phyllanthus emblica	*Pongamia pinnata*	*Prosopis cineraria*
Sesbania sp.	*Shorea robusta*	*Syzygium cumini*
Tamarindus indica	*Tectona grandis*	*Terminalia arjuna*
Terminalia bellerica	*Zizyphus mauritiana*	

Fig. 6.4 Pit plantation technique

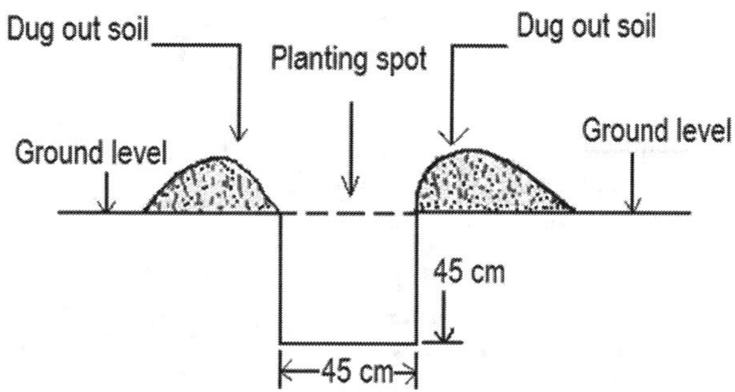

Table 6.5 Pollution-tolerant tree species for revegetation programme in coal mine spoil (Maiti 2010)

Pollution-tolerant plants	
Botanical name	Common name
Acacia arabica Willd	Kateria babul
Aegle marmelos Correa	Bel
Ailanthus excelsa Roxb.	Mahaneem
Albizia lebbeck Benth	Siris
Alstonia scholaris R.Br	Chitwan
Azadirachta indica A. Juss.	Neem
Bougainvillea spectabilis Willd.	Bougainvillea
Dalbergia sissoo Roxb.	Shisham
Ficus benghalensis L.	Bargad
F. infectoria Roxb.	Pakar
Lagerstroemia flos-reginae	Jarul
Leucaena leucocephala Benth.	Subabul
Madhuca indica J.F. Gmel.	Mahua
Mimusops elengi Sieber ex A. DC.	Bakul, Maulsri
Phoenix sylvestris Roxb.	Khajur
Phyllanthus emblica L.	Amla
Pithecellobium dulce Benth.	Jangal jalebi
Polyalthia longifolia Benth. & Hook.	Ashok
Tamarindus indica L.	Imli
Zizyphus mauritiana Lam.	Ber
Anthocephalus cadamba Miq.	Kadamb
Delonix regia Raffin.	Gulmohar
Bauhinia variegata	Kachnar
Cassia fistula L.	Amaltas
Morus alba L.	Shahtoot
Mangifera indica L.	Aam
Litchi chinensis Sooner.	Lichi

6.13 Techniques of Tree Plantation

Trees are introduced as partly grown plants (by nursery) as pit plantation, though direct seeding is also practised (e.g. *Acacia, Prosopis, Leucaena*).

1. *Pit plantation*: The pit size is generally 45 cm × 45 cm and filled with OB/topsoil, 4:1, with 10–15 kg of FYM. 10 g BHC is recommended for sandy area where termite attack is envisaged (Fig. 6.4).

2. *Notch planting*: A notch is made in the erosion-prone area, unlevelled area and rock area with a spade; the notch can be a single slit, to T- and L-shaped. The roots are inserted into the notch, spread out, and the spade withdrawn. Finally, the notch is closed firmly with the foot. Notch planting is suitable for seedlings/sapling/bigger size seed (Fig. 6.5).

3. *Pocket planting*: Suitable for rocky overburden dump plantation. It is just like a pit planting. A hole or depression is created by excavating or moving the rocks. The hole is filled with manures (one basket—15 kg), 200 g of slow-releasing fertiliser and 150 g limestone (required for low pH spoil material) and watered. A small sapling is plated into the pocket. Instead of manures, topsoil could be used (Fig. 6.6).

4. *Contour Terracing and Vegetation in Overburden Dump Slope and Vegetation*
   In overburden dumps, trenches on the slope contours to detain water and sediment transported by water or gravity down slope need to be constructed. These contour terraces or contour furrowing shall be lined with geotextiles or filled with rock and used as erosion control structure.
   The main purposes of contour trenches are to break up the slope surface, to slow run-off and

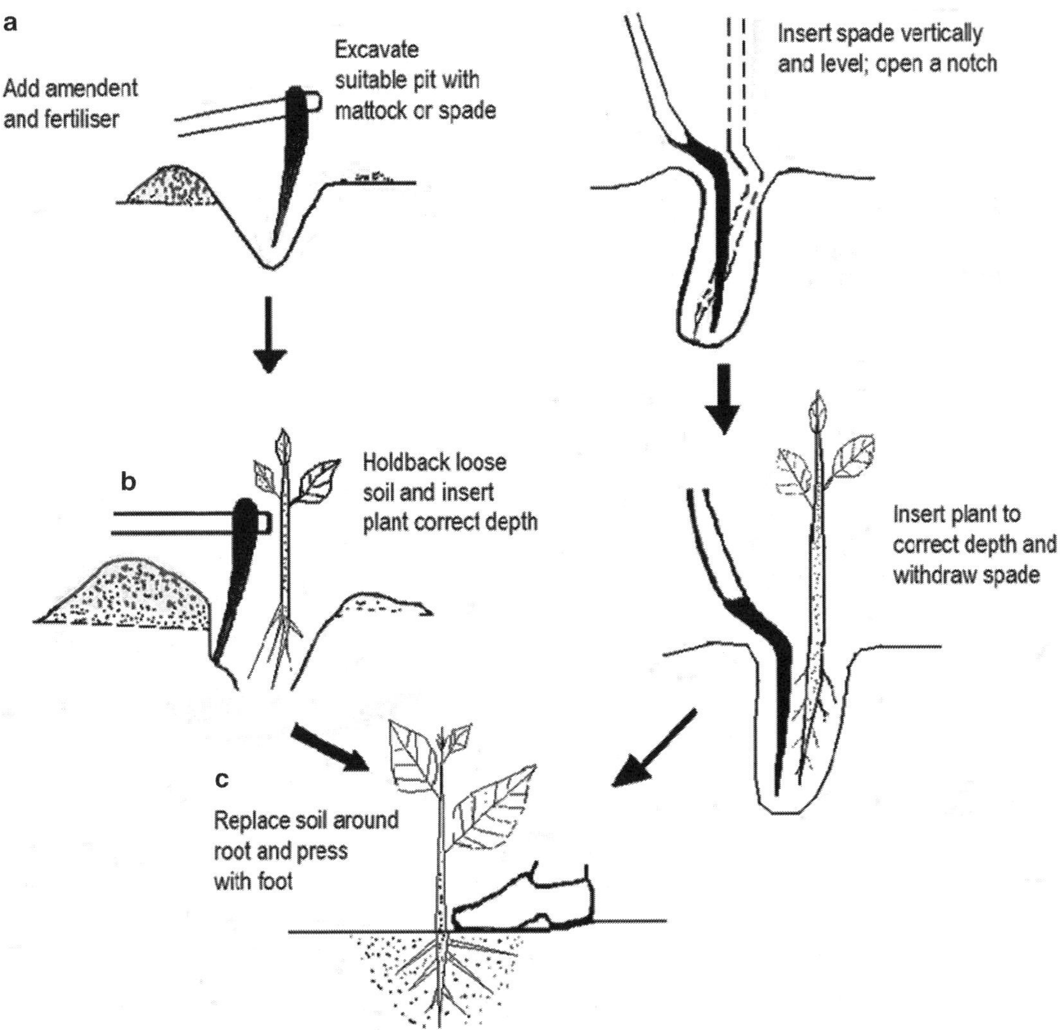

Fig. 6.5 Notch planting techniques for sapling plantation (Coppin and Bradshaw 1982)

allow infiltration and to trap sediment. Rill formations can be stopped by the trenches. Trenches or terraces are often used in conjunction with seeding. They can be constructed with machinery (deeper trenches) or by hand (generally shallow). Width and depth vary with design storm, spacing, soil type and slope. Trenches trap sediment and interrupt water flow, slowing run-off velocity. They work best on coarse granitic soils. When installed with heavy equipment, trenches may result in considerable soil disturbance that can create problems.

Trenches must be built along the slope contour to work properly. Trenches have high visual impact when used in open areas but tend to disappear with time as they are filled with sediment and covered by vegetation. A trench about 10 ft long, one foot wide and one foot deep is very effective. Detailed layout of contour terracing along with vegetation scheme is shown in Fig. 6.7.

6.14 Case Study I: Plantation Activities in KDH Project (CCL Area)

The KDH project is located in the North Karanpura (NK) area, under Central Coalfield Limited

Fig. 6.6 Method of pocket planting on coarse rocky spoil (Coppin and Bradshaw 1982)

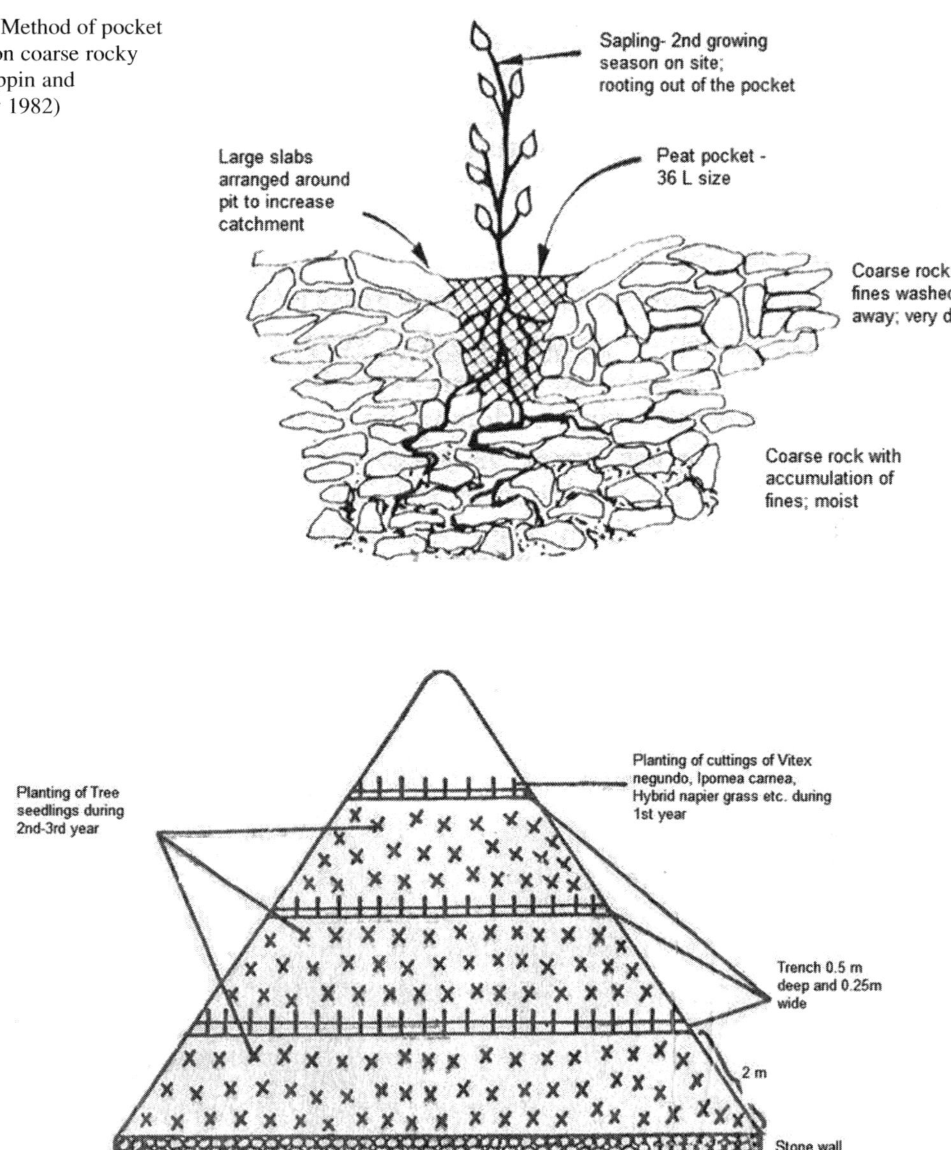

Fig. 6.7 Contour trenching on slope and vegetation establishment in overburden dump

(CCL) in Ranchi district, Jharkhand (India). The field analysis revealed that 12 types of tree species were planted in the overburden dumps (Table 6.6). The plantation work has been done by State Forest Department, Jharkhand. Out of 12 trees species, 2 trees, namely, *Gmelina arborea* (white teak or gamar) and *Dalbergia sissoo* (shisham) constituted more than 55% of tree popula-tion in the dumps. Later phase of afforestation, *Acacia harecia* (*Acacia mangium*) was planted. *A. harecia* grows well in the dumps, and it has good self-regeneration capacities and evergreen in nature. However, *A. harecia* withered after 5–6 years. Field sampling revealed that out of all the plants, *D. sissoo* and *Cassia siamea* grow very well in the acidic dumps (Maiti and Ghose 2005).

Table 6.6 Plantation at KDH external OB dump during 2002 monsoon (Maiti and Ghose 2005)

Sl. No.	Types of trees	No. of trees	% of species
1.	*Gmelina arborea* (Gamhar)	3,250	32.5
2.	*Dalbergia sissoo* (Shisham)	2,500	25.0
3.	Bamboo (*Bambusa* ssp.)	1,500	15.0
4.	Samol (*Bombax ceiba*)	500	5.0
5.	Chakundi (*Cassia siamea*)	525	5.25
6.	Acacia harecia (*Mezium acacia*)	400	4.0
7.	Karanj (*Pongamia pinnata*)	300	3.0
8.	Jacaranda (*Jacaranda mimosifolia*)	300	3.0
9.	Bakain (*Melia azedarach*)	275	2.75
10.	Imli (*Tamarindus indica*)	250	2.50
11.	Neem (*Azadirachta indica*)	200	2.0
	Total no. of trees	10,000	100

6.15 Case Study II: Plantation Activities of Chirimiri Area (SECL)

The reclamation strategies adopted in the Kurasia opencast mines in Chirimiri area of SECL was studied during July 2005 and April 2006 (Maiti 2007). The overburden dumps and mined-out areas were reclaimed by plantation of tree species. The plantation was carried out by Chhattisgarh Van Vikas Nigam at the rate of Rs.50/tree sapling, which includes digging of pits and overall maintenance for 3 years. To assess the biological reclamation aspects, different aged reclaimed overburden dumps were selected on the basis of age of plantation, and spoil samples were also taken from the rhizosphere of different tree species to study the amelioration of dump physico-chemical and notional properties.

- Tree species composition in young vegetation stand (2 years old, plantation year: 2003–2004).
- Tree species composition in intermediate age of 4 years old, plantation year 2001.
- Tree species composition in old reclaimed dumps (6 years, plantation carried out in 1999).

6.15.1 Tree Species Composition in Young Vegetation Stand

Tree plantation was carried out by pit digging followed by sapling plantation. The following tree species were found growing satisfactorily on the 2-year reclaimed overburden dumps of Kurasia OCP, which was the youngest reclaimed OB dump of the area. The details of tree species composition are Arjun (*Terminalia arjuna*), Gamar (*Gmelina arborea*), Subabool (*L. leucocephala*), Sissoo (*Dalbergia sissoo*), *A. auriculiformis*, Sitaphal (*Annona squamosa*), Karanja (*Pongamia pinnata*), Bakayan (*Melia azedarach*), Silver oak (*Grevillea robusta*), Chakundi (*Cassia siamea*) and Vilayati imli (*Inga dulcis*). Under the planted tree species, natural succession of herbs has been observed and dominated by gigantic herb *Hyptis suaveolens* (Ban tulsi), *Calotropis procera* (Akand), *Lantana camera* (Chotra) and grasses. Other common small herbs were *Scoparia dulcis, C. sparciflorum*, etc.

6.15.2 Tree Species Composition in Intermediate Age

Eight tree species comprising both deciduous and evergreen types were planted for the reclamation of the OB dumps. Average aerial heights of the trees were measured between 3.0 and 4.0 m. Out of eight trees, two tree species *Inga dulcis* and *Psidium guajava* were fruit tree, which attract avifauna. Generally, deciduous tree yields higher amount of litter fall that ameliorates dump properties by increase in moisture-holding capacity, organic matter,

Photo 6.1 (a) Row plantation of *Grevillea pteridifolia* on reclaimed coal mine overburden dump of Kurasia OCP, Chirimiri areas, SECL (Photo Maiti 2003). (b) Showing growth and decaying of grasses and litter accumulation under the *Grevillea pteridifolia* on reclaimed coal mine overburden dump of Kurasia OCP, Chirimiri areas, SECL (Photo Maiti 2003)

nutrients and microbial activities. Maximum litter fall observed underneath of *Casuarina equisetifolia* (common Casuarinas) and *Tectona grandis* (Teak). In these reclaimed OB dumps, the afforestation of tree species has been done systematically, for example, similar type of tree species was planted in each row. One row of *Cassia siamea* (chakundi, aerial height 2.5–3 m) is followed by one row of *Grevillea pteridifolia* (aerial height 4–4.5 m), one row of *C. equisetifolia* (guava: 2–2.5 m), one row of *Gmelina arborea* (aerial height 3.5–4 m) and one row of *Inga dulcis* (Vilayati imli: aerial height 3.5–4 m) and again *Cassia siamea* (chakundi). In every 5th row, *Cassia siamea* (chakundi) was planted. Other plants like *Tectona grandis* (teak), *Acacia catachu* (khair) and *Pongamia pinnata* (karanja) were found growing satisfactorily. This dump remains green even in the month of summer (May–June) due to the evergreen nature of tree species planted like *Cassia siamea*, *Grevillea pteridifolia* and *Casuarina equisetifolia*. The growth of *Inga dulcis*, *Terminalia arjuna*, *L. leucocephala* and *Agave sisalana* in the reclaimed area was seen satisfactory. The row plantation of *G. pteridifolia* is shown in Photo 6.1a. Invasion of common shrubs, herbs and grass species was observed in these 5-year-old reclaimed dumps due to *natural succession*. Vigorous growth of *Lantana camera* (shrubs) in the dumps was noticed during the survey.

Table 6.7 Details of floristic survey in the afforested OB dumps of Chirimiri area, SECL

Name of plants	Nos.	Aerial height (m)	DBH (cm)
Cassia siamea	7	5.5–6	30–32
Pongamia pinnata	7	5–5.5	20–24
Acacia auriculiformis	5	5.5–9.2	25–32
Grevillea robusta	3	7–7.3	35–40
Psidium guajava	2	2	–
Acacia catachu	1	4	15
Total	25		

Natural growth of grasses and litter production by tree species (*G. pteridifolia*) are essential for the amelioration of mine spoil properties on dump (Photo 6.1b).

6.15.3 Tree Species Composition in Old Reclaimed Dumps

Major dominant planted tree species was *C. siamea* (chakundi) attaining a height of 2.5–4 m. Other plants consist of *A. auriculiformis*, *P. pinnata* (karanj) and *C. equisetifolia* and fruit trees like guava and jamun were also planted. In an old reclaimed area, quantitative tree survey has been conducted, and details are presented in Table 6.7. The density of stem was found 2,500/ha. Maximum aerial height and DBH were observed in *G. robusta*.

References

Ashby WC, Vogel WG (1993) Tree planting on Mined Lands in the Midwest – a handbook. Coal Research Center, Southern Illinois University, Carbondale, p 115

Coppin NJ, Bradshaw AD (1982) Quarry reclamation. Mining Journal Books, London

CPCB (2000) Guidelines for developing green belts, Programme objectives series: PROBES/75/1999-2000. Central Pollution Control Board, New Delhi, p 203

Donahue RL, Miller RW, Shickluna JC (1990) Soils – an introduction to soils and plant growth, 5th edn. PHI, New Delhi

Grewer G (1988) Direct seeding of trees. In: The Esperance Tree Book. Esperance Soil Conservation District Advisory Committee, Esperance WA, pp 37–48

Jha AK, Singh JS (1992) Influence on microsites on redevelopment of vegetation on coalmine spoils in a dry tropical environment. J of Env Manage 36: 95–116

Jha AK, Singh A, Singh AN, Singh JS (1999) Tree canopy development in young plantations raised on coalmine spoil affects the growth of herbaceous vegetation. Indian For 125(3):305–307

Maiti SK (1995) Some experimental studies on Ecological aspects of reclamation in Jharia coalfield. Ph.D dissertation, Indian School of Mines, Dhanbad

Maiti SK (2002) Ecological environment. In: Saxena NC et al (eds) Environmental management in mining areas. Scientific Publishers, Jodhpur, pp 110–141

Maiti SK (2003) MoEF report on "An assessment of overburden dump rehabilitation technologies adopted in CCL, NCL, MCL and SECL mines" (No. J-15012/38/98-IA II (M)). MOEF, New Delhi

Maiti SK (2006) Ecorestoration of coalmine OB dumps – with special emphasis on tree species and improvements of dump physico-chemical, nutritional and biological characteristics. MGMI Trans 102(1&2):21–36

Maiti SK (2007) Minesoil properties of different aged reclaimed coal mine overburden dumps of Korba, Gevra and Kusmunda area of SECL, India. MINE-TECH 28(2&3):93–98

Maiti SK (2010) Revegetation planning for the degraded soil and site aggregates in Dump sites. In: Bhattacharya J (ed) Project environmental clearance. Wide Publishing, Kolkata, pp 189–228

Maiti SK, Ghose MK (2005) Ecorestoration of acidic coalmine overburden dumps – an Indian case studies. Land Contam Recl 13(4):361–369

Maiti SK, Reddy MS (2003) Nutrient accumulation in reclaimed overburden dumps of Ramagundam OCP-1, SCCL, India. In: Srivastava BK (ed) Proceedings of "Environmental Management in Mines" (SEMMI 2003). Mining Engineering Department, BHU, Varanasi, 17–18th Jan 2003, pp 249–256

Maiti SK, Shee C, Ghose MK (2007) Selection of plant species for the reclamation of mine degraded land in the Indian context. Land Conta Recla 15(1):55–66

Miller S (1999) Successful tree planting techniques for drastically disturbed lands: a case study of the propagation and planting of container-grown oak and nut trees in Missouri. Online document: http://www.mcrcc.osmre.gov/PDF/Forums/Reforestation/Session%204/4-6.pdf

Mine Rehabilitation (2006) Leading Practice Sustainable Development Program for the Mining Industry, Gov of Australia. http://www.dmp.wa.gov.au/documents/mine_rehab.pdf

SER (2004) The SER International Primer on Ecological Restoration. www.ser.org & Tucson: Society for Ecological Restoration International

Vogel WG (1987) A manual for training reclamation inspectors in the fundamentals of soils and revegetation. USDA, Northeastern Forest Experiment Station, Berea, KY

7

Contents

7.1 Introduction

A typical seed includes three basic parts: (1) an embryo, the baby plant; (2) an endosperm – a source of supply of nutrients to the embryo and (3) a seed coat. The seed coat (testa) develops from the integument of the ovule. It can be a paper-thin layer (e.g. peanut) or something more thick and hard (e.g. Gulmohor, Teak, *Gmelina*). The seed coat helps in protecting the embryo from mechanical injury and from drying out. Seeds serve several functions for the plants, amongst them are nourishment of the embryo, dispersal to a new location, and dormancy during unfavorable conditions. Seeds are fragile, living organisms, and the self-life of the seed is affected at the beginning of the plant life cycle. Generally, seeds are stored for more than 1 year, hence it is important that after harvesting of seeds, they should be processed and stored properly. Most important factors that effect the seed viability and vigour are time of harvesting, methods of extraction and cleaning, transportation and storage. It is easy for seeds to become damaged at any of these stages.

7.2 Seed Collection

Seeds may be collected from the naturally growing mother plants or directly purchased from the market. During the natural collection, following points should be taken into consideration:

S.K. Maiti, *Ecorestoration of the Coalmine Degraded Lands*,
DOI 10.1007/978-81-322-0851-8_7, © Springer India 2013

- It is better to establish seed-orchards for species those are rare, produce limited number of seeds; seeds are difficult to collect, or seeds are not easily available.
- Seeds should be collected only from well-grown and healthy trees. Too old and too young trees should not be used for this purpose.
- Collect information well in advance when (i) a particular species produces flower, (ii) time taken for maturation of seeds and (iii) identify the mother plant from where seeds are to be collected.
- Collect seeds only when they are mature. Some seeds are fall quickly, so collect as soon as possible for these species (i.e. Margosa seeds). Some seeds are hanged on long period, and in such cases, it may be collected later on (i.e. *Peltophorum, Cassia fistula, Legistromea* etc.).
- Seeds should be thoroughly cleaned and free from impurities.
- Seeds should be stored in clean gunny bags or in suitable containers. The store should be free from damp, excessive heat, pests and insects. *The container should not be air tight as the seeds need oxygen.*
- Each container should have the following information written and enclosed in it:
  (a) Name of the tree and variety
  (b) Date and place of collection
  (c) Weight of the seed

Purchasing of seeds from market
- Purchase from reputed seed merchants
- Arrange for long-term contract for seed supply
- Get information about the date of seed collection and condition of storage
- Check the rate of germination

7.3 Seed Processing (Pods, Capsules, etc.)

When seeds are ready to be processed, the entire seedpod, capsule or seed head will become brown and dry. During the maturation process, the ripening pods and capsules change colour from green, to yellow-green, to yellow, to light brown, to a darker brown or dark grey. Ripening and maturation may be uneven within the pod or capsule, uneven on the plant or uneven within the stand of plants. For this reason, the pods of many plants are harvested individually. Seeds of legumes often develop a split along one side of the pod. This is the best time to collect the seed, before the pods start to open and scatter their seeds. After harvest, seeds are thrashed to remove the seeds from the surrounding plant material. A period of air-drying is important before seeds are thrashed.

7.4 Drying of Seed

Drying is a normal part of the seed storage process. Seeds should be dried as quickly as possible after washing. Reduction of moisture content of the seed is important in order to maintain seed viability and vigour. If seeds are not dried properly, quality will be lost due to mould (fungus) growth. Slow drying may result in mould growth or premature sprouting of the seed. Some seeds must dry down to minimum moisture content before they can germinate. Low seed moisture content is a prerequisite for long-term storage and is the most important factor affecting longevity. Seeds lose viability and vigour during processing and storage mainly because of high seed moisture content. Silica gel is the most effective desiccant (moisture absorbing material) for drying seeds.

Methods of drying of seeds:
Two methods of drying of seeds are practised: (a) sun drying and (b) artificial (forced) air drying. Sun drying method is cheap. Generally, drying of seeds to 4–6% moisture content decelerates the rate of seed deterioration substantially.
- Seeds are extracted from hard fruits by drying the capsules or pod in the sun and remove by thrashing.
- Seed should be cleaned before storage.

- Clean seed should be stored in dry containers after mixing with insecticides.
- Loss of seed viability during storage is common (ascertain the viability of seed by germination test).

7.5 Seed Storage and Longevity

The general effect of temperature on longevity is that lower the temperature, higher the longevity of seeds. This is true for 'orthodox' seeds, that is, most seeds that follow some general 'rules of thumb' regarding longevity during the storage life of the seeds. The relationship between temperature and seed longevity is that for each 5.6°C (10°F) decrease in temperature, longevity doubles (Harrington 1972). This rule applies to seeds stored between temperatures of 0 and 50°C. For short-term storage, a simple thumb rule can be used between temperature and relative humidity. The storage temperature (in deg. F) plus relative humidity should be equal to 100. For example, 60°F (15.6°C) at 40% RH or 55°F (12.78°C) at 45% RH or 65°F (18.33°C) at 35% RH. For long-term storage of seeds, temperature does not exceed 10°C (50°F), and the RH is not higher than 50%; most species of seeds will retain their full initial viability for a few years

7.6 Seed Testing (Germination Rate)

The following conditions are necessary for germination:
1. *Water or moisture*: No seed germinates if they are kept in a perfectly dry condition. Water is essential to bring about vital activities like metabolism of dormant embryo to dissolve various salts, hydration of organics stored in the cotyledons and softens seed coat, which helps the embryo to come out.
2. *Supply of oxygen or air*: Seeds fail to germinate in absence of air (oxygen). It is observed that if seeds are immersed under water they do not germinate. Thus, constant supply of air is required for the respiration in germinating seeds.
3. *Suitable temperature:* Seeds fail to germinate in very low temperature, and germination is retarded even water-soaked seeds kept under high temperature. To carry out the vital activities of protoplasm, optimum temperature is essential. For tropical climate, 20–40°C is optimum, and in temperate climate, it may be low. Most seeds fail to germinate below 0°C and above 50°C.
4. *Light*: It has not direct effect on germination, as germination does take place even in night. But light facilitates and hastens embryo to grow perfectly, so light indirectly influence germination.

Changes during seed germination:
- Intake of water through micropyle (a tiny hole in the seed coat) and imbibitions to the seed causes increase in volume.
- Pressure exerted due to swelling will rupture the seed coat.
- Dry seed coat, which was impermeable to gases in dry conditions, now becomes permeable because of imbibitions of water into seed.
- The imbibe water helps in digestion of reserve food and converts it into simpler form which is used up by embryo.
- Due to assimilation, growth of embryo takes place and ultimately seedling with root and shoot develops.

Seed germination, generally expressed as percentage, measures the number of seeds in a lot that can be expected to germinate and grow into a healthy plant. Seed vigour is defined by normal seedling morphology and the rate at which seeds germinate and grow in the early stages. Strong seed vigour has many advantages because it can fight against diseases, weeds and insects than the weak seeds.

Seed testing for germination and purity is normally undertaken by the seed testing laboratory. Seeds are tested under their optimum conditions of light, temperature and humidity with required pretreatment to overcome dormancy. Seed test can be performed by placing some seeds on moist absorbent paper (e.g. blotting paper or filter paper) in a covered Petri dish. The numbers of seeds germinated are counted at 7, 14 and finally 21 days and expressed as a percentage of total seeds sown. A real value (RV) or pure live seed (PLS) count can be calculated for any seed lot by the following formula:

$$\text{RV or PLS } (\%) = \frac{\%\ \text{purity} \times \%\ \text{germination}}{100}$$

Seed that has low purity or germination due to long storage, for example, can be retested, and sowing rates is increased to compensate for the reduced PLS count.

In the laboratory, seed germination test are conducted under ideal conditions and may not reflect the actual harsh field conditions. Therefore, it may be preferable to conduct seed tests under less ideal conditions, such as high temperate, reduce moisture, etc. Intermediate counts of germination at 7 and 14 as well as 21 or 28 days (for grass and legumes, woody seeds) will give an indication of the speed of germination (germination vigour).

Germination vigour is a useful concept when quick establishment, particularly of grass seeds, is considered. Germination vigour is simply the number or proportion of the seeds that have germinated at the first (7 days) count during the seed test. Germination vigour could also be expressed as the length of time taken, say 50% of the total or pure live seed to germinate.

7.7 Causes of Seed Dormancy

A large number of ripen seeds fail to germinate even after a lapse of particular time period is over in spite of providing all favourable environmental conditions. Seed dormancy is defined as a seed fails to germinate under environmental conditions optimal for germination, normally when the environment is at a suitable temperature with proper soil moisture. This temporary suspension that is definitely due to some internal conditions within seeds that inhibit germination is known as dormancy. Seed dormancy introduces a temporal delay in the germination process that provides additional time for seed dispersal over greater geographical distances. It also maximises seedling survival by preventing germination under unfavourable conditions. The most important impression regarding the dormancy is that it has a "positive value to the plant".

The important factors that cause dormancy are-

1. *Impermeability of seed coat to water*: At maturity, a large number of legumes develop seeds with hard thick seed coats which are completely impermeable to water. Germination cannot take place until water penetrates through the seed coat.

2. *Mechanically resistant seed coat:* In a number of plants, the seed coats are made with some stony cells which are critical in mechanically limiting the enlargement of embryo. Seeds of a number of plants retain a considerable period of dormancy as their coats are strong enough to prevent any expansion of embryo.

3. *Seed coat impermeable to oxygen*: The seeds of a number of grasses and many members of Compositae family have their seed coat, which limits the gaseous exchange and causes a prolonged dormancy.

4. *Growth inhibitors*: Seed coats and pericarp contain relatively high concentrations of growth inhibitors that can suppress the germination of embryo. They are very common and wide spread in the nature, such as coumarin, parasorbic acid, ammonia, phthalides, ferulic acid, dehydracetic acid and abscisin II (Taiz and Zeiger 1998). Abscisic acid (ABA) is a common germination inhibitor present in seeds, and repeated washing (*leaching*) removes dormancy (12–24 h soaking is sufficient). Seeds with hard seed coats can be soaked in hot water to break open the impermeable cell layers that prevent water intake.

7.8   Seed Treatment to Overcome Dormancy

Most grasses and legume seeds (herbaceous and forage categories) have little seed dormancy and do not require treatment to increase germination. Many woody seeds have some inbuilt dormancy. Some species have a hard, impermeable seed coat that prevents the seed imbibing sufficient water to germinate. This seed coat dormancy will gradually overcome as the coat is damaged or soften over time, and this can be hasten by a simple scarification treatment to break down the seed coat.

Embryo dormancy is more difficult to overcome. The seed needs to undergo some predetermined sequence of events before it will germinate, such as wetting and dry cycles (i.e. Teak seed). Some species require a period of after-ripening of the seed for the embryo to mature.

Scarification of seeds with hard seed coats:
- *Mechanical scarification*: The seed is tumbled in sandpaper-lined drums or mixers and depending upon the species the time will vary. However, around 5–10 min is usual and optimum time of scarification should be determined beforehand for different seeds. Over-scarification may lead to exposure of the embryo and seed damage.
- *Acid scarification:* The seeds are soaked in a solution of hot (60 °C) H_2SO_4, at 5 N concentration, approximately for 5 min. Again, the length of time varies with species.
- *Hot water scarification*: This is the simplest and easiest method but gives variable results. Simply pour 4–5 times of volume of boiling water over a seed batch and stand to cool for 6–12 h. Seed can be re-dried for short-term storage or use directly.

The ideal seed treatment process should be (1) very effective against seed-borne pathogens; (2) relatively non-toxic to animals and plants, even if misused; (3) effective for a long-term seed storage; (4) easy to use and (5) economical. The hot water treatment method meets many of these criteria and easy to use than the chemical treatment methods, and it is more effective and non-toxic.

7.9   Calculation of Quantity of Seed

The required quantity of seed is determined by weight. For this, one should take a standard unit weight of seeds (100 g, 500 g or 1,000 g) and count the number of seeds. In case of smaller seeds, lesser amount of weight (10–15 g) can be taken for counting of seeds. Higher unit weights can be taken when the seeds are large in size. Once the number of seeds per unit weight and the total number of seeds required are determined, the quality of seeds required is calculated using the following formula:

$$\text{Quantity of seeds required (kg)} = \frac{\text{Total number of seeds}}{\text{Numbers of seeds per unit weight}}$$

For example, assume that there are 120 seeds in 100 g and the requirements are 34,041 seeds of a particular fruit tree, than, applying the above formula, the quantity of seeds to be ordered is calculated as

$$\text{Quantity of seeds required (kg)} = \frac{34,041}{(120/100)}$$

$$= 28,367 \text{ g} = 28.37 \text{ kg}$$

Details of some common seeds (number of seeds/100 g) collected from the trees growing at the Indian School of Mines, Dhanbad, are given in Table 7.1.

7.10   Seed Record

Seed record provides information on the time of collection and sowing of seeds for the most common fruits and non-fruit trees (Table 7.2). Columns are provided for entering information on number of seeds/100 g and germination percentage. This table will serve as a ready reference for nursery raiser.

Table 7.1 Quantity of seeds of some common leguminous plant (number of seeds/100 g)

Sl. no	Name of tree	Length of pod (cm)	Number of seeds/pod	Number of seeds/100 g
1	*Albizia lebbeck* (Siris)	19.5 ± 4 (15–27)	8 ± 1.5 (4–11)	973
2	*Cassia siamea* (Chakundi)	22 ± 3(14–27)	16 ± 4(12–24)	6,500
3	*Peltophorum pterocarpum* (Copper pod)	6.5 ± 1.7(6–10)	1–3	1,450
4	*Delonix regia* (Gulmohar)	31 ± 5(26–40)	13 ± 3(10–20)	313
5	*Cassia fistula* (Amaltas)	41 ± 16 (30–53)	53 ± 20(40–60)	553
6	*Leucaena leucocephala* (subabul)	17.5 ± 2(14–21.5)	19 ± 3 (12–15)	2,000
7	*Jatropha curcas* (Jatropaha)	–	–	176

Table 7.2 Time of seed collection sowing time and germination percentage

Sl. no	Species	Seed collection time	Sowing time	Germination percentage
Fruit trees				
1	Amla	Nov–February	July	40
2	*Wood apple (bel)*	Apr–June	June–July	–
3	Ber	Oct–March		
4	Imli	Feb–April	June–July	60
5	Jamun	Jan–July	April–July	80
6	Mango	June–July		85–90
Fuel, fodder, timber				
1	Arjun	Feb–May	June–July	–
2	Babul	April–June	June–July	50
3	Bakain	Nov–Dec	Feb–July	70–80
4	Cassia siamea	May–June	Mar–July	60
5	Jangle jalebi	May–June	Mar–July	90
6	Kachnar	Dec–Jan	May–July	65
7	Kala siris	June–July	Jun–July	–
8	Mahua	June–July	June–July	65
9	Neem	Jan–Feb	June–July	70
10	Peltophorum	Nov–Jan	June–July	40
11	Shisham	Oct–Dec	June–July	50
12	Subabul	May–June	May–July	85

The approximate number of seeds/100 g should be determined by the field person in order to estimate the amount of seeds to be purchased

References

Harrington JF (1972) Seed storage and longevity. In: Kozlowski TT (ed) Seed biology, vol III. Academic, New York, pp 145–245

Taiz L, Zeiger E (1998) Plant physiology, 2nd edn. Sinauer Associates Inc., Sunderland

Contents

S.K. Maiti, *Ecorestoration of the Coalmine Degraded Lands*,
DOI 10.1007/978-81-322-0851-8_8, © Springer India 2013

8.1   Preamble: Nursery Development

For biological reclamation of mined-out areas, overburden dumps and subsided areas and development of greenery in and around mining areas, suitable value-added plants are required to be planted. There are large varieties of plants ranging from evergreen to deciduous, foliage species to flowering plants, etc., that are required to be raised in the nursery. Hence, judicious selections of plant species are to be made before any plantation works. This will only be successful provided that desired saplings are easily available in proper time and at right place. Hence, raising of on-site nursery is essential to achieve the goal of biological reclamation/development of greenery in mining areas. There are several techniques available for raising greenery:

- Sowing of seeds on prepared sites
- Planting of root cuttings (rhizomes) and shoot cuttings (e.g., *Vitex negundo*)
- Nursery raising—the general principle is that well-growing seedlings conditioned and rearing in the nursery will be in a better position to resist the extreme environmental conditions in fields. The methods generally applicable for nursery raising of plants are value-added trees, that is, container-growing saplings (e.g., Teak, Eucalyptus, etc.)
- Planting naked seedlings
- Planting seedlings with a ball of earth

For nursery raising of different tree species different techniques are used.

(a) Nursery development of common forest trees, where no special seed treatment is required. A case example of raising *Eucalyptus* saplings are discussed in detail.

(b) Nursery development methods of raising leguminous saplings, where seed treatment is required, i.e, inoculation of plant-specific *Rhizobium* bacteria in need. Now a days, vesicular-arbuscular mycorrhizal (VAM) fungi are inoculated in the saplings in nursery itself. The culturing of VAM inocula, inoculation techniques and evaluation of effectiveness of VAM fungi are discussed

in the Chapter 11 (Biofertiliser (Mycorrhiza) Technology in Mine Ecorestoration).

(c) Nursery raising technique of bamboo saplings from seeds is discussed.

(d) Nursery raising technique of bamboo saplings from cuttings is also discussed.

But there are large varieties of plant saplings required to be developed in nursery. For most of the plants, nursery techniques and general requirements are same, but there are some plant-specific differences in terms of collection of seed material, its storing and treatment. Lastly, nursery practices for important plant species are discussed in brief. Those species are *Acacia, Azadirachta indica, Albizia lebbeck, Cassia fistula, Dalbergia sissoo, Delonix regia, Tectona grandis, Melia azedarach*, etc.

8.2   Nursery Techniques of Common Tree

Site selection: Selection of suitable nursery site is an important factor determining the *quality* of the planting stock and consequently the success of plantation. Therefore, the question of *temporary* and *permanent* nursery still remains a debatable point.

- *Temporary nursery* is favoured to reduce transport cost and reduce damage cost to the planting stock during transport. Generally, if saplings are required for one or few years, temporary nurseries could be developed in the plantation site itself. The temporary nursery could be used in at most 4–5 years.
- *Permanent nursery* has advantages where large supplies of plants are wanted annually and where communications are good for getting the stock quickly to the plantation site. It could be developed in the area office itself.

8.2.1   Criteria for Selection of Nursery Site

The area requirements for the nursery depends upon the number of plant saplings to be raised at a time, types of species to be raised and age of saplings at the time of planting out. The area calculated in terms of these considerations should be increased by 50% to have paths, roads and for irrigation facilities. Normally the nursery area constitutes 0.5% of the plantation area. For economic viability, the nursery should have a capacity to raise at least 25,000 seedlings at a time. Following general guidelines can be used for selection of nursery site:

- Preferably along the road site.
- Soil should be fertile, deep and sandy loam.
- Availability of good quantity of irrigation water.
- Easy availability of sufficient quantity of farmyard manure (FYM).
- Site selected should offer scope for further extension to meet future requirement.
- Effective fencing of the nursery is necessary. Four-strand barbed fencing supported on wooden or angle iron poles spaced about 3-m apart is normally provided.
- Buildings to be provided in nursery should include residential houses for supervisory staff, stores for seeds, implements, fertiliser, fungicides, insecticides, machinery shed, shelters for labours, etc.

8.2.2 Protection of Nursery

The nursery site should be well protected from the domestic animals especially sheeps and goats. Therefore, effective fencing of the nursery is necessary. Four-strand barbed-wire fencing supported on wooden or angle iron poles spaced about 3-m apart is preferred for a permanent nursery site. This is combined with trenches for a third method of protection. If nurseries are located in forest areas, they should be protected from wild animals too.

(a) *Protection Trench*

A cattle-proof trench having a top width of 1.2 m, base width of 1 m and depth of 90 cm is dug all round the nursery area. The exca-

vated earth is piled on the inner side of the trench at a distance of 15 cm from the edge and arranged in the form of a ridge. This is sown with species like *Prosopis, babul or Pithecellobium dulce* (Jungle jalebi/Vilayati imli). Sowing is carried out in June and the hedge germinates during the rains. This live hedge becomes effective in 6 months time.

(b) *Live Hedges*

Where pressure from cattle is not much, live hedge or a fence around the nursery alone is sufficient. This is done by planting trees or bushes along the boundary which are not graze by cattle. Any tree, which does not branch profusely, can be used for live hedges. If bushes are planted, select those that are quick growing and profusely branching. A proper combination of trees and bushes is essential for the development of live fencing in a nursery site. They are periodically trimmed down to proper shape. Hence, thorny bushes are unsuitable for live fencing. Trimming of live fences provides a lot of biomass for composting and mulching in the nursery beds.

(c) *Barbed Wire*

Fencing with barbed wire is costly but is not time consuming as in live fencing, nor is it laborious as in digging trenches.

The choice of any protection measure is made on the basis of cost, labour availability, time at one's disposal and the level of effectiveness of protection.

8.2.3 Timing of Nursery

If tall saplings are required for planting on roadsides when there is fear of damage from animals, the nursery work should be started from *September to October* and sapling will be ready for planting in *June–July* (during monsoon). Practically 10-month saplings are needed.

For planting in other areas, the nursery work could be started from *February to March* and sapling would be ready for planting in *June–July*. In this case, about 6-month saplings are generally raised.

Purpose	Start of nursery	Sapling ready for transplantation	Time (months)
Roadside plantation	September–October	June–July	10
Other area	February–March	June–July	6

8.2.4 Common Resources Needed for Nursery Development

1. *Land with shading arrangement.* Twigs are generally used for shading of the nursery bed.
2. *Seeds*: About 200 g/bed of 10 m^2. Seeds can be collected from
   - The nearest forest range officer
   - The mature tree standing nearby
   - Commercial store
3. *Fine sand*: About 200 g/bed
4. *Manure*: 25 kg of compost/bed
5. *Shading arrangement*: Bamboo/wooden pole
6. *Container*: Perforated polythene bag (150 gauge) 10 × 15 cm
7. *Irrigation facility*: Water resources nearby—well, tube-well, canal, pond, etc.

8.2.5 Tools Needed for Nursery Development

The following tools are necessary for manual nursery development works. However, for large nurseries (*permanent nursery*), mechanisation is necessary. Tools required for temporary nursery (smaller size, i.e. *temporary nursery*) are given below:

Tools	No. needed
Pickaxe (gaiti)	1 per bed
Hoe (kodal)	1 per bed
Garden harrow/rake	1 per bed
Rose can	1 per bed
Measuring tap	One
Scraper (khurpi)	1 per bed
Sickle (da)	Two

8.2.6 Laying Out and Preparing Bed

1. *Selection of site*: The site should be selected, where the soil is light textured (sandy loam), well drained and rich in organic matter. Heavy clay soil should be avoided.
2. *Shape*: In plains, the nursery bed should be rectangular and divided into blocks by permanent path (inspection path) about 1-m width. The long side of beds should run *east–west* for convenience of shading.
3. *Demarcation of area*: The area to be used for nursery bed is demarcated. The size of the area depends on the number of saplings needed to be raised. A minimum size would be to have capacity to raise 10 nursery beds, each bed measuring 10 m × 1 m. In each bed, working space of 0.5 m is provided all around. The effective area of 10 nursery bed is 100 m^2.
4. A unit area of 1 m^2 can accommodate 100 polythene bags with seedling. The bed of 100 m^2 could accommodate *10,000* saplings. *Practically, the 100* m^2 *nursery bed area will be sufficient to supply saplings for 4* ha (*density 2,500 plants/ha*). Accordingly, if the area of afforestation is known in advance, the nursery bed preparation activity could be started well in advanced. By thumb rule, *for 1* ha *area to be afforested = preparation of 25* m^2 *nursery bed is require*. For raising 1,00,000 sapling, the calculation of shading area requirement of 2,500 m^2 is given below:

Calculation of demarcation area	
(a) Size of one nursery bed = 10 m × 1 m	= 10 m^2
(b) Saplings in one nursery bed	= 1,000. (100 saplings/m^2 of bed)
(c) Total no. of nursery bed req.	= 100
(d) Effective area of total bed = 10 m^2 × *100*	= *1,000* m^2
(e) Inspection path and *1* m surrounding periphery	= *1,500* m^2
(f) Demarcation area	= *2,500* m^2

5. *Fencing*: The demarcated areas are fenced with locally available fencing materials.

6. *Soil preparation*: The soil in the nursery bed is dug out by pickaxe up to a depth of 30 cm. All stumps are removed and stones and weeds are picked out. Hoe the area thoroughly to get the soil powdered (well tilled).
   - If the soil is clayey (heavy), about 10-basketful fine sand is added per bed and spread it all over the bed uniformly. In addition, 25 kg of compost is added in each bed and also spread uniformly.
   - Both mixed compost and sand (if it has been added) are spread upper 10 cm of soil layer of the bed and surface is levelled.
7. *Collection of seeds*: For example, *Eucalyptus* flowers during September–October, and the seeds are ripe in December–January. The seeds are collected from the fruits when ripe, from the trees themselves. Another method of collection of seed is to tie a cloth or canvas supported by stakes, around the tree, at about 1.5m height, and the seeds dropping from the dehisced fruits are collected. The seeds store well when dried before storage and stored in a sealed tin. Seeds are viable for long time. The seeds are very small and there are over 20,000–30,000 in an ounce (28.35 g).
8. *Preparation of seed bed*: Shallow furrows of about 1-cm width, across the length of the bed with 5-cm space in between is done by gentle running of fore finger on the prepared bed. The furrow should be shallower so that the germination of the seeds is not hampered.
9. *Sowing of seeds*: To determine the quantity of seed to be sown in each bed, preliminary knowledge of the number of seed per kg weight and probable germination percentage is required. The reference list is available for most of the trees. For smaller sized seeds, equal quantity of fine sands are mixed (seed to sand = 1:1). Then *sand-mixed seeds* are sown on the furrows in a continuous line so that the entire quantity covers the nursery bed uniformly. For example, in the case of *Eucalyptus*, 20 g of seeds are mixed with 20 of fine sand and sown. About 3,000–10,000 seeds are sown per m^2 of germination bed area.
   - *Treatment of seed*: Hard-coated seeds should be softened by placing them in boiling water for a few minutes, and then allowing the water to cool.
10. *Seed covering*: After sowing, the bed is covered by spreading the thin layer of straw/hay. The soil, if dry, must be well moistened before the sowing is done. *Protection*: It is sometimes necessary to cover the newly sown beds with thorns to keep off birds and rodents
11. *Watering*: Watering is done by using a rose can. Sprinkling should be uniform over the bed so as to make at least 12–15 cm of the topsoil layer moist. *Precautions*: Only a very fine rose can should be used; otherwise, seeds will be washed out. The watering should be done every day, but flooding must be avoided, as it results wash out of seeds.
12. The nursery beds are to be carefully irrigated in the morning in preference of evening, because photosynthesis activity is at its highest at noon. *Damping off*, a disease, is caused due to excessive accumulation of water in the root zone. Watering should be avoided on the rainy days. *Growth of moss* on the surface of the soil usually gives an indication of over watering, and is often accompanied by an unduly yellowish colour in the leaves.
13. *Germination of seed*: The straw cover is removed after germination has started. For example, *Eucalyptus* seeds start germinating 5–6 days after sowing (sometimes after 7–10 days or even 2 weeks) and are completed in about 2–3 weeks. The seedlings attain 12–15 cm heights in about a month and ready for transplanting. Each bed yields about 12–15 thousands seedlings.
14. *Weeding*: Nursery beds should be kept scrupulously clean of weeds. Good watering and soil aeration are as important as irrigation. The tendency to stress the importance of irrigation at the expense of weeding and soil working results in more weeds than plants.

Estimation of total shaded area: The total shaded area requirement is estimated to accommodate the needed number of seedlings in polythene bags as per following requirements.

(a) Polythene bags with seedling should be kept to gather as a unit arrangement in 10×10 and area requirement is 1 m $\times$ 1 m.

(b) A working space on 0.5 m should be kept all around to facilitate watering and taking care of seedling.

(c) *Shading area:* A unit area of 1 m^2 can accommodate 100 polythene bags with seedlings. About 2,500 m^2 shade area will be sufficient to accommodate about one lakh seedling transplanted in polythene bags filled with soil and compost mixtures.

(d) The area is shaded with tree or bush twigs supported by bamboo/wooden post, adjacent to the nursery bed.

15. *Preparation of soil mixture*: The soil mixture is prepared by mixing sufficient quantity of soil with compost or farmyard manure (FYM) in *4:1* ratio. If the soil is heavy, soil is mixed with fine sand in the ratio of soil to fine sand to FYM = 4:2:1.

16. *Filling of polythene bags*: The perforated polythene bags are filled up to the brim with the soil mixture.

17. *Transplanting*: The seedlings are picked out, those that are about 7 cm in height (4-leaf stage), with the help of the tip of sickle. A hole is made with a stick and seedling is put into it and pressed from all sides and well packed. The bigger saplings are transplanted first in to the polythene bags. As a rule, the smaller the plant, the lesser it suffers from the actual shock of being uprooted and replanted. Very small plants are however more difficult to handle and more liable to fatal damage in various ways, so the most common size of plant used is about 8 cm.

Aftercare of transplanted seedlings in polythene bag:

(a) The polythene bags with transplanted seedlings are arranged in the shades in groups of 100 (10×10). Perforation of polythene bags should be ensured before transplanting to avoid waterlogging. The polythene bags are arranged in sunken beds.

(b) The seedlings are watered immediately after transplantation, which is just enough to moist the entire soil in the polythene bags. The watering is repeated every day, but excessive watering is avoided.

(continued)

Aftercare of transplanted seedlings in polythene bag:

(c) The shade is removed as soon as seedlings become erect indicating that they have survived.

(d) The seedlings that do not survive are replaced.

(e) The polythene bags with seedlings are shifted once in a week to avoid penetration of the roots into the ground. The tap roots pierce the bottom part of the polythene bag. If any weeds are grown in the polythene bags, that must be removed. If any insect pest is noticed, it should be hand-picked and destroyed.

- The *transplanted* seedlings are maintained from the day of transplantation to the days of transportation to the planting site.

- The seedling should be carefully transported to the plantation site so as not to disturb the root system of the plant or damage the polythene bags. About 1.5- to 2.0-m-tall plants are preferable for plantation. The plant obtained from September–October sowing attains this height by July–August.

- The plants are to be transported to the reclamation site in an advance. The pit should be sterilised with 10% BHC solution or gamaxine powder, and then the planting is done. Before plantation, the polythene bags are removed, and saplings along with soil is put into the pit and filled with soil/excavated loose material. Pits of 30–60 cm^3 in size are dug depending upon the site conditions.

8.2.7 Important Nursery Activities

1. *Preparation of Nursery Bed*:
   - Measurement of bed
   - Soil working
   - Mixing of manure
2. *Aftercare*
   - Watering of bed
   - Removal of straw after germination
   - Picking and transplanting in polythene bags
   - Attention to rearing in polythene bags
   - Percentage of mortality (should not exceed 5%)

8.3 Raising of Subabul Plant (*Leucaena leucocephala*) Saplings

Timing: The nursery work should be started from February–March; the saplings will be ready for planting in June–July (during monsoon).

8.3.1 Requirement of Resources

Along with the common requirement as discussed in Sect. 8.2.4, the following additional resources are required as given below:

(a) *Culture*: In addition, extra *Rhizobium* culture is needed, which can be easily collected from any agricultural university. The application rate is 1 packet per kg of seeds.

(b) *Seeds*: The quantity of seeds should be doubled the sapling needed (800 seeds weigh approximately 100 g).

(c) *Manure*: Compost or FYM.

(d) *Container*: Polythene bags—150 gauge, 10 cm × 15 cm, number as per requirement.

(e) *Shade*: Bamboo/wooden poles and twigs to erect the shade.

8.3.2 Procedures of Nursery Development

1. Demarcate the area where subabul seedling will be raised (an area of about 2,800 m^2 will accommodate 1,00,000 seedlings).

2. During February-March, erect the shade with wooden/bamboo poles and twigs, allowing penetration of sunlight to some extent.

3. Procure the polythene bags as required, plus 5% more to compensate for wear and tear.

4. Select a patch of land where good soil (preferably sandy) is available. Collect the soil and transport it to nursery site. Mix the soil with compost and fine sand (if the soil is heavy) in 4:2:1 ratio. Approximately, 1/2 kg of soil mixture may be needed for filling one polythene bag. So, the total quantity of soil, compost and sand (if required) needed may be calculated accordingly.

- *Treatment of subabul seeds*:
During the evening hours, boil water in an earthen container. The required quantity of seed is put in the boiling water for 2–3 min. After 2–3 min, the seeds are taken out and strained through a piece of cloth and kept overnight inside the cloth. In the next morning, the treated seeds are mixed with *Rhizobium* culture at 1 packet/kg of seeds as per the instruction contained in the *Rhizobium* culture packet.

5. The next morning, two treated seeds are sown in each polythene bag filled with soil mixture, about 1-cm deep. The sown polythene bags are arranged in groups of a hundred in a square. The water is sprinkled over the sown bags with a rose can to make the soil just moist.

6. The watering is done in sown bags every alternate day till the seedlings are transported to the planting site or otherwise disposed of in the planting season (June–July). On rainy days, watering should be avoided.

7. The germination of seeds will be observed on 7th or 8th day. If germination fails, re-sowing should be done, following the procedure. The shade is removed after 10–15 days of sowing if germination is found be satisfactory, that is, 80–90%.

8. *Shifting*: Once a week, the polythene bags with seedlings, are to be shifted to avoid the penetration of tap root of the seedling into the ground by piercing the bottom of the polythene bags.

9. *Weeding*: The weeds, if any, grown along with the seedling in the polythene bag are to be removed. The visible insect and pest should be hand-picked and destroyed.

8.4 Raising of Bamboo (*Bambusa arundinacea* and *Dendrocalamus strictus*) Saplings

8.4.1 Introduction

There are several species of bamboo and the most important for reclamation of dumps are two: *Dendrocalamus strictus* and *Bambusa arundinacea*.

Dendrocalamus is found in well-drained sandy loam soils of our forest. The clumps of bamboos contain two or more and up to 50 shoots called 'culms' and are developed annually from the rhizomes and attain there full size in one season only. The culms are 10–15 m in height and of 15–20 cm in girth.

Bambusa arundinacea grows in moist soil and localities and along perennial streams. This is a thick bamboo reaching 15–20 m in height, with girths up to 30–45 cm.

Life cycle of bamboo: Bamboos flower once in their lifetime and thus called cyclic flowers. After flowering and seeding, the clumps dried up. The life of the bamboo plant is about 40 years, but with repeated felling and damage due to grazing, fire, etc., the flowering can occur much earlier. The general season of flowering is January/February, and seed can be collected in March.

Planting: Bamboos can be planted:
- Along the field boundaries
- Along the fence
- In the form of a plantation

For planting along the field boundaries and fence, a spacing of 4 m is maintained. In case of plantations, the spacing of 5 m by 5 m is sufficient. In an area of 1 ha, about 400 plants can be planted.

Site Preparation for Area to Be Planted
- Area to be planted must be clean.
- About 2 months in advance, a pit of 0.3 m^3 is dug up.
- After planting, the pit is provided with a 1-m-diameter saucer for storage of water, forming a small mound round the plant.

Aftercare: Saucer pit should be weeded for every month or once in 2 months in the first year. Causalities should be replaced with every weeding and soil working till September. In the second year, two weeding and one soil working are sufficient. The soil working should preferably be done in January–February before the summer season. The depth of soil working should be 15–30 cm.

Timing: The nursery work should be started in September–October, and the sapling will be ready for planting after 10 months, in June–July (during the monsoon).

Resources to Be Utilised
- *Same as earlier.*
- Bamboo seeds collected and stored for a period of more than 2 years should not be used as they lost the viability.

8.4.2 Requirements of Materials

(a) *Seeds*: 1 kg of seeds (with husk on) per bed. In the case of *Dendrocalamus strictus*, the seed count is about 20,000 per kg, and in the case of *Bambusa arundinacea* the seed count is 69,000–70,000 per kg.
(b) *Fine sand*: About 10 basket per bed (if soil is clayey).
(c) *Manure*: 25 kg of compost or FYM per bed.
(d) *Straw/hay*: About a bundle per bed.
(e) *Container*: Perforated polythene bags 150 gauge, 25 cm × 15 cm; 15,000 bags per bad of *Dendrocalamus strictus* and 45,000 for a bed of *Bambusa arundinacea*.
(f) *Shade*: Bamboo/wooden poles and twigs as per requirement.

8.4.3 Nursery Procedure

1. Select a site preferably where soil is light textured, well drained and rich in organic matter. Demarcate the area to the maximum extent for preparing five nursery beds for raising *Bambusa arundinacea* and two nursery beds for raising *Dendrocalamus strictus*, each bed measuring 10 m × 1.5 m with a working space of 0.5 m all around.
2. Fence the demarcated area with locally available fencing material. Dig the soil with a pickaxe over the whole area, up to at least 0.3-m deep. Hoe the area thoroughly to get the soil powdered (good tilth). Remove the weeds, pebbles and clods with the help of a garden harrow/rake.

3. Prepare sunken beds of 15-cm depth each, with a dimension of 10 m × 1.25 m. This can be done by removing the upper soil layer of the bed area and spreading it over the working space. If the soil is clayey (heavy), add about 10 basketful of fine sand per bed and spread this uniformly all over the bed.

4. Add 25 kg of compost to each bed and spread uniformly. Mix compost and sand (if it has been added) with the upper 15 cm of soil layer of the bed and level the surface of the bed. Prepare shallow furrows of about 1-cm width across the length of the prepared bed with a space of 5 cm in between, by gently running your forefinger across the soil. (The furrow should be shallow so that the germination of the bamboo seed is not hampered).

5. The seeds are sown with the husk in the furrows, in a continuous line. Sow 1 kg of seeds per bed, uniformly in the furrow. The sown bed is covered by spreading the straw/hay thinly over it.

6. Sprinkle water from a rose can uniformly over the bed so as to moisten at least 15 cm of the topsoil layer. Water the beds twice or thrice a day but avoid flooding and waterlogging.

7. Remove the straw cover after germination has started. Germination generally starts within 10 days of sowing and continues till nearly 50–60% germination is achieved. *Bambusa arundinacea* will give nearly 40,000 and *Dendrocalamus strictus* nearly 10,000–12,000 seedlings per bed. Water the beds twice daily with a rose can once the germination starts, but avoid waterlogging as well as flooding

8. Construct the shade area with tree or bush twigs supported by bamboo/wooden posts adjacent to the nursery beds. The total shade area needed to accommodate the required number of seedling in polythene bags will be as per the following estimation:
   (a) 100 polythene bags with seedlings should be kept together as a unit arranged in 10 by 10
   (b) A working space of 0.5 m should be kept all around to facilitate watering and taking care of the seedlings. (*A unit of* 2 m × 2 m *will accommodate 100 polythene bags with seedlings. About* 3,000-m^2 *area will be required to keep one lakh transplanted seedling in polythene bags*).

9. Prepare a soil mixture by mixing sufficient quantity of soil with compost of farmyard manure in 4:1 ratio. If the soil is heavy, mix the soil with fine sand and compost manure in 4:2:1 ratio. Fill the perforated polythene bags with soil mixture up to the brim.

10. Pick out the seedlings of about 7-cm height with the help of the tip of the sickle, the bigger ones first, and transplant 2–3 seedlings in each polythene bag. Leave a few seedlings in the nursery bed and maintain them by regular watering and removal of weeds to meet the eventual mortality in the polythene bags. Make sure that the collar zone (the junction of the root and shoot portion) of the seedlings is not buried under soil. Arrange the polythene bags with the transplanted seedlings under the shade in groups of 100.

11. Water the seedlings immediately after transplantation, just enough to moisten the entire soil in the polythene bag. Repeat watering ever day but avoid excessive watering. Watering should be avoided on rainy days. Remove the shade after 4–5 days as soon as you notice that the seedlings have survived

12. Shift the polythene bags with seedlings once a month to avoid penetration of the roots into the ground through the polythene bag. Remove the weeds, if any, growing with the seedlings. If any insect or pest is noticed, it should be hand-picked and destroyed.

13. Re-transplant in the polythene bags if previously transplanted seedlings have not survived after 3–4 months of transplantation, that is, in March–April. Maintain the seedlings from the day of transplantation to the

day of transportation to the planting site. (Note: *Seedlings are to be maintained in the nursery for a minimum period of 1 year and a month before they become ready for planting.*)

8.5 Raising of Bamboo Cutting (*Bambusa vulgaris*) in the Nursery

Nursery work should be started in June–July (at the onset of the monsoon). The pre-sprouted cuttings will be ready for planting after 2 years during the monsoon (June–July).

8.5.1 Resources to Be Utilised

(a) Land to prepare for the nursery beds, preferably near to plantation site. One nursery bed to raise 100 bamboo cuttings will measure $25\ m^2$ (5 m × 5 m) plus a working space of 0.5 m all around

(b) The required number of bamboo cuttings can be obtained either from a farmer's land nearby or the forest department.

(c) Irrigation facility: Water should be easily available in the nursery site, for example, nearby well, tube-well, canal, pond and river.

8.5.2 Required Materials and Tools

(a) *Bamboo cuttings*: Culm cuttings of *Bambusa vulgaris*—100.

(b) *Fine sand*: The fine sand is needed in case the soil is clayey; *25 baskets per bed.*

(c) *Compost/FYM*: 50 kg per bed.

(d) *White oil paint*: 1 L for 100 bamboo cuttings.

8.5.3 Nursery Procedure

1. *Site selection*: Select a site preferably where the soil is light textured, well drained and rich in organic matter for preparing the nursery bed, although bamboo grows best in clayey or other heavy soil in field conditions. Its water requirement is somewhat high. So it is excellent for growing near water courses.

2. *Demarcation of area*: Demarcate the area to the minimum extent of preparing at least one nursery bed measuring 5 m × 5 m with working space of 0.5 m all round. Fence the demarcated area with locally available fencing material.

3. *Soil preparation*: The soil is dug with a pickaxe over the whole area up to 0.30 m. Hoe the area thoroughly to powder the soil (good tilth). Remove weeds, pebbles and big clods manually/mechanically. Keep 0.5 m of working space surrounding the nursery bed. Mix the soil of each nursery bed with 25 baskets of sand (if the soil is clayey) and 50 kg of compost manure and level the surface of the bed.

4. *Collection of bamboo cuttings*: On a cool, cloudy or rainy day, cut with a sharp-edged axe (to avoid splitting while cutting) full length, living, matured bamboos as required from available standing bamboo clumps. White oil paint should be applied immediately to those portions of each bamboo smeared due to the removal of branches.

5. *Precautions*: See that while cutting, pruning the side branches and transporting, the vegetative buds located at each node of the bamboo culm are not damaged. At no stage of cuttings should the bamboo culms split. So, it is again emphasised that to avoid splitting, a sharp-edged axe should always be used. Prepare bamboo cuttings from the collected bamboos on the same day as per the details given below:

   Cut each bamboo into approximately three equal lengths and apply oil paint at the upper tip of every piece. Keep the middle portions in one lot and the upper and lower portions in another separate lot. From the lot containing the middle portions of culms, prepare cuttings, each with two nodes, and apply oil paint on the upper tip of each cutting. From the other lot, prepare cuttings, each with 3 nodes, and apply white oil paint on the upper tip of each cutting.

6. *Planting*: While planting the cutting in the nursery bed, the following details should be adopted:
   (a) Insert the cuttings by hand.
   (b) Keep the painted ends of the cuttings up.
   (c) Bury the other end of the cuttings in the ground so that one of the nodes is buried at a depth of about 5 cm.
   (d) The cuttings should be planted obliquely and not vertically, with all the cutting pointing in one direction, preferably away from direct midday sunlight.
   (e) The space between cuttings should be 50 cm.
   (f) Flood/irrigate the nursery bed just after planting.
7. *Precautions*:
   (a) The cuttings and collection of bamboos from the field, preparation of cuttings and planting of cuttings should be done on the same day, which should be in a cool, cloudy or rainy day.
   (b) Repeat flood/irrigation once every fortnight, except when the rains are heavy and the soil is fully saturated with water. Continue this practice uninterruptedly for a year.
8. *Weeding*: Remove weeds regularly from the nursery bed.
9. *Hoeing*: Loosen the soil of the bed by hoeing once a month, particularly after the monsoon period. *Insect pests*, when noticed, should be hand-picked and destroyed (note: *Generally, 25% mortality has been observed in the nursery. So, to get 100 successful pre-sprouted cuttings for planting in the field, about 125 cuttings should be reared in the nursery bed*).

8.6 Nursery Practices and Raising of Important Plants

Acacia species *(Leguminosae, Mimosoideae)*. Almost all *Acacia*'s flowering during January/February and pods (fruits) remain on the trees for long. The seeds can be collected in April/May, when the pods are greyish white colour. The seeds are hard and can be stored for 1 year in gunny bags or container. About 5 species of *Acacia* are grown for mine reclamation. These species are
1. *Acacia arabica* (Babul): deep brown bark
2. *Acacia leucophloea* (Safed babul): white bark
3. *Acacia catechu* (khair)
4. *Acacia mangium*
5. *Acacia auriculiformis*

8.6.1 *Acacia nilotica* (Babul, Kikar, Gum Arabic)

Acacia nilotica (L.) Willd. Ex. Delile. (syn. *Acacia arabica* (Lam.) Willd.) (Family: Fabaceae—Mimosoideae), commonly known as Babul/Kikar (gum arabic tree), is naturally widespread in the drier areas. The species is used as a pioneer species in land rehabilitation and as a barrier to desertification. This species is used extensively on degraded saline/alkaline soils, growing on soils up to pH 9, with a soluble salt content below 3%. It also grows well when irrigated with tannery effluent and colonises easily in coal mine degraded lands.

8.6.2 *Acacia auriculiformis* (Australian Wattle, Akashmoni)

Acacia auriculiformis (A. Cunn. ex Benth.) (Mimosaceae), commonly known as Australian wattle (English), Akash moni (Hindi, Bengali), is a fast-growing, evergreen and one of the very common tree seen in reclaimed overburden dumps and other mine degeraded sites in India. Its spreading, superficial and densely matted root system makes *A. auriculiformis* suitable for stabilising eroded land (Photo 8.1). Its rapid early growth, even on infertile sites, and tolerance of both highly acidic and alkaline soils make it popular for stabilising and revegetating mine spoils. It has been reported that plantations of *A. auriculiformis* improve soil physio-chemical properties such as water-holding capacity, organic C, N and K through litter fall. *A. auriculiformis* can fix nitrogen after nodulating with a range of Rhizobium and Bradyrhizobium strains. It also has associations with both ecto- and endo-mycorrhizal fungi.

Photo 8.1 *Acacia auriculiformis* (showing flattened leaf stalks known as phyllodes, function as leaves, and has tinny yellow flowers)

Seed treatment: As seeds have hard seed-coat, to obtain good germination, older seeds are generally sown. The seeds can also be immersed in boiling water before sowing.

Direct sowing: Pretreated (boiling water treated) seeds can be directly sown where good soil cover is available (excavated cattle-proof trench's edge); live vegetative protection, side of roads, that is, where dense vegetation of Acacia is desirable.

Nursery raising: Very useful for container-grown seeding. Acacia can be developed in polythene bag (20 × 10-cm size, 180 gauge). The pretreated seed is normally shown in each bag. The depth of sowing is 1.5 cm; regular watering is necessary (but not excessive). Seedling becomes plantable within three months. This technique is good for far-off nursery, when seedlings have to be transported. The polyethylene bags must be removed at the time of planting.

Ball of earth: The pre-treated Acacia seeds are shown directly in standard nursery bed (10 m × 1 m). The seedling raised in beds are taken out and transported with a ball of earth to the planting site. Seedlings of 6 months to 1 year are used. It is very useful for temporary nursery.

8.6.3 *Acacia catechu* (Khair)

Acacia catechu Willd. belongs to family Fabaceae (mimosae) commonly known as cutch tree (Eng), khair (Hindi) or khayer (Bengali) is a small thorny tree, native to India, hardy, drought resistant and grows up to 15 m in height. The stem is dark brown to black; the flowers are white or pale yellow; the dark brown seed pods are 5–10-cm long and contain 3–10 seeds, which are dark brown in colour, flat, 5–8-mm diameter. The tap root and branches may go up to 2-m depth. The seeds are good source of protein. Kattha (catechu), an extract of its heartwood, is used as an ingredient to give red colour and typical flavour to paan. Branches of the tree are quite often cut for goat fodder and are sometimes fed to cattle. Propagated may be through by seeds or container-growing sapling. It is the very commonly used tree species for coal mine dump reclamation (Photo 8.2).

Photo 8.2 *Acacia catechu* (*khair*) plant growing in coalmine overburden dumps of KD Heslong project, CCL; *close view* of inflorescence

Pre-treatment of seed: Due to hard seed coat, pre-treatment is necessary for seed germination, which is done either by (i) soaking in water for 24–48 h, (ii) soaking in boiled water for 6 h then cool, and (iii) treatment with conc H_2SO_4 for 2–5 min.

8.6.4 *Acacia mangium* (Mangium Acacia)

Acacia mangium, Willd. (Family: Fabaceae—Mimosoideae), commonly known as Black Wattle (English) or Mangium Acacia, is a fast-growing medium-sized tree which tolerates pH levels between 4.5 and 6.5. It can achieve a mean annual diameter increment of up to 5 cm and a height of up to 5 m in the 1st year. It is reported to grow 3 m tall in the 1st year, and it reached an average height of 8 m and diameter 9.4 cm after a further 2 years. However, growth declines rapidly after 7 or 8 years. Provenances from Papua New Guinea consistently show better growth in height and diameter and superior form. Optimal growth of trees is achieved most effectively if vesicular-arbuscular mycorrhizal fungi such as *Glomus fasciculatus* and *Gigaspora margarita* are present in combination with *Rhizobium*.

It demands full light for growth; in shade, growth is stunted and spindly.

Trees are renowned for their robustness and adaptability, which makes them good plantation species. Survival after planting out is high: 60%. Plantation canopy cover occurs after 9 months to 3 years, depending on soil fertility; plantation with an initial spacing of 3 m × 3 m, the canopy closed in 1 year. In the 1st year, the plantation should be protected from livestock.

As trees have a tendency to produce multiple leaders from the base, singling is carried out at 4–6 months after planting. The productivity of trees has been found to be closely related to 'total' soil potassium levels (accounting for 50% of the variation in data) and phosphorus levels.

Tree flowers precociously, and viable seed can be harvested 24 months after planting; from the onset of flower buds to pod maturity takes about 6–7 months. The tree is a hermaphrodite and pollinator. Pollinators are generally insects. After planting, it takes about 18-20 months for flowering and production of seeds. Mature fruits occur 3–4 months after flowering period; flowers are present in May and the seeds mature in October–December and fruits mature in July (Photo 8.3).

Photo 8.3 *Acacia mangium* growing coalmine degraded land

8.6.4.1 Propagation Methods

Tree can be propagated from seed (direct sowing or in the nursery); seeds are pretreated before sowing by immersing them in boiling water (100°C) for 30 s then soaking them in cold water for 24 h; alternatively, they may be manually scarified.

The germination rate is high, generally 75–90%, and germination is rapid, occurring within 1 month. Seeds may be sown in seedbeds and pricked out 6–10 days after sowing; recovery rate with this method is about 37%. Sowing in germination trays (wet towel method) and pricking out the seedlings 6–10 days after sowing when the radicle emerges gives 85% recovery. Another option is direct sowing in containers (polythene bags, open-ended hanging pots called 'root trainers' or other permanent pots) followed by pricking out to maintain 1 seedling per container. There are no specific requirements for the type of substrate; mixtures of topsoil, peat, old sawdust, rice husks, sand and vermiculite are used. Even a pure mixture of peat vermiculite with a pH of 3.1 presented no problems.

Nitrogen–phosphorus–potassium fertilisers are generally applied in the nursery, but fertilisation is stopped when '*hardening off*' the plants by reducing watering and exposing them to full sun-light. The appropriate height for transplanting is 25–40 cm, when the seedlings have been in the nursery for 9–16 weeks.

8.6.5 *Azadirachta indica* (The Margosa Tree, Neem)

Azadirachta indica (L.) (Meliaceae) or neem tree is also known also known as Indian lilac (Photo 8.4). Flowering in March and fruits in April/May. Seeds are collected manually but have low viability. Therefore, fresh seeds should be used for sowing, as the seed does not store well, seeds should not be used after 12 months of its collection. No pre-sowing seed treatment is required. The plant will grow in full sun to partial shade, growing best in a well-drained soil mix. The neem tree is noted for its drought resistance. Neem can grow in many different types of soil, but it thrives best on well-drained deep and sandy soils. This tree is of great importance for its anti-desertification properties and possibly as a good carbon dioxide sinks.

Seeds and sowing: Neem propagates easily by seed without any pretreatment. Seeds are collected from June to August and remain viable for 3–5 weeks only, which necessitates sowing

Photo 8.4 *Azadirachta indica* (the Margosa tree) (**a**) close view of leaves and flowers (**b**) external appearance of the tree as a whole

within this short time. Seeds may be de-pulped and soaked in water for 6 h before sowing. Seeds are sown on nursery beds at 15 × 5 cm spacing, covered with rotten straw and irrigated. Seedlings can be transplanted after 2 months of growth onwards either to poly bags or to field. For field planting, pits of size 50–75 cm^3 are dug 5–6 m apart, filled with topsoil and well-rotten manure, formed into a heap, and seedling is planted at the centre of the heap. *Manuring:* FYM is applied at 10–20 kg/plant every year. Chemical fertilisers are not generally applied.

8.6.6 *Albizia lebbeck* (Siris Tree)

Albizia lebbeck (L., Benth.), commonly called as Siris tree, is a deciduous tree, growing to 30 m tall (Photo 8.5). It is a nitrogen-fixing tree; the extensive, shallow root system makes it a good soil binder and suited to soil conservation and erosion control. It is grown on well-drained soils of moderate to high fertility; it will grow on less fertile soils, but is not adapted to heavy clay or waterlogged soils. It is adapted over a wide range of pH from acid to alkaline and also tolerates moderate soil salinity.

It can be raised either by container-grown seedlings or seeding. Fruit is pod, 15–30 cm long, contains 6–12 seeds. Seeds (pods) are collected from January to March. Pods should preferably be collected from the trees and should not be swept from the tree floor. Seed stores well and should preferably be stored in a sealed tin container. Seeds are not hard, germinate rapidly without any treatment. Seed inoculation is not necessary and about 40g/m^2 required for nursery bed. Best seedling development is obtained in full sunlight.

8.6.7 *Cassia fistula* (Indian Labrum)

Cassia fistula Linn. belongs to family Fabaceae (Caesalpinioideae), also known as golden shower (English), amaltas (Hindi) and the Indian labrum (Photo 8.6). This plant flowers during April-May and is nearly leafless during flowering. The flowers are yellow in colour and fruits are very long pod, about 1–2 ft in length and 3/4 in. in diameter. This plant is suitable for parks, garden and avenues.

Propagation: It is commonly propagated from seeds. It is advisable to plant more seed, due to poor germination capacity of seeds. Germination is hastened by boiling the seeds for 5 min before sowing to soften the hard seed coat. Studies indicated soaking the seeds in concentrated H$_2$SO$_4$ results in highest germination; puncturing the seed coat proved to be the simplest, most effective method to break dormancy. Seedlings,

Photo 8.5 *Albizia lebbeck* (Siris tree) (**a**) view of pinnately (bipinnate) compound leaves and inflorescence consisting of axillary cluster of 40–50 pedicellate flowers (**b**) Pod (flat and oblong) and seeds

Photo 8.6 *Cassia fistula* (Amaltas tree) (**a**) showing evenly pinnate compound leaves, golden yellow flowers (terminal, drooping racemes) and fruit which is long, black colour, hanging indehiscent pod.
(**b**) showing pods and seed embedded in the black pulp

planted in plastic bags containing 7 kg soil, survived transplant quite well. Other studies reported that *Cassia fistula* seeds when soaked in concentrated H_2SO_4 for 5–20 min and then soaked in water for 24 h resulted in 84% germination. Soaking in water alone for 24 h failed to germinate.

8.6.8 *Cassia siamea* (Kassod Tree, Chakundi)

Cassia siamea Lamk. (subfamily Caesalpinioideae, Fabaceae) (Syn. *Senna siamea*) also known as kassod tree or chakundi, is a medium-sized evergreen tree, leaves pinnate, pinnate oblong-elliptic, young leaves with a slight brownish tinge, flowers yellow (Photo 8.7). Leaves are not browsed by cattle. It has been reported that hot water treatments and a manual scarification treatment followed by soaking in water for 24 h are the easiest ways to break dormancy. It is also reported that while hot water treatment is more practical for large-scale plant production, manual scarification should be used in small nurseries.

8.6.9 *Dalbergia sissoo* (Sissoo)

Dalbergia sissoo Roxb. ex DC. (subfamily Papilionoideae, Fabaceae) is also known shisham (Hindi), shishu (Bengali) and Indian rosewood (English), is an important fuelwood, shade, shelter and timber tree in India. This plant can be grown in nursery. The seeds are collected in December/January. It is difficult to separate the seed from the pods. Seed (in broken pods) can be stored for 6 months without loss of viability and contain 13,000–53,000 seeds/kg. Seeds should be stored in sealed tin containers. Storage for 1 year does not affect the germination behaviour of the seed. Rate of seed germination is high (70–100%) and takes about 15 days for complete germination of seeds.

Photo 8.7 *Cassia siamea* (Chakundi tree) (**a**) close view of numerous thin young pods (strap shaped, dehiscent), along with flowers and compound leaves (**b**) showing alternate, pinnately compound leaves, with terminal racemes and mature pod turns to black colour

Photo 8.8 *Dalbergia sissoo* (Shisham tree) (**a**) showing imparipinately compound leaves, alternately arranged with 5-leaflets, leaf size is broad ovate with acuminate tip along with fruits (pods are thin, oblong to flat, containing 1–3 seeds) (**b**) close view of old pods broken into pieces containing one seed

The plant species has root suckers and runners that make it useful for erosion control in gullies. Sissoo are the first trees to come up on freshly exposed ground and newly deposited alluvium. They have special nodules on their roots that add nitrogen to the soil and improve fertility. It has been found that once sissoo is established, they improve soil, add nutrients and control temperature and winds and thus help more advanced vegetation to grow (Photo 8.8).

Nursery practices: Sowing in nursery is done in February–March. While seeds may be sown without pretreatment, it is recommended that they be soaked in water at room temperature for 24–48 h, inoculated with *Rhizobium* after soaking and sown immediately. Seeds or broken pieces of pods are sown in lines about 25-cm apart. Irrigation at regular interval is necessary for germination of seeds and good growth of seedlings. Thinning of the seedlings to a spacing of about 5–10 cm in lines is necessary to ensure good growth. The seedlings become plantable in July–August, are lifted from nursery bed and are put in polythene container, as usual.

8.6.10 *Delonix regia* (Gulmohar)

Delonix regia (Boj. ex Hook.) Raf. (Caesalpiniaceae, Fabaceae) (Syn. *Poinciana regia*

Photo 8.9 *Delonix regia* (the Gulmohar tree) (**a**) showing the flame-colored flowers formed in dense clusters and (**b**) isolated seeds taken out from pod

Boj. ex Hook.), commonly known as Gulmohar tree, seems to be native of Madagascar. The Gulmohar is a fast-growing tree, produces a spreading umbrella-like canopy, suitable for avenues where both flowers and shades are desired (Photo 8.9). Tree is 5–10-m tall, adapted to variety of soil conditions. The roots are shallow and spreading. It grows well on rocky soil, also. It is an ornamental plant and grows well in deep fertile soil, raised by planting out nursery seedlings.

Propagation methods: Gulmohar is easily propagated from seeds which have a very hard, woody testa and take a long time to germinate. They may lie for 2–3 years in the soil without germinating. Seed pretreatment is mandatory to break dormancy, where seeds are boiled in hot water, then allowed to soak for 24 h. The pretreated seeds are sown in unshaded nursery beds, where seeds are germinated within 5–10 days, with a germination rate of up to 90%; subsequent growth in the nursery is quite fast.

Otherwise, the seeds can be directly sown in polythene bags, containing 4–5 seeds/bag. Seedlings are watered and weeded regularly and are planted out in the rainy season, with total time required in the nursery being 3–5 months. Keeping the plants for more than 9 months is not desirable, as they become too tall to handle, but seedlings can be transplanted even when 20–25-cm high. Natural regeneration is common. Young plants are not fire resistant and should be protected from grazing.

8.6.11 *Eucalyptus citriodora* (Lemon Gum)

Eucalyptus citriodora Hook. (Myrtaceae), commonly known as lemon-scented gum, has a clean silvery green trunk and strongly scanted narrow leaves. The plant is coppice at a height of 90 cm to 1 m to get a larger yield of foliage, which contain aromatic oil. On an average, the foliage contains up to 0.8% of oil, of which 60–80% is of citronellal that imparts to it the lemon-like odour.

Seed: May be sown during early spring either in polythene bags in the nursery or directly in the field. The small seeds are surface sown on coarse sand and peat mixture and kept moist. Seeds germinate in 30–45 days at 22°C.

Transplantation: About 10–20-cm-tall seedlings are transplanted in the field (without exposing their roots) during rainy season and planted at 45 × 30-cm spacing.

Fertiliser: 50 kg of phosphorous and potassium is added per ha during planting and supplemented with about 100 kg of nitrogenous fertiliser/ha.

8.6.12 *Gmelina arborea* (Gamhar)

Gmelina arborea Roxb. (Verbenaceae) is a medium- to large-size deciduous tree with a straight trunk. It is wide spreading with numerous branches forming a large shady crown. Wood is one of the best and most reliable timbers of India (Photo 8.10).

Photo 8.10 *Gmelina arborea* (Gamhar)

Seeds and sowing: The best method of propagation is by seeds. Seed formation occurs in May–June. Seeds are dried well before use. Seeds lose their viability after a year of storage. They are soaked in water for 12 h before sowing; however, for quick germination, the seeds should be soaked for 48 h. The seeds germinate within 20–50 days under ideal conditions; the average rate for a healthy seed lot is 60%. Seeding rate is 3 kg/ha. Seeds are sown in nursery beds shortly before rains. Seeds germinate within 1 month. Seedlings are transplanted in the first rainy season when they are 7–10-cm tall. Pits of size 50 cm^3 are made at a spacing of 3–4 m and filled with sand, dried cow dung and surface soil, over which the seedlings are transplanted.

After cultivation: 20 kg organic manure is given once a year. Irrigation and weeding should be done on a regular basis.

Plant protection: The common disease reported is sooty mould caused by *Corticium salmonicolor*, which can be controlled by applying 1% Bordeaux mixture. Stem borer caterpillar is seen infesting in some areas.

8.6.13 *Grevillea robusta* (Silver-Oak)

Silver-oak (*Grevillea robusta* A. Cunn.) (Proteaceae) also often called silk-oak or Silver oak is a medium to large tree commonly planted for eco-restoration of overburden dumps. It has been established as a forest tree in some countries and shows promise as a fast-growing timber tree (Photo 8.11).

Seed Production and Dissemination: Silk-oak is a prolific seeder. Seeds are about 10 mm (0.4 in.) long, flattened and surrounded by a membranous wing. There are reported to be 64,000–154,000 seeds per kilogramme. Because of their relatively large wing, the lightweight seeds are widely disseminated by wind. Seed is viable for a short

Photo 8.11 *Grevillea robusta* (silver-oak)

period, unless they are dried and stored under refrigeration. Under refrigeration, seed will last 2 years or more. The seeds, if kept at 10% or less moisture content, can be stored for as long as 2 years at −7° to 3°C (20° to 38°F) with little loss in germinability.

Germination of fresh, unstratified seeds requires about 20 days. After 48-h water soak, substantially increases germinative capacity of seeds that have been stored.

Seedling Development: Germination is epigeal. Seedlings are grown in flats or containers in nurseries. Methods vary among the countries where silk-oak is grown. In some countries, 4–6-week-old wildings are lifted and potted and later replanted. Elsewhere plants are grown to 45-cm (18-in.) heights in large baskets so that they can compete when outplanted. In India, seedlings in individual containers can be grown to a plantable size of 20-cm height and 4-mm calliper in 12–14 weeks.

8.6.14 *Heterophragma adenophyllum* (Katsagon)

Heterophragma adenophyllum (Wall. ex G. Don) Seem. ex Benth. & Hook. belongs to family Bignoniaceae, is a moderate-sized deciduous tree (Photo 8.12). Leaves are compound, large 0.3- to 0.6-m long, usually 5–7 leaflets per leaf. The flowers are large, brownish yellow, flowers in November. The fruit is a large capsule (30–90 cm long), cylindrical ribbed and twisted, maturing between January and February.

It is an intolerant tree that requires full sunlight to develop to a mature tree. It grows well in moist situations on deep soils that are well drained. It requires a precipitation zone of at least 800 mm/year. It prefers a subhumid, tropical climate. It has no known insect or pest problems.

It can be reproduced from seed. The seeds are viable only when fresh. It grows approximately 1 m in height every 2 years.

8.6.15 *Leucaena leucocephala* (Subabul)

Leucaena leucocephala (Lam.) De Wit belongs to family Fabaceae (Leguminosae); subfamily Mimosoideae (Mimoseae), commonly known as Leucaena (Eng), koo babul, ku-babul and subabul (Hindi). Tree grows up to 18 m and is forked when shrubby and highly branched

Photo 8.12
Heterophragma adenophyllum (Katsagoan)

Photo 8.13 *Leucaena leucocephala* (subabul)—fruits and seeds

tree. Leaves are bipinnate with 4–9 pairs of pinnae, variable in length up to 35 cm. Flowers are numerous, in globose heads with a diameter of 2–5 cm. Pod is 14–26 cm × 1.5–2 cm, pendant and brown at maturity. Seeds are 18–22 per pod, 6–10 mm long and brown (Photo 8.13). It is highly valued as ruminant forage and as a fuelwood, is grown in dense rows as a living fence and has popularly been used as a reclamation species following mining in India.

There are 15,000–20,000 seeds/kg. Seedlings and direct sowing are recommended methods of propagation; seed must be scarified to break the impermeable testa. Seeds can be stored without special considerations for several years and maintain its viability. Pretreatment of seed with a water soak will speed up germination process.

Photo 8.14 *Melia azedarach* (Bakain) (**a**) leaves with long petiole, 2 or 3 times compound (odd-pinnate); leaflets 5, serrate margin, (**b**) long inflorescence, flowers white to lilac growing in clusters

Hot water treatment involves soaking of seeds for 2 min. Mechanical scarification, using coarse sandpaper (for small seed lots) or abrasive-lined rotating drum scarifiers, is now preferred.

Small areas can be planted using either seed or seedlings. Seedlings are normally raised in poly bags for plug planting at 3–4 months old. Seedlings can also be raised in beds and removed for planting, normally grown as a hedgerow with grasses or crops grown between hedgerows. *L. leucocephala* is highly palatable to most grazing animals.

8.6.16 *Melia azedarach* (Bakain)

Melia azedarach, L (Meliaceae), commonly known as Bakain (Hindi), Ghora nim (Bengali) and Persian lilac (English), is a fairly fast-growing deciduous tree. The crown is spreading and rounded (Photo 8.14). This tree is native to India but is now grown in all the warmer parts of the world. The tree develops spreading crown and the branches in a week. Bakain is generally raised by planting nursery-raised seedlings. It requires a precipitation zone of 600–1,000 mm/year or more and relatively insect and disease free.

Seed collection is done in January/February. The seed stores well and approximately 70% of the seed will be viable. The seeds can be stored for approximately a year without loss of viability. The sowing in the nursery is done in March–April in lines about 15-cm apart. About 2- to 3-month-old seedlings are transplanted in the field. Entire transplants are planted out with a ball of the earth.

8.6.17 *Peltophorum pterocarpum* (Copper-pod)

Peltophorum pterocarpum (DC.) Backer ex Heyne, belongs to family Fabaceae (Caesalpinioideae), commonly known as Copper-pod (English) and Radhchura (Bengali, Hindi), a large deciduous tree usually reaching a height of 15 to 24 m and a diameter of 50 cm (Photo 8.15). The bark is smooth, grey in colour, leaves are bipinnately compound with wide spreading dense crown. The fruit is pods, flat and thin, contains 1–4 seeds, dark red when ripe (copper colour) and turning black. The pods are remained in the tree for long times. The yellow flowers are borne in long bunches (spikes) at the top of the tree crown. The dense green foliage accents the flower's colour. Flowers develop in May through August, while fruit and seed are produced in the autumn. The tree can be propagated by seeds, grafting or branch cuttings. Seedlings are best raised in the nursery for 1 year before transplanting to the field.

Used during coal mining reclamation due to its fast-growing nature and little effort is needed to maintain plantations. It is a widely appreciated shade tree, due to its dense spreading crown and used in shelter belts. It has the ability to fix nitrogen and is used as source of green manure.

8.6.18 *Pongamia pinnata* (Indian Beech, Karanj, Karanja)

Pongamia pinnata (Linn.) Pierre (Syn. *P. glabra* Vent., *Derris indica* (Lam.) Bennet.) (Papilionaceae, Fabaceae), also known as Indian beech

Photo 8.15 *Peltophorum pterocarpum* (Copper-pod) (**a**) close view of terminal panicle inflorescence with yellow flowers (**b**) distance view of blossom

Photo 8.16 *Pongamia pinnata* (Indian Beech; Karanj; Karanja) (**a**) shows alternate, pinnately compound leaves, consist of 5 or 7 leaflets, arranged in 2 or 3 pairs with a single terminal leaflet. Pods are elliptical contain a single seed. (**b**) mature tree with old brown colour pods

tree, with drooping branches, shining green leaves laden with lilac or pinkish white flowers (Photo 8.16). The plant comes up well in tropical areas with warm humid climate and well-distributed rainfall. Though it grows in almost all types of soils, silty soils on riverbanks are most ideal. It is tolerant to drought and salinity. The tree is used for afforestation, especially in watersheds in the drier parts of the country. *Pongamia pinnata* is one of the few nitrogen fixing trees (NFT) and seeds contain 30-40% oil.

Seeds and sowing: It is propagated through seeds and root suckers. Seeds remain viable for 1 year. Seed setting is usually in November. Seeds are soaked in water for few hours before sowing. Raised seedbeds of convenient size are prepared, well-rotten cattle manure is applied at 1 kg/m^2 and seeds are uniformly broadcasted. The seeds are covered with a thin layer of sand and irrigated. One-month-old seedlings can be trans-

planted into poly bags, which after 1 month can be planted in the field. Pits of size 50 cm^3 are dug at a spacing of 4–5 m, filled with topsoil and manure and then planted.

Manuring: Organic manure is applied annually.

8.6.19 *Phyllanthus emblica* (Aamla, Amloki)

Phyllanthus emblica L. (syn. *Emblica officinalis* Gaertn.) (family: Euphorbiaceae) known as Indian gooseberry (English), Aamla (Hindi), Amloki (Bengali), is a small to medium size deciduous tree, normally reaching a height of 18 m; shedding its branchlets as well as its leaves. Leaves are alternate, pinnate and oblong; small, inconspicuous flowers are borne in compact clusters in the axils of the lower leaves (Photo 8.17). The fruit is round or oblate, indented at the base, and smooth, though 6–8 pale lines, sometimes faintly evident as ridges,

Photo 8.17 *Phyllanthus emblica* (amla)

extending from the base to the apex, giving it the appearance of being divided into segments or lobes. The fruit has one of the highest concentrations of vitamin C, a strong antioxidant, and is used in many medicinal and cosmetic products, especially those for hair such as hair oils and tonics.

It is considered the best of the Ayurvedic rejuvenative herbs, in the treatment of diverse ailments associated with the digestive organs, jaundice, dyspepsia and coughs.

Fruit is considered diuretic and laxative; the dried fruit yields ink and hair dye and, having detergent properties, is sometimes used as a shampoo; fruit is also used in Indian cooking mainly as pickles or as mouth fresheners.

Soil improver: The branches are lopped for green manure. They are said to correct excessively alkaline soils. The hard but flexible red wood used for minor construction, the foliage furnishes fodder for cattle, and wood serves also as fuel.

Each pinna has a single pair of ovate-oblong leaflets that are about 2- to 4-m long. The flowers are greenish-white, fragrant, sessile and reach about 12 cm in length, though appear shorter due to coiling. The flowers produce a pod with an edible pulp. The seeds are black, dispersed via birds that feed on the sweet pod. It is drought resistant and can survive in dry lands, suitable for cultivation as a street tree. *Pithecellobium* is usually propagated by seeds. This tree often used for land rehabilitation, by reason of its soil hardiness and drought tolerance; it is also used as shelter, wind break and defensive live hedge.

The genus includes several other important species—*P. arboreum, P. unguis-cati, P. flexicaule, P. jiringa* and *P. parviflorum*. Seed viability is long under dry cool storage. No pretreatment is necessary for seeds to germinate, although nicking may improve and hasten the process. Germination occurs quickly, normally in 1–2 days.

8.6.20 *Pithecellobium dulce* (Manila Tamarind)

Pithecellobium dulce (Roxb.) Benth. belongs to Fabaceae (Mimosaceae) commonly known as Manila tamarind (English), Vilayati babul, Jungle jalebi (Hindi) is a medium to large tree that reaches a height of about 10–15 m (Photo 8.18). Its trunk is spiny and its leaves are bipinnate.

8.6.21 *Tectona grandis* (Teak)

Tectona grandis L. (Verbenaceae), also known as sagwan (Hindi), Indian oak, teak tree and teak wood (English), is a large, deciduous tree reaching over 30 m in height in favourable conditions (Photo 8.19). Crown opens with many small branches. Natural regeneration is particularly abundant in forests exposed to fires and often

Photo 8.18
Pithecellobium dulce
(Manila Tamarind)
(**a**) branches showing
paripinnate leaves with one
single pair of pinnae and
one single pair of leaflets
per pinna, (**b**) fruit is
greenish pods, coiled

a

b

Photo 8.19 *Tectona grandis* (Teak)

occurs in patches. Teak grows well in deep sandy loam soils. It needs a well-drained soil and does not withstand waterlogging or salinity. Teak flowers in July/August fruit appear in November/January and ripe for collection in February/March.

Seeds: It is propagated from seeds. Seeds collected from the forest floor are generally used to establish plantations. It is recommended that seeds be collected from trees over 20 years old. The seed coat is very hard and can be stored for about 1 year or even more. Germination percentage declines with storage for more than one and half years. There are about 1,500 seeds per 1 kg. The collection is done by hand from below the good mother trees. Because of hard seed coat, seed pretreatment is needed before sowing. It consists of following steps.

Pretreatment of teak seeds: The graded clean seed is spread on a hard sloping ground or on cement floor to a thickness of about 15–30 cm. A thin layer of hay is spread on the seed to provide humidity. The spread seed is watered twice in the day, morning and evening. Thus, the seed is alternatively wetted and dried. The above watering and drying are continued for about a fortnight (15 days) until the seed develops a white radicle, then it is ready for sowing.

The nursery bags are sown with well-treated seeds in April, about 8 kg per bed (standard bed

Photo 8.20 *Putranjiva roxburghii* (Putranjiva) (**a**) old mature tree with dense canopy (**b**) fruit is ellipsoidal drupe with pointed tip

12 m × 1.5 m). After sowing, they are covered with a light cover of hay and watered daily. When germination sets in, the hay cover is removed. The beds are watered daily till the rainy season sets in; watering is also done when rains cease or there is a long gap in rains.

8.6.22   Other Trees

8.6.22.1 *Putranjiva roxburghii* (Putranjiva)

Putranjiva roxburghii Wall. (family: Euphorbiaceae) commonly known as the lucky bean tree (Eng), Putranjiva (Hindi and Bengali), a medium-sized evergreen tree with drooping foliage, distributed throughout India (Photo 8.20). The tree grows up to 13 m; leaves are simple, alternate, dark green and shiny; and it has a large shady head composed of innumerable expanding branches, is a very good avenue tree giving a close and pleasant shade and belongs to one of the most graceful trees. Fruits are ellipsoidal or rounded drupe contains normally one seed.

8.6.22.2 *Mimusops elengi* (Bakul)

Mimusops elengi L (family: Sapotaceae), commonly known as Spanish cherry (Eng), Maulsari (Hindi) and Bakul (Bengali), is an evergreen, medium to large tree, reaches about 16 m height, and provides dense shade during the month of April to July. Fragrant flowers bloom from January to March and starts bearing fruits from January to May. The fruit is a berry, yellow, ovoid 2.5-cm long (Photo 8.21).

8.6.22.3 *Polyalthia longifolia* (The Mast Tree)

Polyalthia longifolia Sonn. (family: Annonaceae) is a lofty evergreen tree, native to India, commonly planted due to its effectiveness in alleviating noise pollution. It exhibits symmetrical pyramidal growth with willowy weeping pendulous branches and long narrow lanceolate leaves with undulate margins. The tree is known to grow over 30 ft in height. *Polyalthia longifolia* is sometimes incorrectly identified as Ashoka tree (*Saraca indica*) because of very close resemblance of both trees. *Polyalthia longifolia* var. *pendula is* one of the prime choices of landscape designers (Photo 8.22).

8.6.22.4 *Bombax ceiba* (Cotton Tree)

Bombax ceiba L (Syn. *Bombax malabaricum* DC., *Salmalia malabarica*) (family: Malvaceae) is commonly known as Cotton tree (Photo 8.23). This tropical tree has a straight tall trunk and its leaves are deciduous in winter. Red flowers with 5 petals appear in the spring before the new foliage. *B. ceiba* grows to an average of 20–30 m. The trunk and limb bear numerous conical spines particularly when young, but get eroded when older. The leaves are palmate with about 6 leaflets radiating from a central point, an average of 7–10-cm wide and 13–15 cm in length.

Photo 8.21 *Mimusops elengi* (Bakul) (**a**) twig showing arrangement of simple leaf alternately with flowers and fruits, (**b**) close view of twig with green colour fruits (ellipsoidal berry) with pointed tip, turns yellow to red when mature

Photo 8.22 *Polyalthia longifolia* (the mast tree)

8.6.22.5 *Tamarix aphylla* (Athel Tree)

Tamarix aphylla (L.) Karst. (family: Tamaricaceae), commonly known as Athel tree and Laljhar (Hindi), is a fast-growing, moderate-sized evergreen tree, grow up to 18 m high, with 60–80-cm DBH with many stout spreading purplish brown and smooth branches, twigs drooping and a deep and extensive root system, about 10 m vertically and 34 m horizontally. Leaves are bluish-green, alternate, reduced to tiny scales ensheathing wiry twigs and ending in points, hairless, often with epidermal salt glands each forming a joint along the twig. Flowers are many, nearly stalkless, tiny, whitish-pink, in racemes 3–6-mm long, 4–5-mm broad at end of twigs, drooping. Fruit is a small capsule, many, narrow, pointed, 5-mm long, splitting into 3 parts. Seeds are many, 0.5-mm long, brown, each with tuft of whitish hairs 3-mm long. The specific name means without leaves.

Photo 8.23 *Bombax ceiba* (cotton tree)

Erosion control: The species is highly valued for stabilising waste dumps, due to its fast growth and deep and extensive root system.

Shade or shelter: An important tree for shade. Very useful for obtaining temporary shelter as quickly as possible, which can be removed once the adjacent longer-term shelter belt has attained sufficient size. *Soil improver*: The tree sheds leaves and twigs abundantly forming a compact litter that improves water-holding capacity of the sand. However, it is reported to have a high water output through transpiration.

Propagation methods: Closed capsules do not contain any fertile seed, therefore, only capsules which are just opening or have partly opened should be collected. Not usually propagated from seeds as they lose viability rapidly (maximum 1 week). There are 100,000–286,000 seeds/kg. Easily propagated from cuttings, cuttings of 10 cm length, stripped of foliage, stored in moist sand for 10 days to develop root buds are planted in the nursery with 1.5 cm exposed above the soil. Young plants require watering, especially in dry periods to facilitate good establishment. Saplings are often planted close together.

8.6.22.6 *Prosopis cineraria* (Khejri Tree)

Prosopis cineraria (L.) Druce (family: Leguminosae, subfamily: Mimosoideae) is commonly known as Khejri which is a small thorny, irregularly branched evergreen tree, 5–10 high; forms an open crown; and has thick, rough grey bark with deep fissures. The tree prefers a dry climate and the most important areas of its distribution are characterised by extremes in temperature and grows on a variety of soils. Khejri is a nitrogen fixer, which means it improves soil quality by making nitrogen in the soil more available to other plants. Growth above the ground is slow but below the ground the roots penetrate deeper and deeper for the subsoil water. Very deep roots help in securing firm footing and in obtaining moisture supplies from deep soil layers.

Propagation: Natural regeneration through seed is confined to moist places, not in dry situations, since the tree regenerates itself by root suckers. The seeds need scarification and soaking in water before sowing; 24 h is recommended as a pre-germination treatment. Germination rate is about 65%. Seeds (25,000/kg) remain viable for decades in dry storage and establish well with 80–90% germination. About 1-year-old nursery plants are planted in the field. Seedlings are raised in a nursery and transplanted when 2–3 months old at the onset of the rainy season. Trees can be planted in close lines as a hedge with 1-m spacing between trees, but tree densities of 50–100/ha are recommended.

8.6.22.7 *Holoptelea integrifolia* (Indian Elm, Chilbil Tree, Papri Tree)

Holoptelea integrifolia (Roxb.), Planch (family: Ulmaceae) is medium-sized to large well-spreading deciduous tree about 18–25 m in height. It is mainly known as roadside tree and found to naturally established and grow well in coal mine degraded lands. Leaves are simple, alternate, elliptic and entirely glabrous with cordate base. Flowers are greenish yellow. Male and bisexual flowers are mixed in short racemes near leaf axils. Fruits are suborbicular samara with membranous wings. It grows fast and is suitable for gravely soil; propagated from seeds and cuttings.

8.6.23 *Vetiveria zizanioides* (Khus)

Vetiveria zizanioides (Linn.) Nash (family: Poaceae) commonly known as the KhasKhas, Khas or Khus grass in India. It is a tall perennial grass and had spongy and much branched fine root system. It is a densely tufted grass with the culms arising from an aromatic rhizome up to 1.5–2-m tall; the roots are stout, dense and aromatic.

Cultivation: The cultivation procedure adopted is also very simple. About 15–20-cm-long rooted slips are transplanted. In the field, it planted in 45 cm apart in the ridges during monsoon. About 10–12 tonne/ha of FYM or compost is added during planting. Roots are dug out after 12–18 months during dry month. The aerial parts are cut at a height of 15–20 cm and removed.

8.6.24 *Cymbopogon citratus* (Lemon Grass)

Cymbopogon citratus (DC.) Stapf. is a perennial grass which grows well on poor marginal lands. It is a hardy drought-resistant plant and is grown under a wide range of climatic conditions. Plant prefers full sun and fast-draining soil. Leaves yield an aromatic oil containing 75–80% citral. This important grass that is grown on about 90% of the produced is exported.

Propagation: This grass is propagated by cuttings, and herbage contains 0.35% of oil on fresh weight basis. The grass yield ranged from 18 to 25 tonne/ha.

Establishment of Grass and Legume Cover

9

Contents

9.1 Introduction

Grass and legume mixture is primarily used for erosion control and minimisation of run-off volume. It is a technique for quickly covering the surface of a disturbed or degraded site. Native revegetation is different from temporary reseeding, which a practice is used to provide short term cover for a site scheduled for future disturbance.

Purpose: Seeding with native grasses, legumes (nitrogen-fixing plants) and forbs (broad-leaved herbaceous perennials and not a grass) is an inexpensive method to quickly cover a site. Native grasses and forbs are adapted to regional conditions of climate and disease and so require relatively low maintenance.

Temporary seeding controls run-off and erosion until permanent vegetation or other erosion control measures can be established. In addition, it provides residue for soil protection and seedbed preparation and reduces problems of dust pollution from bare soil surfaces.

Materials: Grass, legume and forbs seeds; mulch and/or netting.

The grass-legume mixtures are introduced as seeds in degraded site. They can be established in two ways: (1) either use natural seed bank (i.e., topsoil), or (2) from commercial seed sources. If no commercial quantities of seeds are available, topsoil is used as seed bank, which can be collected either from undisturbed grassland or

S.K. Maiti, *Ecorestoration of the Coalmine Degraded Lands*,
DOI 10.1007/978-81-322-0851-8_9, © Springer India 2013

forestland. The seed in surface soil is found in the top 20–30 cm or even in thin layer of 10 cm only. The seeds of grass–legumes mixture can be collected from agriculture university, nursery and research institute.

Characteristics of grass–legume mixtures:
- *Grass*: Local, perennial grass, fibrous root system, high binding capacity
- *Legume*: Forage, perennial, good nitrogen fixer, creeping type

9.2   Benefits of Grass–Legume Mixtures

The sowing of grass–legume (GL) mixture seeds acts as pioneering species at the initial stage of revegetation in the degraded site. The GL mixture being a fast grower provides erosion control, enhances in situ moisture conservation, supplies organic matter and promotes nutrient cycling, fixes atmospheric nitrogen (i.e. *Alfalfa*—225 kg/ha), supplies fodder and provides conditions congenital to the tree species. Figure 9.1 explains the importance of GL mixture towards the development of self-sustaining forest cover through ecological succession approach.

9.3   Important Legumes

The use of legume and grass mixture for the initial establishment of vegetation cover on mine spoil has been practised by many coal companies abroad (Maiti 2002). As commonly known, legumes are members of a plant family Leguminosae (Fabaceae); ranged from small herbs (*Desmodium*), undershrubs (*Trifolium, Medicago, Stylosanthes*), to trees (*Acacia, Dalbergia sissoo*, etc.); and fixed atmospheric nitrogen by bacteria (*Rhizobium*) nodulated on the roots. Different legumes need different species of bacteria to produce maximum amounts of usable nitrogen (Lyle 1987). Legumes grown on mine spoil enhance soil-forming process, increase soil nitrogen as mine spoil is low in organic matter and nitrogen. The rate of nitrogen fixation by different legumes is as follows: alfalfa

(*Medicago sativa*)—224 kg/ha, red clover—129 kg/ka, kudzu—123 kg/ha, soya bean—112 kg/ha, cowpea—101 kg/ha (Donahue et al. 1990). Nitrogen released by decomposition of legume foliage and sloughed roots quickly becomes available to trees.

Types of Legumes

There are two types of legumes that can be use in biological reclamation as an initial coloniser:
- Forage legume (herbaceous) like *Stylosanthes* sp.
- Pulses—peas, *Cajanus*, etc.

9.3.1   *Stylosanthes humilis* Kunth. (Stylo)

Stylosanthes humilis, also known as townsville lucerne (English), is a profusely branched annual plant, usually less than 50-cm tall, and the leaves are trifoliate with prominent veins. They have prostrate stems (creeper), which in contact with moist soil develop adventitious roots away from the taproot. The plant has very good regenerating capacity, acts as fodder (2–2.5 t/ha) and is a good N fixer. Seeds are sown just before the rainy season @ 2–3 kg/ha. The natural growth of *Stylosanthes humilis* is shown in Photo 9.1.

Establishment: Fresh Stylo seeds can have >90% embryo dormancy, which lasts about 4 months. Seed does not soften during normal storage. Germination of commercial seed can be improved by mechanical scarification (kept in hot water at 80°C for 10–15 min, then cool and dry). *S. humilis* is fairly promiscuous in its *rhizobial* requirements, but seed can be inoculated with CB 82, CB 756 *Bradyrhizobium*, or their equivalents, to ensure best results. Seedlings are only moderately vigorous, but rapid root development provides tolerance to dry conditions and overcomes the competition from associated species. The hooked seeds are readily spread by adhering to livestock, wind and water movement. It is one of the best suited legumes for mine spoil reclamation (Box 9.1).

Dry matter: Dry matter yields range from as little as 1 t/ha to a high of 7 t/ha, depending on soil and

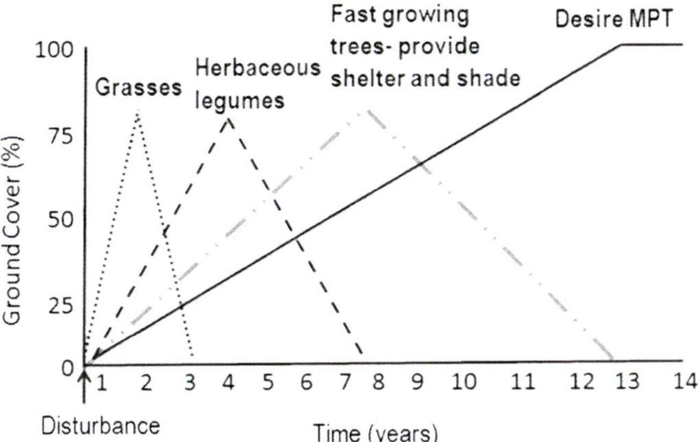

Fig. 9.1 Ecorestoration in mine-degraded site seeks to stimulate natural succession processes leads to forest. All vegetation types are established at the initial stage (grass and legume seeds are sown in the interspacing of tree rows) of reclamation. As time passes, grass and legume covers enhance yield of organic matter and nitrogen for the fast-growing trees and desire multipurpose trees (*MPT*), which gradually mature and develop forest (Modified after Burger and Zipper 2002)

Photo 9.1 *Stylosanthes humilis*

climatic conditions. Yields are depressed by the presence of taller grasses.

Seed production: Average yield is about 330 kg/ha, but yields up to 1,100 kg/ha have been obtained under good conditions.

Advantages of S. humilis:
- Adapted to low-fertility soils
- Tolerates high soil manganese and aluminium
- Free seeding, self-regenerating and tolerates heavy grazing

9.3.2 *Stylosanthes hamata* (L.) Taub.

Stylosanthes hamata, commonly known as Caribbean stylo (English), is a herbaceous annual to short-lived perennial, semi-erect, reaches 30-75 cm height and sometimes prostrate. short-lived much-branched herbaceous perennial; semi-erect, mostly 30–75 cm, sometimes prostrate. Stems are fine soft green, leaves trifoliate, seeds occurred in pod differing from *S. humilis* in having only fine white hairs down one side, but no bristles as in *S. humilis*.

Box 9.1 Field Experimental Study of *Stylosanthes humilis* (Maiti 1995, 1997; Maiti and Saxena 1998)

A field experiment was conducted for 3 years to study ability of *Stylosanthes* legume to enhance the organic matter and nitrogen in the bare coal mine spoils in Kusunda opencast project (BCCL).

Results showed that organic carbon was increased to 1.9% after 3 years (initial level was 0.44%). The rate of increments were 141 and 79% for second and third years, respectively. Initial rapid increment of OC level was due to accumulation and incorporation of legume biomass (leaves) on the surface.

The study reported that the rate of N accumulation on coal mine spoil was 267 kg/ha/year, which is much higher than the reported values in the range of 50–150 kg of N/ha/year for most leguminous plants. The higher value of N accumulation is attributed to higher accumulation and decomposition of leaves of legumes, N fixation by legume itself, increase in moisture content due to massive growth of legumes and climatic conditions. The easily mineralisable N and total N were found to increase up to 225 and 1,640 ppm, respectively.

We observed that below 10-cm depth there was very little accumulation of nitrogen and source of nitrogen accumulation was due to symbiotic nitrogen fixation. The N accumulation rate in 7 years old coal spoil (control plot) was found 53 kg/ha/year. It clearly shows that N accumulation on spoil depends on the nature of spoil also. The interrelation between nitrogen level, organic carbon and time are significantly correlated. Improvement of C:N ratio was observed 48:1 to 25:1 to 11.6:1 after 2 and 3 years, respectively.

Establishment: Seeds are hard and embryo dormancy is very high. Level of hard seed can be reduced by dry heat treatment, followed by cooling. High soil surface temperatures have a similar effect of breakdown of hard seed coat. Seed can also be scarified using hot water treatment. Seed is best sown at the end of the dry season at the rate of 1-4 kg seed/ha.

9.3.3   Compatible Trees

Selection of compatible trees is essential for the development of "tree-compatible ground cover" to minimize competition with tree seedlings. The list of some trees compatible with specific grasses are given below:

Trees	Grasses
Babel (*Acacia tortilis*)	Anjan (*Cenchrus ciliaris*)
Babul (*Acacia arabica*)	Guria (*Chrysopogon fulvus*)
Sirsi (*Albizia lebbeck*)	Sain (*Sehima nervosa*)
Subabul (*Leucaena leucocephala*)	Dennanath (*P. pedicellatum*)

9.3.3.1 Fertiliser

For biological reclamation purpose, 7–10 t/ha of FYM and 50 kg/ha of nitrogenous fertiliser are recommended.

9.4   Inoculation Techniques of Leguminous Seed

Inoculation of legume seed is an efficient and convenient way of introducing effective rhizobia to soil and subsequently the rhizosphere of legumes. Deaker et al. (2004) reviewed the legume seed inoculation technology and reported that a

Table 9.1 Legume inoculation techniques (After Deaker et al. 2004)

Technique	Description
Seed inoculation	
Dusting	Peat inoculant is mixed with the seed without re-wetting
Slurry	Seed is mixed with a water solution of peat often with the addition of an adhesive
Lime or phosphate pelleting	Seed is treated with a slurry peat inoculants followed by a coating of calcium carbonate (superfine limestone) or rock phosphate
Vacuum impregnation	Rhizobia is introduced into or beneath the seed coat under vacuum
Soil inoculation	
Liquid inoculation	Peat culture mixed with water or liquid inoculant applied to the seedbed at the time of sowing (liquid inoculants may also be applied to seed)
Granular inoculation	Granules containing inoculum sown with seed in seedbed

Photo 9.2 *Vetiveria zizanioides* (*khus* grass)

minimum number of rhizobia per seed is require for successful inoculation of legume seeds. Factors like, desiccation, temperature, seed size, seed coat toxicity and loss of viability during inoculation influence the survival of rhizobia in the field. The efficacy of inoculation varies depending on several factors, all of which affect the number of viable rhizobia available for infection of legume roots. Rhizobia may be introduced to legumes by inoculation of the seed or soil. Seed may be inoculated immediately prior to sowing or custom inoculated by local seed merchants with coating facilities to be sown within a week. Alternatively, legume seed may be commercially inoculated and stored prior to its sale (Table 9.1). This product is commonly referred to as pre-inoculated seed.

9.5 Important Grasses

9.5.1 *Vetiveria zizanioides* (Linn.) Nash (Khus Grass)

Vetiveria zizanioides (Linn.) Nash is popularly known as *Khas Khas*, *Khas* or *Khus* grass in India. Khus is a perennial grass with thick fibrous adventitious roots which are aromatic and highly valued. It is a densely tufted grass, found throughout the plains (Photo 9.2). Two species of vetiver are found in India, of which *V. zizanioides* is the common source of the well-known oil of vetiver, which is used in medicine and in perfumery. Harvesting is usually done by uprooting the whole plants and then cutting the roots

Photo 9.3 *Cymbopogon citratus* (lemon grass)

and cleaning mud and other parts of the root system. In areas where Khas is systematically cultivated, the roots are harvested at the age of 10–12 months. Khus is used for erosion control for its tuft-forming habit and thick root system. Khas is recommended by World Bank in India and Southeast Asian countries for growing as live bunds which gives good economic returns. The vetiver grass can also be used as bioremediation of Cu tailings (Das and Maiti 2009).

9.5.2 *Cymbopogon citratus* (DC.) Stapf. (Lemon Grass)

Cymbopogon citratus, commonly known as Lemon grass or Citronella grass, is a perennial, evergreen, fast-growing aromatic grass. It forms dense clumps, growing to about 1 m (3 ft) high with long, thin leaves (Photo 9.3). It produces a network of roots and rootlets that rapidly exhaust the soil. The bright bluish green leaves release a citrus aroma when crushed. The leaves are distilled to extract lemon grass oil. The lemon grass plants rarely produce flower. It is used as cover species during ecorestoration and also used for stabilisation purposes. It has been reported that lemon grass produced more biomass than vetiver grass, and Cu tailings amended with 5%(w/w) chicken manure produced significantly higher shoot and total plant biomass. Lemon grass is

recommended for bioremediation (phytostabilisation) of toxic Cu tailings (Das and Maiti 2009).

9.5.3 *Pennisetum pedicellatum* Trin. (Dennanath Grass)

Pennisetum pedicellatum Trin. (Dennanath grass) is a tall, annual, bunch grass (in discrete tufts or clumps), grow up to 1 m height, branched from the base and leafy (Photo 9.4). In Jharkhand (India), it grows between June to September, produces enormous quantity of seeds and dispersed by wind (anemochory). It has good drought tolerance and spreads rapidly and regenerates naturally in each year. In India, in the beginning, propagation is done by broadcasting of seeds, preferably in a row of 45-cm apart sown just before the rainy season (May–July in India), @ 12.2 kg seeds/ha. In the growing season, several times it is cut and used as fodder. Approximate seed yield is 2 t/ha and annual maximum yield of 7.5–8 t/ha has also been reported.

9.5.4 *Dichanthium annulatum* (Forssk.) Stapf. (Marvel Grass)

Dichanthium annulatum (Forssk.) Stapf., commonly known as Marvel grass, is a tufted perennial grass, 60–100 cm high, culms are erect, leaf

Photo 9.4 *Pennisetum pedicellatum* (Dennanath grass)

Photo 9.5 *Dichanthium annulatum* (marvel grass)

blades are linear (3–30 cm long and 2–7 mm wide), roots grow up to 1m deep and widely used fodder grass (Photo 9.5). Along with marvel grass, anjan grass is grown in the interspacing and additional green matter yield of 2.0–2.5 t/ha has been reported. It is often grown in low-fertility soil, prefers neutral to alkaline conditions, even found to grow in pH 5.5; not considered to be shade tolerant and flowers throughout the growing seasons. It is a good spoil stabiliser also. Marvel grass is one of the best grasses for soil erosion control and ground cover and helps binding the soil even on 200 slopes (FAO 2010).

Establishment: Vegetatively, the grass can be established from rooted slips, planted at 60-cm distance in staggered rows 60-cm apart. Commercial seed is rarely available. Optimum temperature for germination is 32°C, although germination can be achieved between 15 and 40°C. It can grow with the company of legumes *Medicago sativa* and *Stylosanthes hamata*. Dry matter production was reported in the order of 2–6 t/ha.

9.5.5 *Cenchrus ciliaris* L. (Anjan Grass)

Cenchrus ciliaris L (Syn. *Pennisetum cenchroides* Rich.; *P. ciliare* (L.) Link) is commonly known as Buffel grass (Australia) and Anjan grass (India), very drought resistant. It is a tufted perennial grass growing up to 50 cm tall, leaf blades linear, excellent for spoil stabilizer and has good fodder yield of 3.0-3.5 t/ha (Photo 9.6). *C. ciliaris* spreads well by seed where the soil pH ranged from pH 7–8. Aerial sowing is common, and under ideal conditions, it is surface sown and lightly harrowed.

Photo 9.6 *Cenchrus ciliaris* L

It is best sown just ahead of the expected rainy season @ 0.5–4 kg seeds/ha depending on quality of seed supplies, costs and expected ground coverage. Number of seeds per kg is around 450,000–703,000 nos. Seed yield is around 10–60 kg/ha of clean seed per harvest. Seed remains viable for 2–3 years.

9.5.6 *Saccharum munja* Roxb. (Munja grass) and *S. spontaneum* L. (Kansi, Kas grass)

Saccharum munja and *S. spontaneum* are known as Munj and Kas grass respectively in Hindi, grow up to 1.8–2 m, found in arid areas and along river banks in India. The flowering season of these tropical grasses ranges from September to November in India (Photo 9.7). They are found to be naturally colonised and grow in fly ash lagoons and overburden dumps. Both the species have 25–27% root wt., which are fibrous and relatively more important for preventing the movement of spoil.

9.5.7 *Eulaliopsis binata* (Retz.) CE Hubb. (Sabai Grass)

Eulaliopsis binata (Retz.) is locally known as Babui grass or Sabai grass mostly grown in the eastern part of the country like West Bengal, Bihar, Jharkhand and Odisha. Thin and long leaves of the plant with high-quality fibre constitute a major raw material for paper industries; also the flexibility and strength properties of the leaves are utilised for making ropes and other rope-based utility items. *E. binata* is a perennial fibre yielding grass that is economical to grow, has a growing market and eco-friendly as it checks surface run-off and erosion. It possesses drought and infertility resistance, thus can easily be grown in almost every climatic condition. It can quickly establish a grass cover, commonly reaching a height of 1.5–2 m, and has capacity to retain surface water and soil. The most important use of this grass is to make ropes for domestic purposes.

9.6 Planting Procedure

Before sowing of grass–legume mixture, complete the grading operation, prepare favourable seedbeds and install all necessary erosion control practices, such as dikes, waterways and basins. It is advisable to avoid steep slopes, or reduce the steep slope to 3:1, because they make seedbed preparation difficult and increase the erosion hazard. If soils become compacted during grading, loosen them to a depth of 6–8 in. (15–20 cm) using a ripper, harrow or chisel plow. Adequately

Photo 9.7 *Saccharum spontaneum*

Photo 9.8 A typical cultipacker is used to cover the GL mixture in levelled ground

prepare the seed bed. A good seedbed is well-pulverised, loose and uniform.

Revegetation, as distinguished from temporary seeding, is used for the long-term restoration of a degraded site. Once established, native annuals herbs and shrubs will reseed the area themselves, although they may require protection from exotic or invader species. If left unmanaged, natural succession will usually result in the invasion of shrubs and trees. Some species of native grasses, legumes, and forbs are relatively easy to grow.

In case of steeper slopes (i.e, >3:1), prepare groove or furrows on the contour before the sowing of grass-legume mixture. Select an appropriate seeding species or mixture (many commercial mixes are available). Evenly apply seed using a cyclone seeder (broadcast), driller or hydroseeder. Use seeding rates as given by the supplier. Small seeds should be planted not more than 25 mm (1 in.) deep, and grasses and legumes not more than 13 mm (1/2 in.) deep. Broadcasted seed must be covered by raking or chain dragging and then lightly firmed with a roller or cultipacker (Photo 9.8).

Where hydroseeding method is used, the surface may be left with a more irregular surface of large clods and stones. Apply lime as required and soils with a pH of 6 or higher need not be limed.

Hydroseeded mixtures should include wood fibre (cellulose) mulch, cow dung (Farmyard

Box 9.2 Key Points in the Maintenance of GL Mixture in the Reclaimed Site (After Maiti and Banerjee 1993)

- Select appropriate GL mixture for the mine spoils and climatic condition.
- Innoculate with correct *Rhizobium* strain.
- Maintain near neutral pH (by liming) or select legume species which tolerate/adapt to the acidic pH (pH > 5 or higher is most suitable for nitrogen fixation).
- Maintain available phosphate concentration (may be adding P fertiliser).
- Ensure adequate available field moisture in the spoil (initially).
- Control grass growth by cutting.
- Regularly monitor the increase in accumulation of organic matter, nitrogen, phosphorus and moisture.
- Choose the GL mixture which has the higher self-replication capabilities.

manure) and seeds. Mulching is necessary when seeding in on slopes steeper than 3:1, during excessively hot or dry weather and adverse soils (shallow, rocky or high in clay or sandy). If the area subjected to high rainfall, such case of grass-legume seed mulch should be anchored with netting.

9.6.1 Additional Information of Grass–Legume Mixture

9.6.1.1 Quality Check of Seed Stock

Many suppliers advertise native grass and legume seeds. Before placing purchase order, check the suppliers' reputation. Purchase a mixture appropriate for the conditions at the site. Restorationists frequently use hand-collected seed to obtain particular species and to ensure that the stock is native to the area. Processing

(presoaking, scarification, etc.) of some species is necessary for successful germination.

Site preparation, addition of amendments and chemical treatment is a common practice to restored native species. Aggressive exotic species has to be removed (like *Lantana, Eupatorium*) from the site completely. In some habitats and for some species, seed may need to be added for multiple years, since conditions for successful seed set, germination and survival may not occur each year. Mowing should be repeated whenever competing species reach 1–1.5 ft (304–457 mm) height.

9.6.1.2 Seeding of Grass–Legume Mixture

When interseeding, it is best to sow seeds in fall or early spring. Legume seeds are generally inoculated with nitrogen-fixing bacteria to enhance growth. Seeds may be broadcasted by hand. In large areas, machine seeding can be used.

On steeper slopes, hydroseeding may be advisable. After sowing, seeds should be incorporated into the soil using hand raking, harrowing, disking or drilling. A variety of seed drills is available from different manufacturers. As a rule of thumb, seeds should be covered by a layer of soil twice their thickness. An alternative to burying the seeds is to sow in late fall.

While site is completely clear prior to seeding, it is best to start as early in spring as the ground can be conducive to work. A layer of mulch will help to reduce water loss and enhance vegetation, but it is difficult to mulch large sites. The key points for the maintenance of GL mixture is given in the Box 9.2.

References

Burger JA, Zipper CE (2002) How to restore forests on surface-mined land. Reclamation guidelines for surface mined land in Southwest Virginia, Pub 460–123. ser. org/sernw/pdf/VSU_COOP_reforest_surface_mine.pdf

Das M, Maiti SK (2009) Growth of *Cymbopogon citratus* and *Vetiveria zizanioides* on Cu mine tailings amended with chicken manure and manure-soil mixtures: a pot scale study. Int J Phytoreme 11(8):651–663

Deaker R et al (2004) Legume seed inoculation technology—a review. Biol Biochem 36:1275–1288

Donahue RL, Miller RW, Shickluna JC (1990) Soils – an introduction to soils and plant growth, 5th edn. PHI, New Delhi

FAO (2010) http://www.trc.zootechnie.fr/ node/4260

Lyle ES Jr (1987) Surface mining reclamation manual. Elsevier, New York

Maiti SK (1995) Some experimental studies on Ecological aspects of reclamation in Jharia coalfield. Ph.D. dissertation, Indian School of Mines, Dhanbad

Maiti SK (1997) Nitrogen accumulation in Coalmine spoils by legume (*Stylosanthus humilis*). Environ Ecol 15(3):580–584

Maiti SK (2002) Ecological environment. In: Saxena NC et al (eds) Environmental management in mining areas. Scientific Publishers, Jodhpur, pp 110–141

Maiti SK, Banerjee SP (1993) Coal spoils reclamation with legumes and grass – an experimental field study. In: Banerjee SP (ed) Minerals and ecology. Oxford/IBH, Calcutta, pp 111–120

Maiti SK, Saxena NC (1998) Biological reclamation of coalmine spoils without topsoil: an amendment study with domestic raw sewage and grass-legumes mixture. Int J Sur Min Reclam Environ 12:87–90

Contents

10.1 Mulching

Primary use of mulching is to control of erosion on disturbed areas. Additionally, it increases soil moisture, increase infiltration and promote germination of planted seeds. Mulching and matting protect the soil surface from the forces of raindrop impact and overland flow. Mulch and mats foster the growth of vegetation, reduce evaporation and insulate the soil. Spoil surface is stabilised by the application of mulches, stabilisers and binders, and amendments. Before application of any stabiliser, spoil pH should be corrected to around 6.0–7.5 by addition of lime (acidic spoil) or gypsum (alkaline spoil), as the case may be.

Spoil surface can be covered with various organic mulches, for example, straw, saw mill waste, hay and cellulose mulches (bark, wood chips, wood fibre) (Photo 10.1). Surface mulches are important because these usually improve conditions at the surface of the spoil.

Mulching can help in establishing plant growth by
- Preventing erosion both due to water and wind
- Facilitating water infiltration
- Improving soil moisture conditions by reducing evaporation
- Spoil temperature amelioration by its colour and insulating properties
- Being compatible with plant development, improve germination conditions and protecting seedlings

S.K. Maiti, *Ecorestoration of the Coalmine Degraded Lands*,
DOI 10.1007/978-81-322-0851-8_10, © Springer India 2013

Photo 10.1 Different types of mulching material

- Reinoculating microorganisms into spoil
- Preventing of soil crust formation
- Helping soil structure formation and enhancing nutrient supply

In addition, it suppresses herbaceous vegetation growth and thus eliminates competition between undesirable weeds the trees. The effectiveness of different mulch materials in soil erosion and vegetation establishment will be dependent on:

(a) Type of mulch material used
(b) Mulch morphology, for example, corn stalks are more effective in erosion control than corn leaves
(c) Application rate
(d) Method of application (surface versus incorporated)
(e) Soil type
(f) Slope
(g) Climatic characteristics

Decisions on what type of mulch is to be used are usually based on local availability and cost. Often crop residues are used for livestock feeding, fuel or thatching, so they may not be available for mulching. Also, in areas of high fire hazard, certain combustible mulches may not be appropriate. Other problems associated with mulches include harbouring of diseases and pest and the creation of favourable habitats for rodents.

10.1.1 Durability of Mulch Materials

The durability of different mulch materials is important, as this will affect their effective life span. Mulches composed of residues with low carbon/nitrogen (C/N) ratios, such as legumes, will decompose quickly, whereas straw and corn-stalk are longer lasting as they have relatively higher C/N ratio. Decomposition rates are also affected by whether the mulches are surface laid, incorporated or covered with soil.

Straw and hay mulches are usually applied @ 1–2 tonne/acre. Table 10.1 shows the effects of different straw mulch rates, slopes and soil texture on soil erosion. It is probably best to apply 1 tonne/acre or more on slope of over 10%. The 1 tonne/acre would leave about 25% of the soil surface exposed.

Table 10.1 Different straw mulch rates, slopes and soil texture on soil erosion

Straw mulch rate (tonne/acre)	Soil surface coverage (%)	Loam, gradient 15% (tonne/acre)	Silt loam, gradient 5% (tonne/acre)	Silt, gradient 10% (tonne/acre)
0.0	0	27.8	12.4	19.3
0.25	33	9.0	3.2	7.5
0.50	50	8.7	1.4	4.4
1.0	75	5.1	0.3	4.2
2.0	90	1.1	0.0	–
4.0	95	0.7	0.0	–

A thick mulch can prevent or retard plant establishment in two ways: It can act as a physical barrier to seedling emergence, or it can prevent the soil from reaching a temperature high enough for seed germination. Thick mulching also obstructs the light. Non-uniform application can result from many factors such as defective equipment, unskilled operators, poor mulching material or poor weather condition. Table 10.2 describe the mulching materials and application rates.

Application: Mechanical *mulch blower* is used to apply hay materials. The mulch blower separates the stems, chops them into shorter lengths and blows them through a nozzle that can be moved vertically and horizontally. Straw or hay mulch can be held in place by pushing parts of the mulch into the soil with discs or crimpers.

10.2 Soil Amendments

The addition of organic amendments (coal combustion by-products, biosolids, poultry manure, sewage sludge, pepper mill sawdust or wood residue) can ameliorate the impoverished mine spoils drastically. These amendments can alleviate the adverse spoil conditions by improving soil fertility and enhancing plant growth. These organic amendments can decrease soil bulk density, increase water-holding capacity, improve aggregate stability and enhance availability of plant nutrients.

Organic matter is an excellent ameliorant, since it contains nutrients; improves the water-holding capacity and cation exchange capacity of the sandy or stony soils; improves aeration and drainage in heavy soils. It provides the basis for soil structure and beginnings of nutrient cycle. The most commonly used amendments are farm manures, compost, sewage sludge and municipal garbage.

For the improvement of soil structure and fertility, amendments like sawdust (25 t/ha), fly ash (1 t/ha), gypsum (3 t/ha), farmyard manures (FYM; 50 m³/ha) and pressmud (50 m³/ha) are applied.

Sawdust

This makes the spoil more porous. On decay, sawdust develops into a good substructure for holding the moisture and giving a good soil texture in course of time. It also acts as a plant protection chemical for controlling the nematode multiplication.

Fly Ash

It adds micronutrient for soil with limited application. Fly ash has the potential to improve mine spoil and soil quality for establishment of sustained vegetative cover. The major benefits are improved physical properties and neutralization of acidity. Most fly ashes will increase the water holding capacity of coarse textured, high rock fragment spoils. Alkaline fly ashes contain significant alkalinity and act as effective liming materials for acid soils and spoils.

Pressmud

The waste from sugar cane factory contains NPK and calcium, which helps in the plant growth. It also helps in multiplication of microorganisms, improving the soil fertility and its texture.

Stabilisers and Binders

Soil stabiliser falls into two basic types:

Chemical glues which bind the fine soil particles are originated from natural sources and have capacity to retain moisture/absorbed moisture.

Table 10.2 Mulching materials and application rates

Material	Rate per acre	Notes
Straw	1–2 tonnes	From wheat or oats, spread by hand or machine, should be tacked down
Wood chips	5–6 tonnes	Treat with 12 lbs. Nitrogen per tonne; apply with mulch blower, chip handler, or by hand; not for fine turf
Wood fibre	0.1–1 tonne	May be hydroseeded; do not use in hot weather
Bark	35 cubic yards	Apply with mulch blower, chip handler or by hand; do not use asphalt tack
Jute net	Cover area	Withstands water flow, best if used with organic mulch
Fibreglass	Net cover area	Withstands water flow, best if used with organic mulch
Wood fibre	Net cover area	Withstands water flow
Fibreglass roving	0.5–1 tonne	Apply with compressed air ejector. Tack with emulsified asphalt at a rate of 25–35 gals/100 sq. ft

Source: NRCS Planning and Design Manual, Washington State Dept. of Transportation, An Introduction to Water Erosion Control, Alberta Agriculture, Food, and Rural Development (2003)

Examples of these are lignosulphonates, resinuous adhesives and different polymeric substances.

10.3 Geotextiles

Netting includes lightweight plastic, cotton, jute, wire or paper products which leave much of the underlying surface exposed. It is mainly used to hold mulch in place. Mats are organic, synthetic or combination materials which blanket the surface and perform the roles of both mulch and net.

Application of geotextile has the following advantages:
- Enhance infiltration
- Improve drainage, because it increases local hydraulic conductivities
- Surface erosion control
- Slope stability and reinforcement
- Amelioration of site conditions for vegetation establishment and growth

Geotextiles used for soil erosion control can be classified by their:
- Composition (natural or synthetic, which in turn affects their durability on site)
- Mode of installation (surface or buried)

Geotextiles made from natural, often vegetative, materials will bio- or light degrade in time, so their durability is temporary. In theory, as they decompose, the natural vegetation will establish and develop sufficiently to control erosion. Geotextiles will rot in about 2 years.

Two popular geotextile materials commonly used in coal mine spoil reclamation are:
- Jute, a commonly used raw material in surface-applied (Photo 10.2a)
- Coir mat, made up from coconut fibre

A comparison of Jute net and coir net is given below:

Jute net	Coir net
Raw material for jute netting is derived from the jute plant, which grows in India and Bangladesh, and is a 100% natural and renewable resource	Coir erosion control blankets, also known as coconut blankets or coir blankets, are used extensively for slope stabilisation, landslide stabilisation, riverbank protection purposes, landscaping, vegetation establishment and sediment control purposes
Jute netting is the best and most common form of erosion control material	Coir blankets are well known for superior performance compared to other organic blankets
It is safe, biodegradable and non-toxic to both soil and plants	Coir is naturally resistant to rot, mould and moisture
It allows air and water to move freely through to the soil and instals easily	Hard and strong, it can be spun and woven into matting

(*continued*)

Jute net	Coir net
Effectively controls erosion	It also provides adequate strength and durability to protect slopes from erosion while at the same time allowing vegetation to flourish
High water absorption capacity	They are easily transported and deployed, can be secured in place and can eventually become part of the soil structure
Decomposes after vegetation is established (2–3 years)	They are environmentally friendly and do not pose potential contamination risks
Moulds easily to landscape	Over a period of time, coir, which is ecofriendly and biodegradable, completely disintegrates leaving only humus

(continued)

Jute net	Coir net
Plant through and hydroseeding compatible	Coir fibres have an advantage over synthetic fibres
Creates a favourable microclimate for new plant development	They are highly water absorbent while retaining their physical properties
Biodegradable and non-toxic	They store up water and build an ideal microclimate for the seeds underneath
Inexpensive	Of all natural fibres, coir has the greatest tensile strength

Some artificial tests into geotextiles durability were carried in laboratory. Four types of fibre, namely, cotton, jute, sisal and coir, were tested. After 1 year of extreme test coir, textile had degraded least. It has been reported that coir takes 15 times longer than cotton and 7 times longer that the jute to degrade, especially in a wet

Photo 10.2 (**a**) Application of Geojute nets in Samleswari project (MCL) for reclamation of steep slope, (**b**) coir mat rolls available, (**c**) close view of coir mat rolls showing the coconut fibre, (**d**) view of coir mat applied in a steep slope (Photo: Author)

Table 10.3 Selected geotextiles used in soil erosion control

Geotextile	Type	Composition	Natural/synthetic	Surface applied/buried
A. Temporary geotextiles				
Geojute/soil saver	B	Open-weave jute mat	Natural	Surface
Fine Geojute	*B*	Open-weave fine jute mat	Natural	Surface
Enviromat/excelsior	*B*	Mat of wood chips in a light-sensitive mesh	Natural/synthetic	Surface
Coir oven mesh	*B*	Open-weave coir mat	Natural	Surface
Covamat	*B*	Pre-seeded, coir, straw and cotton waste in a light-sensitive mesh	Natural/synthetic	Surface
Eromat	*B*	Coir fibre and straw in a light-sensitive mesh	Natural/synthetic	Surface
Cocomat	*B*	Coir fibres in light-sensitive mesh	Natural/synthetic	Surface

B blanket

environment, with very fertile soils. The result of sisal was not reported. Detailed of the geotextiles materials are given in Table 10.3.

Fibre type mat, which forms an interlocking suitable cover over the spoil surface. Geojute and coir netting were used in steep slope of Mussoorie region, Dehradun. Geojute is an open mesh of 2- to 5-mm-thick jute yarn with about 10-mm-size apertures. Jute mesh is gives an area of approximately 60–65%. Coir netting consists of netting of coir with square-shaped opening of 1.5–2.5 cm size, available in roll of about 1–2 m width and 50 m length (Photo 10.2b,c). Both the fibres are biodegradable and they retain tensile strength of about 3–5 years.

Other synthetic geotextiles are *geocell, enviromat, green fix and netlon.*

10.3.1 Installation of Nets and Mats

1. Apply lime, fertiliser and seed prior to laying net or mat. However, grass-legume mulch seeds can be sown on the surface of the coir-mat, then loose soil is spread above the mat.
2. Start laying the mat or net from the top of the channel or slope and unroll it down the grade.
3. Allow netting to lay loosely on the soil without wrinkles. Do not stretch.
4. To secure the net, bury the upslope end in a slot or trench at least 6-in. (150 mm) deep, cover with soil and tamp firmly. Staple the net every

12 in. (300 mm) across the top end and every 3 ft (0.9 m) around the edges and bottom. Where two strips are laid side by side, the adjacent edges should be overlapped 3 in. (75 mm) and stapled together. Each strip of netting should also be stapled down the centre every 3 ft (0.9 m). Do not stretch. To join two strips, cut a trench to anchor the end of the new net. Overlap the end of the previous roll 18 in. (380 mm) and staple every 12 in. (300 mm), just below the anchor slot (Photo 10.2d).

10.4   Super Absorbent

Super absorbent polymers (SAPs) are compounds that absorb water and swell into many times of their original size and weight. It has been reported in the literature that SAPs are capable of withholding water/moisture by absorbing it as high as 100–1,000 times of its weights. They are lightly cross-linked networks of hydrophilic polymer chains. The network can swell in water and hold a large amount of water while maintaining the physical dimension structure. It keeps this absorbed water stored in a readily available form to the plant roots (Photo 10.3). Once plant roots absorb water, it sinks. Super absorbent can handle several cycle of sink and swell. However, in a typically growing environment, super absorbent average life is taken as 0.5–1 year.

Photo 10.3 Application of superabsorbent in the root zone. (**a**) Application of super absorbent in the root zone (Maiti 2010). (**b**) Super absorbent is used for establishment of tree species in drastically degraded land (Maiti 2010)

Fate of Super Absorbent

They are made up of starch, which is act as food for microbes, hence degraded easily. Thus, super absorbent without starch has higher life as it will not be attacked easily by microbes.

Benefits of Super Absorbent

- Reliable and efficient water supplier: It traps waters and stores in the root zone.
- Improve aeration and drainage.
- Nutrient availability: It increases nutrient retention capability of growing media. Nutrients are not allowed to leach. Increase CEC and enhance the effectiveness of fertiliser resulting in lower quantity requirements.

10.4.1 Application of Super Absorbent

- During pit plantation
- *Hydromulching*: Super absorbent mixed with mulch, water, seed and other additives that promote seed germination in hydroseeding application. It also acts as lubricant and improves the fluid flow. Super absorbent also acts a binder by holding soil particles in place forming a thin crust cover over the soil surface. The rate of application is about 80 kg/ha and above.
- *Seed germination*: Superabsorbent (terra-sorbers) is broadcast prior to seeding of about 14 g/m^2; enhance germination rate up to 75% as compared to 15% of the area untreated.

References

NRCS (2003) Planning and Design Manual, Washington State Dept. of Transportation, An Introduction to Water Erosion Control, Alberta Agriculture, Food, and Rural Development. ftp://ftp-fc.sc.egov.usda.gov/WSI/UrbanBMPs/water/erosion/mulching.pdf

Maiti SK (2010) Revegetation planning for the degraded soil and site aggregates in Dump sites. In: Bhattacharya J (ed) Project environmental clearance. Wide Publications, Kolkata, pp 189–228

Biofertiliser (Mycorrhiza) Technology in Mine Ecorestoration

11

Contents

S.K. Maiti, *Ecorestoration of the Coalmine Degraded Lands,*
DOI 10.1007/978-81-322-0851-8_11, © Springer India 2013

11.1 Introduction

Coal mine overburden (OB) materials vary widely in their physical, chemical and biological properties than natural soil, which affect the plant establishment, survival and growth. Biological reclamation of these impoverished dumps should be aimed to develop a long-term nutrient cycling system between soil-plant-microbes to make restoration self-sustaining. The long-term plant community stability on OB dumps relies upon the development of a functional soil microbial community. Soil microorganisms are responsible for (a) decomposition of plant litter, (b) mineralisation of essential plant nutrients, (c) nutrient cycling, (e) accumulation of organic matter and (f) beneficial changes of soil physical characteristics. One group of soil microorganisms important to the development of long-term plant community structure is mycorrhizal fungi. The absence of mycorrhiza may account for the poor survival of plant used for OB dump reclamation. To reclaim overburden dumps biologically, several types of biofertiliser are being used. Biofertiliser are defined as 'fertiliser of biological origin'. The biofertiliser can be broadly classified into three categories (Maiti 1997):

- Nitrogen-fixing Biofertliser (NFBF): For legumes (*Rhizobium*); for cereals, blue-green algae, *Azotobacter, Azolla,* etc.
- Phosphorus mobilising Biofertliser (PMBF): Phosphate solubiliser *Bacillus, Pseudomonas, Aspergillus*; Phosphate absorber—VAM fungi (e.g. *Glomus*)
- Organic matter decomposer Biofertliser (OMD BF): Cellulolytic *Cellulomonas, Trichoderma;* Lignolytic *Arthrobacter, Agaricus*

11.2 The Mycorrhizal Association: A Plant Root/Fungus Interaction

Mycorrhiza (*plural mycorrhizae, literally meaning fungus root*) is formed by association between a plant root and a fungus, and by far majority of vascular plants are involved in this association. Fungus roots were discovered by German botanist Frank in the last century (1985) in pine forest. Five types of mycorrhizae are recognised, only two types of mycorrhizae are used in mine waste stabilisation (Norland 1993): endomycorrhiza (aseptate fungi- VAM; septate fungi- arbutoid, monotropoid, ericoid and orchid mycorrhizas) and ectomycorrhiza (almost all septate fungi). The other types of mycorrhiza are "ericod", restricted to some species in the Ericaceae family; "orchid" restricted to some species in the Orchidaceae family; and ectendomycorrhizae, formed by species in families other than Ericaceae, but in the Ericales (Norland 1993).

11.2.1 Vesicular Arbuscular Mycorrhiza (VAM)

Endomycorrhiza are known as VAM (*vesicular-arbuscular mycorrhiza*) fungi which largely, Zygomicotina and Ascomycotina group, do not form sheath. They form *vesicles* and *arbuscules* within the cells (*intracellular*) of the root cortex (Photo 11.1). They appear to serve as *organ of storage and transfer* of carbon compounds and mineral nutrients between the fungal hyphae and host plant. VAM fungi are the most widespread and important root symbionts of all mycorrhizal association. The member of Cruciferae and Chenopodiaceae are devoid of VAM infection. About 80% of all land plants form VAM. Hosts include most families of angiosperms and gymnosperm including Rosaceae, Gramineae and Leguminosae. VAM colonizations are also found in Pteridophyta and Bryophyta. About 150 species of VAM are recognised; all are zygomycetes. Taxonomy of VAM is purely based on spore morphology. Four genera of VAM are recognised. These VAM species are also found on mine OB dumps, namely, *Glomus* spp., *Gigaspora, Scutellospora gregari* and *Acaulospora laevis.*

11.2.2 Ectomycorrhiza

Almost all septed fungi belong to Basidiomycotina and Ascomycotina; mycobionts are *Lactarius, Laccaria, Pisolithus, Boletus, Suillus* and *Rhizopogon.* Host plants mostly belong to woody plants and trees. Ectomycorrhiza are characterised by the formation of a sheath or mantle, which surrounds the roots (Photo 11.2a). They do not penetrate the root cells but simply form a sheath around the root with only intercellular penetration

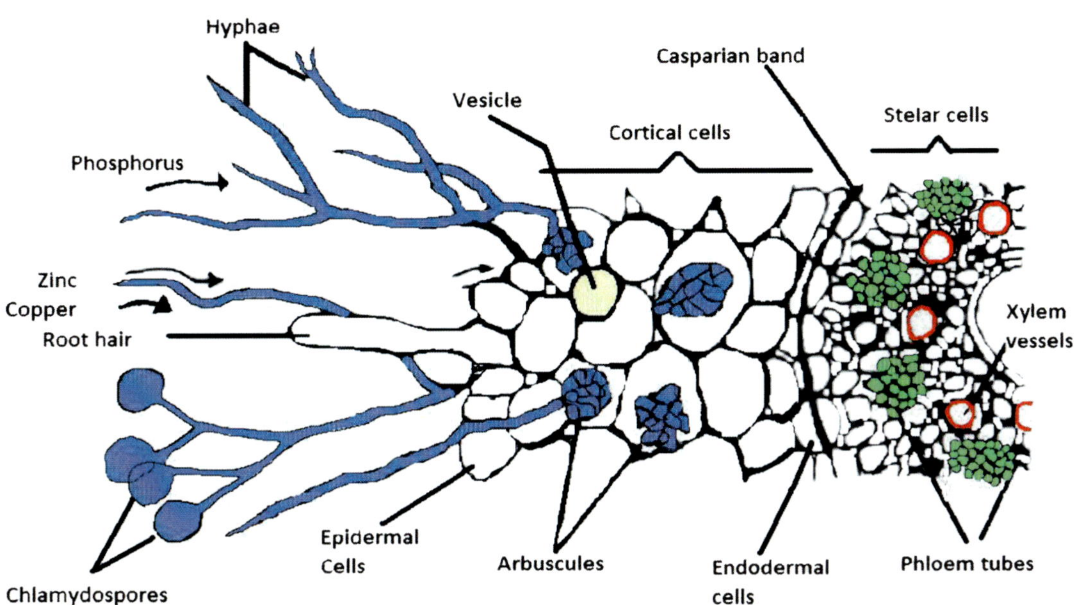

Photo 11.1 Representative cross section through a vesicular arbuscular mycorrhiza (VAM) root. Observe the vesicle and arbuscles in the cortex layer and Exter-nally produced Chlamydospores (They germinate near a plant and germinating hyphae penetrate the root in response to root exudates)

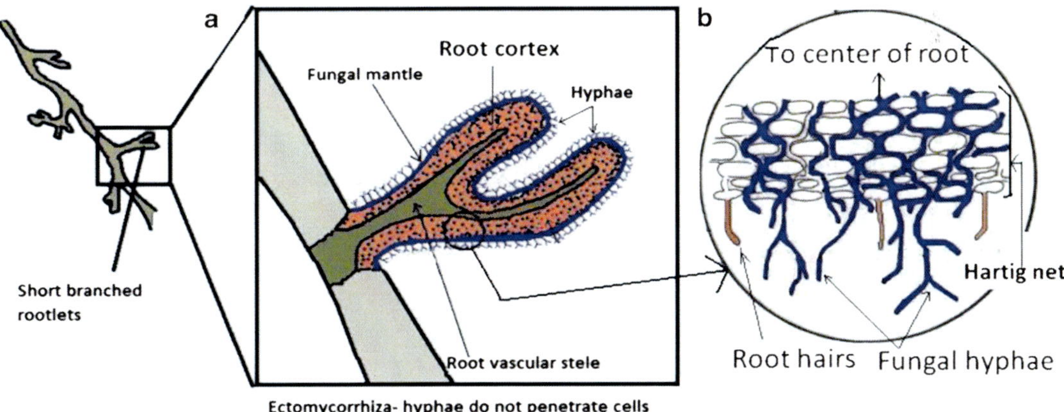

Photo 11.2 (**a**) Representative cross section of an ectomycorrhiza showing fungal mantle or sheath (**b**) enlarge view showing the formation of Hartig net

of root cortical cells (i.e. between the cells of the root cortex) to form the so-called Hartig net (Photo 11.2b). Characteristics of Ectomycorrhiza fungi are

- Hyphae extended outwards into the soil.
- Most ectomycorrhiza are restricted entirely to tree species.
- Ectomycorrhiza have their greatest host range among temperate forest trees, and boreal and high-elevation forest is exclusively ectomy-corrhiza. It is estimated that more than 2,000 species of ectomycorrhiza exist.

- Most are basidiomycetes, but there are some ascomycetes.
- Because of presence of fungal metal on fine roots, it can be seen by magnifying glass.

Ericaceous Mycorrhiza: Both VAM and ectomy-corrhiza are present. These fungi are obligate sym-bionts and have not been isolated in pure culture.

Table 11.1 Some of the main advantages to the plant and microbial partners in the mycorrhiza association (Killham 1966)

Advantages to the plant	Advantages to the VAM fungi
Increased nutrient uptake	A habitat free of competition—sometimes only habitat for growth
Access to organic forms of certain nutrients, for example, nitrogen	A steady supply of photosynthate carbon
Increased rootlet size and longevity	
Protection from pathogen	
Improved water relations	
Drought tolerance	
Enhanced phytohormone activity	
Enhance heavy metal tolerance	

The mycorrhiza association is a symbiotic association, and microbial partner (fungi) requires photoassimilated carbon from its plant host (except orchid mycorrhiza) which is approximately 4% of net photosynthesis. In return, host plants get several advantages, which are given in Table 11.1.

11.3 Benefits of Mycorrhizal Association

The beneficial aspects of mycorrhizal association include the following (Maiti 1997; Killham 1996):

(a) *Increased Nutrient Absorption*: When soil having low nutrient, nutrient absorption is increased by increasing the absorbing surface area of the root system, and the fungal hyphae serve as extension of the root system. Nutrient can therefore be transported beyond the narrow nutrient depletion zone. As mycorrhizal fungi exploiting greater soil volume, the fungal hyphae can often extracts nutrients at lower solution concentration than the uninfected roots. There is clear evidence that some mycorrhiza plants can be connected via hyphae strands and this link provides a means of nutrient transfer from plant to plant. These mycorrhizal pathways facilitate nutrient conservation at the ecosystem level. Mycorrhizal association can also enable the plant host to access nutrients (particularly nitrogen) in an organic form that would otherwise be unavailable. Mycorrhizal structures effectively take up phosphorus from lower concentration at which normal plant roots fail. The Legume –VAM interaction (i.e. *Rhizobia* and VAM) often works synergistically which results better root nodulation, nutrient uptake and plant yield. This interaction is marked when soil has low level of P. This beneficial interaction has been shown in the following legumes: *Stylosanthes guianensis, Centrosema pubescens, Medicago sativa, Phaseolus sp., Glycine max, Arachis hypogaea, Vigna unguiculata, Pueraria sp., Trifolium repens* and *Trifolium subterraneum* (Killham 1994).

(b) *Enhancement of Water Transport Under Water Stress Condition*: Mycorrhiza fungal hyphae may also enhance water transport to trees and drought tolerance due to more hyphal entry point per unit of plant root length that allows more flow of water into the roots and hyphal length and linking of soil surface area with plant roots. It has been reported that (Tisdall 1991) up to 50 m of hyphae length may be present/g of grassland soil. It has been speculated that hyphal diameter of 5 μm and root diameter of 500 μm will lead to 1 m of hyphae having surface area equal to 1 cm of root length. On this basis, it can be estimated that each gram of grassland soil may have hyphae equivalent to 50 cm of fine roots. Improvement of water relation to the host plant is also due to increased cytokinin production by mycorrhiza, which regulates the stomata movement. Sterols are also present at higher concentration in mycorrhizal roots, and this could enhance plant growth under water stress conditions (Smith and Pearson 1988).

(c) *Other Important Benefits Are*

- Increase nutrient mobilisation through biological weathering by breaking down complex minerals and organic substances.
- Serve as a biological deterrent and physical barrier to root infection by soil pathogens.
- Provide tolerance to heavy metal accumulation by restricting the translocation of metals from roots to shoots, as the ions are absorbed on the cell wall of the hyphae in the root.
- Evidence exists that mycorrhizae may provide the host plant with growth hormones such as auxin, cytokinin and gibberellin (Allen and Allen 1980). The mycorrhizal fungi benefit by utilising photosynthates and derivatives from the host plant.

Hence, mycorrhizal status of plants is an important factor in revegetation of severely disturbed sites.

11.4 Importance of VAM Fungi in Coal Overburden Dump Reclamation

Many of the plants that grow on reclaimed coal mine overburden (OB) dumps or naturally colonised invariability have mycorrhiza colonizations which increases the growth and survival rate of these plants. Maiti (1997) reviewed the important mine spoil properties that effects the natural colonisation in the tree species growing on overburden dumps. Recovery of disturbed area in terms of mycorrhizal infection and spore population to be controlled by a number of interaction factors, among them are (a) initial spore count, (b) soil nutrients, (c) texture, (d) moisture, (e) host plant genotype and (f) plant cover age of revegetated site (Norland 1993). In coal OB dumps, the most limiting factors for plant establishment are (a) water stress and (b) nutrients particularly nitrogen and phosphorus (Norland 1993).

The mycorrhiza-mediated drought tolerance to the plant has been discussed earlier. VAM colonization is also influenced by the phosphorus concentration of soil. The degree of colonization was found to be lowest where soil is having high P content. Plant growth is observed to be increased to a greater degree by VAM-inoculation than addition of phosphorus fertilizer (Allen and Allen 1980). The fertilisation of coal spoil with phosphorous fertiliser inhibit mycorrhizal colonisation and sporulation which suggested that during reclamation P fertilisation may reduce VAM colonisation. After 2–3 years of disturbance, percentage of infection and spore count are found to be increased up to 50%. They observed that percentage of infection and spore count are not correlated. Mycorrhizal spores is usually dispershed by wind and animal vector. It has been reported that after 3–7 years of disturbance, the VAM infection is found to be returned to its pre-mining level.

11.5 Factors Affecting Establishment of Mycorrhiza Fungi in OB Dumps

The establishment of mycorrhiza–plant association on OB dumps is influenced by the (a) essential elements, N and P (b) pH (c) organic matter (d) trace elements (e) moisture (f) percentage cover of VAM plants and (g) age of vegetation since disturbance (Norland 1993; Maiti 1997; Mukhopadhyay and Maiti 2009, 2010).

11.5.1 Essential Elements

Soil nutrients particularly nitrogen (N) and phosphorus (P) are primary concern in the formation of mycorrhiza–plant association. High levels of N and P often suppressed the VAM development. The hyphal length was significantly greater at low P than at high P supply. High concentration of soil P has been shown to inhibit mycorrhizae formation. VAM may increase N concentration in plant shoots. Greenhouse and field studies have shown that VAM improve growth, nodulation and nitrogen fixation in legume–Rhizobium symbiosis. Further VAM inoculation enhanced biological N fixation of legumes in a way that is similar to P fertilisation. VAM hyphae can take up and translocate nitrate and ammonium, thereby increases the N uptake by the plants. Studies have shown that calcium, potassium and zinc uptake is increased when plant have mycorrhizal association.

11.5.2 pH

Soil pH strongly controls the mycorrhizal development, altering the bioavailability of nutrients and toxins, and many VAM developments are severely restricted in more acid soils. Root infection by VAM fungi has been to occur over a pH range of 4.2–7.0. *Glomus diaphanum* has been found in abandoned or partially reclaimed surface mine land characterised by acidic and high aluminium content (Duvert et al. 1990). VAM have also been reported in alkaline soil conditions. Optimal pH for spore germination is slightly acidic (5.5–6.5). It has been found that decreasing pH 4.3 decreased spore germination and hyphal growth of *Glomus* species. It has also been reported that spores of *Glomus* species were found only in soil with a pH of 6.8; however, large numbers of *Acaulospora laevis* spores were found in soils with a pH range of 4.5–4.9 and conclude that spoil pH is a major determinant of the distribution of VAM (Norland 1993).

11.5.3 Organic Matter

The number and type of organisms increase with application of organic matter; this can be attributed to the addition of an available energy source and oxidisable carbon. Mining activity typically results in the loss of soil organic matter and reductions in microbial populations. The development of a self-sustaining vegetative cover on OB dumps is dependent on establishment of decomposition and mineralization processes. VAM–plant association may be stimulated by substance produced by organic matter and the properties of organic matter.

11.5.4 Soil Moisture

Soil moisture plays a key role in the formation of mycorrhiza. Excessively high soil moisture is inhibitory when it leads to an anaerobiosis, as all mycorrhizal fungi are obligate aerobes. It has been found that soil moisture characteristics are significantly correlated to propagule level, while soil chemical characteristics are not. Mycorrhiza P supply is likely to be more advantageous for plant growth under arid conditions than under wet conditions since the diffusion coefficient of phosphate in soil is linearly related to soil moisture content. Most aquatic plant and plants growing in wet areas are generally non-mycorrhizal. The lack of VAM formation under saturated conditions has been attributed to low availability of oxygen. Spore germination of VAM fungi is best at soil moisture contents between field capacity and soil saturation. Root infection by VAM fungi is usually most rapid when soil water content is between field capacity and permanent wilting point.

11.5.5 Topsoil Cover

The greatest amount of mycorrhizal infection is found on plots covered with 30 cm of topsoil. Stored topsoil of 1–3.5 years of age is a poor source of VAM inoculation on reclaimed sites, and only 5–10% infection is found.

11.6 Formation of Mycorrhizal Association

Deliberate inoculation of soil and plants with specific mycorrhizal fungi may have one of the following objects (Norland 1993): (i) to increase total inoculums potential or (ii) to establish more efficient fungal symbiosis than those present. Mosse (1981) states that two general principles apply to field inoculation: first, for strongly mycorrhizae-dependent plant species and second, indigenous mycorrhizal fungi are not necessarily the most efficient, particularly when soil conditions are changed or new plant species are introduced. Mycorrhizal inoculums will probably lead to successful mycorrhizal infection in the field when viable inoculum is placed in the root zone of actively growing plants which are not heavily fertilized.

Five basic types of inocula are generally used for bioreclamation of OB dumps in mining areas. These are (a) VAM colonised rhizospheric soil, (b) infected roots, (c) pure cultures of fungi, (d) VAM spores and (e) sometimes by addition of various organic amendments (topsoil) (Norland 1993).

Generally any of the five techniques could be used for mycorrhizal inoculation in OB dumps in mining areas. These are (1) broadcast inoculation, (2) inoculum placement below the seeds at nursery field, (3) slurry dips of sapling in nursery, (4) pelletising seed and (5) mycorrhiza infected seedlings and roots.

1. *Broadcast Inoculation*: A known quantity of inoculum is spread over a given area of spoil surface, and the inoculum is mixed up to a depth of 10–20 cm before seeding.
2. *Nursery Inoculation*: The soil inoculum is placed below the seed in nursery that facilitates the concentration of inoculum near developing roots. This technique is very commonly used for the development of mycorrhizal association.
3. *Slurry Dips*: The slurries of mycorrhiza inoculum are prepared by mixing the inoculum with water and a suitable carrier. Bare roots or container-grown seedlings are inoculated by dipping them into the slurry prior to planting.
4. *Pelletising Seed*: Coating seeds with spores involves incorporation of basidiospores in an external matrix of encapsulated seed. Adhesive such as methyl cellulose has been used to coat seeds with mycorrhiza.
5. *Mycorrhizal Seedlings and Roots*: Transplanting mycorrhizal seedlings is a successful inoculation method. Roots with abundant mycorrhizae could also be selected as a source of inoculum and that can be incorporated. The success rate of mycorrhizal inoculation depends on amount and weight of inoculum used. Field plots are inoculated by placing VAM-infected soil below each sapling. The soil inoculum rates ranged from 2 to 50 g of soil/sapling.

11.7 Bulk VAM-Inoculum Production and Use (Soil Culture Method)

Inoculum of VAM fungi may consist of spores, mycelia and infected root pieces containing vesicles or chlamydospores (Powell 1984). Most common method is soil culture method. In this method, inoculum is produced by growing suitable host plants (onion, clover, sorghum or maize) inoculated with sterilised or unsterilized spores (1–30 nos/plant) in open pots or large bins or sterilised soil or sand. Wet sieving or slurry of inoculum soil containing infected root segments, spores and hyphae is most commonly used. Chopped mycorrhizal roots have also been frequently employed. For pot trials, the inoculum is usually layered below seed at the rate of 0.5–10 g inoculum/plant. In case of field plots, soil is inoculated with mycorrhizal soil on the surface or in the seedbed at the rate of 0.5–2.0 kg/m^2.

11.8 Measurement of Mycorrhizal Growth Response

1. Shoot dry wt. is the most common measure of growth response to mycorrhizal inoculation although shoot fresh wt. is occasionally used.
2. Leaf length and number have also measured, tiller no. in cereals.
3. Root dry wt., fresh wt. or length are often measured and root/shoot ratio is calculated.
4. Assessment of fruit and timber yields.
5. The effect of mycorrhizal inoculation on nodulation and nitrogen fixation of legumes could be assessed visually by the removal and weighing of all nodules.
6. Phosphorus concentration in roots and/or shoots is usually measured in inoculation trials along with other elements (Zn, Cu, N), where necessary.

Pot and Field Trial

1. Plants are grown normally @ 1–10 no per pot of sieved soil/sand mixture on glass house benches.

2. Soils are sterilised with formalin (4.8 L of 2% (w/v) solution/sq. m) or by autoclaving at 121°C for 1 h.
3. Plot sizes for field trials are varied from 0.16 to 500 sq. m. The size of plots should be rectangularly, however, for field trials 6m × 2 m plot are recommended.
4. There should be unplanted guard strip of 1–3 m between plots to prevent mycorrhizal spread, and experiments should be laid out in randomised (square) block design.

11.9 Laboratory Study of VAM Spores

Accurately weight 5 g of soil is taken in a 1,000-mL measuring cylinder and filled with tap water. After mixing the content, wait for 30 min for settlement of heavy particle at the bottom. The suspension is decanted by using only two sieves: (1) 30-mesh (500 μm) to arrest debris and 400-mesh (38 μm) sieves to retain all VAMF spores. The suspension retained on 30-mesh sieve, directly examined under stereo-microscope and observed if any large sporocarp is present. The suspension retained on 400-mesh sieve is carefully removed and transferred along with water to a centrifuge tube. The suspension is centrifuge at 3,000 rpm for 5 min and supernatant is filtered in a marked filter paper as suggested by Gaur and Adholeya (1994). The filter paper is observed under stereo zoom microscope at 6×–80× magnification (Photo 11.3), count number of spore and express in nos/g or nos/5 g or nos/100 g of oven-dried soil. The spores of VAM fungi can be identified using the key of Schenk and Perez (1988).

11.10 Laboratory Study of Root Infection

Approximately 10–15 numbers of root bits (size 1 cm) are taken in a test tube and boiled in 10% KOH for 15–20 min in a water bath (sometimes even 60 min for hard roots, like *Tectona* and *Azadirachta*), washed in tap water and stained in lactophenol following Phillips and Hayman method (1970). For confirmation of infection, the presence of intracellular hyphae, vesicles, arbuscules or both characters is observed under compound microscope (430×–480×) (Photo 11.4). The infected roots are counted by grid line method, and result is expressed as root colonisation (%).

Root colonization (%) =

(Number of root segments colonized × 100)/

(Total number of root segments examined)

Photo 11.3 (a, b)VAMF spores under observed under stereo zoom microscope (magnification 40x; Leixa S6D) (Maiti 2006)

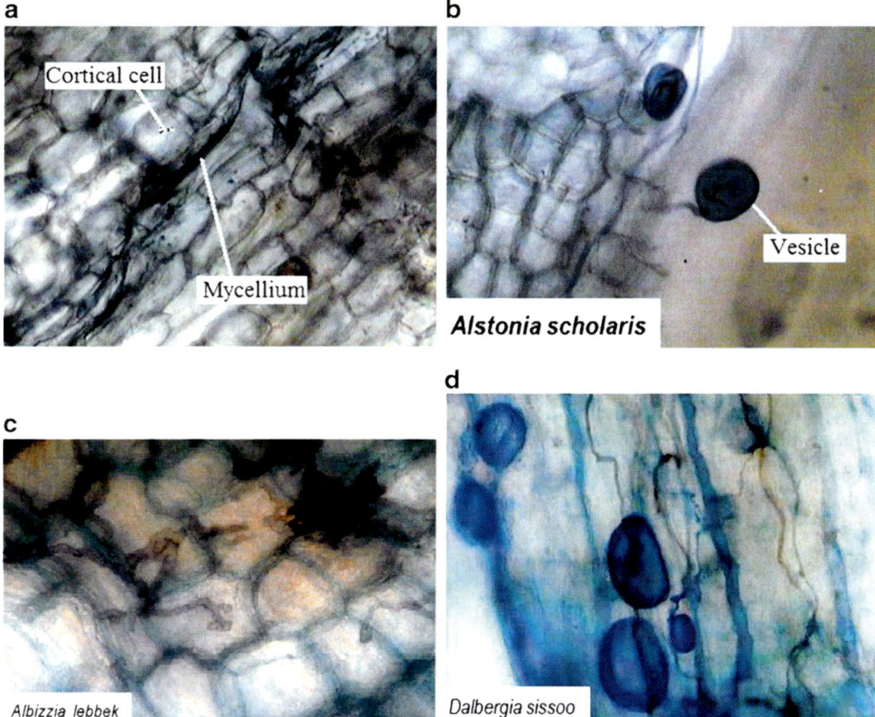

Photo 11.4 Mycorrhiza root infection observed under compound microscope (OLYMPUS BX 60)—430x. (**a**) VAM mycelium in the cortical cells of root, (**b**) mycorrhiza infection in *Alstonia* roots showing vehicles, (**c**) mycorrhiza infection in *Albizia lebbeck* roots and (**d**) mycorrhiza infection in *Dalbergia sissoo* roots (Maiti 2006)

11.11 Application of VAM for Bioreclamation of OB Dumps: Case Studies in India

1. *South Eastern Coalfields Limited (SECL)* (Juwarkar et al. 1994)

Two coal mine overburdens were selected, namely, Chirimari (10.2 ha) and Bishrampur (10.5 ha) projects in SECL. The pit size was 60 cm × 60 cm × 60 cm. The pit treatment was consists of 4 OB (overburden): 1 TS (topsoil) + 15 Kg FYM (farm yard manures). The leguminous plants were infected with *Rhizobium* + VAMF and nonleguminous were *Azotobacter* + VAMF. The VAM association causes profuse root development and three- to five-fold increase in biomass, and survival rate was observed more than 90%. The percentage VAM infection was found in the range of 10–85%. Maximum level of infection was reported in *Tectona grandis* (68%).

The species were consists of timber and fruit bearing trees—*Tectona grandis* (teak), *Grevillea pteridifolia* (silver oak); *Gmelina arborea* (gamar), *Mangifera indica* (mango), *Azadirachta indica* (neem), *Dalbergia sissoo* (sissoo), *Psidium guajava* (guava), *Emblica officinalis* (amla), *Zizyphus jujuba* (Ber), *Dendrocalamus sp.* (bamboo). In Nayveli Lignite spoil, maximum infection was noticed in *Albizia lebbeck* (90%), *Cassia siamea* (60%) and *Tamarindus indica* (45%). Spores from 23 VAM species were found but mostly dominated by *Glomus* sp. In virgin dumps, *G. globuliferum* was found. VAM population depends on age of revegetated site, the degree of surface soil disturbance, amount of topsoil and presence of susceptible root material.

2. *Bioreclamation of OB Dumps of Kamptee and Chandrapur Areas of WCL* (Juwarkar et al. 1994)

Two overburden dumps in Kamptee and Chandrapur areas of WCL were selected for bioreclamation work with the application of mycorrhiza biofertiliser. Total dump area was 20 ha (approx), 12 years old, barren and having low phosphorus (40–95 kg), and pH was acidic (4.0). The biofertiliser was consists of Glomus + Rhizobium + Azotobacter and amended with *sludge from paper mill effluent.* About 56,300 plants were planted consisting of teak, neem, sissoo, shiwan, bamboos, amla, etc. They concluded that survival rate was 78%, and eight- to ten-fold increase in growth was noticed against control plants. The rate of VAM infection was *Dalbegia sissoo, Eucalyptus hybrid* (50%) > *Cynodon dactylon* (40%) > *Tectona grandis* (20%) > *Albizia procera* (15%) > *Delonix regia* (10%). The research concluded that VAM population on OB dumps depends on age of revegetated site, the degree of surface soil disturbance and amount of topsoil and presence of susceptible root material.

3. *Manganese Ore India Limited (MOIL): Dongribuzrug and Gumagaon* (Juwarkar and Malhotra 1994)

Biofertiliser has been used for ecorestoration of overburden dumps Dongribuzrug and Gumagaon mines of Manganese Ore India Limited (MOIL) by NEERI, Nagpur. Five tree species, namely, mango, guava, teak, neem and bamboos, were planted. The treatment at the plantation pits comprises spoil: 1 soil + 100 T/ha sugar mill pressmud + biofertiliser. The biofertiliser consists of *Azotobacter, Rhizobium* and *Glomus* (*VAMF*) species. The enhancement of plant growth was observed 37–59% over control, with a survival rate of 90%.

11.12 Natural VAM Colonisation in Jharia Coalfields

Natural VAM colonisation in the tree species growing the reclaimed overburden dumps of Jharia coalfield was studied by Maiti and Shee (2003), Maiti et al. (2003) and Mukhopadhyay and Maiti (2010). Out of ten tree species studied, roots of *Dalbergia sissoo* contain maximum infection (99%), followed by *Prosopis juliflora* (95%), *A. auriculiformis* and *Tectona grandis* (93%), *Melia azedarach* (83%), *Alstonia scholaris* (78%), *Polyalthia longifolia* (66%), *Cassia siamea* (44%) and *Azadirachta indica* (35%). However, in *Eucalyptus*, no VAM infection was recorded. The percentages of infection in terms of presence/absence of vesicles, arbuscule and fungal hyphae in different plants are shown in Table 11.2.

Table 11.2 Intensity of root infection in the tree species growing in Jharia coalfield (Maiti and Shee 2003)

Name of host plant	Type of infection			
	Vesicle	Arbuscule	Hyphae	% Root colonisation
1. *Azadirachta indica*	−	−	+	35
2. *Cassia siamea*	−	−	+	44
3. *Polyalthia longifolia*	+	−	+	66.6
4. *Alstonia scholaris*	+	+	+	78
5. *Melia azadirach*	−	−	+	83
6. *Tectona grandis*	+	−	+	93
7. *Acacia auriculiformis*	+	−	+	93.5
8. *Prosopis juliflora*	−	−	+	95
9. *Dalbergia sissoo*	+	−	+	99
10. *Eucalyptus sp.*	−	−	−	−

11.12.1 Mycorrhizal Spore Density

A variety of VAM spores were recorded from rhizosphere of different host species in mining area. They are mainly belonging to *Glomus, Gigaspora, Acaulospora, Entrophospora* and *Sclerocystis. Glomus* is the mostly predominant genus found in mine area. But the number of spores found in rhizosphere of mine area of different species was higher (425–600 spores/5 g of soil) than the rhizosphere of non-mining area 300 spores/5 g of soil (garden soil of ISM campus-control area). This is because the plants grown in dumps are always in nutrient and moisture stress condition. Thus, the absorption of nutrient will be effective, if the plants are colonised with VAM fungi. The colonisation of VAM fungi could be easily quantified by studying the density of spores in the rhizosphere of host plant. As the VAM specificity is different, the density of the spore was also different from species to species (Table 11.3).

In the revegetated coal mine OB dumps, about 72% spores were found in between 50 and 100 μm. The distribution of VAM spore size range is given in Table 11.4. The density of spore in mining area (12,320 spores/100 g of soil) was found double than non-mining areas (6,000 spores/100 g of soil). In this case, VAM spore density was found higher than the reported values.

11.12.2 Size Distribution of VAM spores

The size classification of VAM spores was carried out for the mining and non-mining area soil. Spores were classified into 38 categories starting from 30 to 320 μm. The maximum VAM spores were found in 70 μm (17%), followed by 100 (11%), 80 (9%) and 50 (9%). Distribution of VAM spore size frequency of OB spore in Fig. 11.1 shows the distribution of various sizes of spores found in control area. About 87% of the spores were found within 150-μm size more precisely in between 40- and 100-μm sizes. Most of the VAM spores were *Glomus* species. The size of the spores ranged from 41 to 300 μm.

The VAM spores density in afforested coal mining OB was found higher in Jharia coalfield than the other non-coal mining areas. The reasons could be hostile physico-chemical characteristics of OB dumps that promote VAM dependency of host plants. Secondly, spores were quantified by high-power compound microscope, whereas in most of the studies, spore count was done by stereo-microscope. During stereo microscopic counting, small spores or transparent colour spores

Table 11.3 Spore density in different rhizosphere of host plant in overburden dumps (Maiti and Shee 2003)

Name of host plant	Spore density/5 g soil
1. *Azadirachta indica*	425
2. *Cassia fistula*	445
3. *Alstonia scholaris*	456
4. *Melia azadirach*	467
5. *Tectona grandis*	500
6. *Acacia auriculiformis*	585
7. *Prosopis juliflora*	595
8. *Dalbergia sissoo*	600

Table 11.4 Size frequency distribution of VAM spore in mining area and control area (Maiti and Shee 2003)

Range of spore size (μm)	Mining area			Control area		
	No. of spore/ 100 g soil	Percentage (%)	Total no. of spore/ 100 g soil	No. of spore/ 100 g soil	Percentage (%)	Total no. of spore/ 100 g soil
30–50	1,480	12	12,320	1,120	18	6,000
>50–75	3,720	30		1,700	**28**	
>75–100	5,220	42		1,620	**27**	
>100–150	1,680	13		820	13	
>150–200	180	1.4		180	3	
>200–250	0	0		140	2.3	
>250–300	20	0.16		380	6.3	
>300–400	20	0.16		40	0.6	

sometimes overlook. This study will be helpful to identify the efficient mycorrhizal plants for the OB dump reclamation, where available nutrients and moisture contents are always very limiting.

11.13 Natural VAM Colonisation Study in KD Heslong Project of CCL

Natural VAM colonization on tree species in the reclaimed overburden dumps depends age of plantation, types of tree species, nature of mine soils and climatic conditions. Study conducted in KD Heslong project (NK area, CCL) shows that VAM spore density in the rhizosphere of *C. siamea* (27–204 spores/5 g soil) is higher than *Dalbergia sissoo* (17–112 spores/5 g soil) (Table 11.5). Similar species growing in garden soil where moisture and nutrient stress are minimum (i.e. control site: ISM campus garden soil), density of VAM spores is found lower, suggests that stress conditions favor higher VAM colonisation. In control soil, VAM spore density was found higher in *C. siamea* (1,340 spores/100 g soil) than *D. sissoo* (735 spores/100 g soil) (Table 11.6).

This is because the plants grown in dumps are always in nutrient and moisture stress condition. Thus, the absorption of nutrient will be effective, if the plants are colonised with VAM fungi (Maiti et al. 2003). Due to the host specificity of VAM fungi, only few plants have grown and acclimatized in mining area with efficient colonisation of VAM (Habte and Munjunath 1991). The colonisation of VAM fungi could be easily quantified by studying the density of spores in the rhizosphere of host plant. As age of tree species increases, density of VAM also increases (Table 11.7).

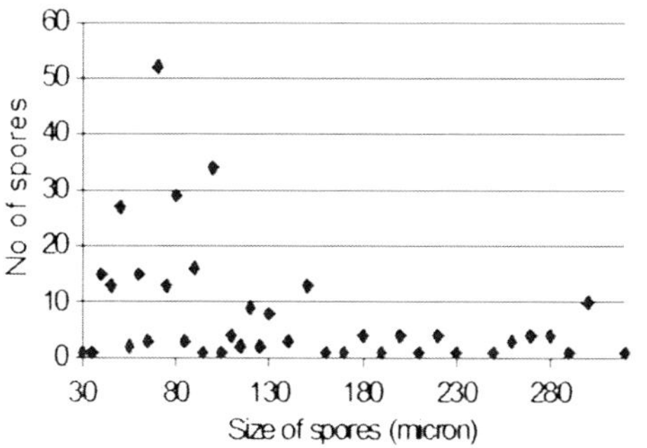

Fig. 11.1 Size distribution of VAM spores in OB dumps

Table 11.5 VAM spore density in the rhizosphere region of *D. sissoo* and *C. siamea* growing on overburden dumps (Maiti 2006)

| Sl. no. | VAMF spore density in rhizosphere, (nos/5 g) | | | | |
	pH	*Dalbergia sissoo*	pH	*Cassia siamea*	Location
1	5.46	112 (87–135)	5.81	204 (175–216)	KDH (1994)
2.	5.35	17 (5–32)	5.34	27 (15–45)	KDH (2000)
3.	na	na	5.15	36 (28–50)	Piparwar (1998)
4.	5.14	86 (73–113)	5.14	184 (151–201)	JCF (1994)
5.	5.92	37 (32–45)	5.91	67 (63–69)	Garden soil[a]

na not analysed
[a]ISM campus (Indian School of Mine campus, Dhanbad, India)

Table 11.6 VAM spore distribution in the rhizosphere of *D. sissoo* and *C. siamea* in garden soil (ISM campus) (spores/100 g soil) (Maiti 2006)

| Name of plant | No. of replicates | | | | |
	R1	R2	R3	R4	Average, spores/100 g soil
C. siamea	1,380	1,360	1,260	1,360	1,340 ± 54.2 (4%)
D. sissoo	640	900	720	680	735 ± 114.8 (15.6%)

Lemon grass—1,940 spores/100 soil

Density of VAM spores reduces as profile depth increases from surface to 45 cm depth. The decrease in VAM spore density with depth attributed due to the fact that VAM fungi are strictly aerobic in nature and as depth increases oxygen/air availability also decreases. Effect of age of plantation, namely, VAM spore density in the rhizosphere of two plants, namely, *D. sissoo* and *Cassia siamea* were studied in KDH and Piparwar dumps and compared with the same plants growing in the OB dumps of Jharia coalfield. In all the places, the VAM spore density in the rhizosphere of *Cassia siamea* was found always higher than *D. sissoo*. This may be one of the reasons why *C. siamea* growth and survival rate always found highest in the OB dumps. In garden soil also (ISM campus), VAM spore density found in *Cassia siamea* was found higher than *D. sissoo*.

11.13.1 Variations in VAM Spores Density Among Tree Species

As the VAM specificity is different, the density of the spore in the rhizosphere of tree species also differs. Density of VAM spores in the rhizosphere of three commonly used tree species, namely, *Dalbergia sissoo*, *Cassia siamea* and *Acacia auriculiformis*, used for biological reclamation of overburden dumps were studied. Samples were collected from the rhizosphere of these three species growing in different aged dumps of KD Heslong, Piparwar and Jharia coalfields. In all the dumps, total VAM spore density was found highest in *C. siamea*, followed by *D. sissoo* and *A. auriculiformis* (Table 11.8).

11.14 Natural VAM Colonisation in the Reclaimed Dumps of SECL

Living VAM spores were analysed in the rhizosphere of five dumps comprising of three tree species, namely, *Grevillea robusta*, *Cassia siamea* and *Acacia nilotica*. As the age of plantation increases, living VAM spore density is also found to be increased. Table 11.9 delineates that *G. robusta* always favours VAM association better than *A. nilotica*. Even in a recent dump (age of plantation is approx 2 years), VAM spore density was found to be 69 spores/10 g of spoil which is better than *C. siamea*.

Table 11.7 No of VAMF spores/5 g soil in different aged overburden dumps (rhizosphere region)

		Year of reclamation				
Sl. no.	Name of plant	2000	1998	1994	1984	Control[a]
1	*Cassia siamea*	27	36	203	186	67
2	*Dalbergia sissoo*	17	25	112	85	36

[a]Control = ISM garden soil

Table 11.8 VAM spore density in rhizosphere of plants growing on OB dumps (spores/100 g soil) (After Maiti and Shee 2003; Maiti and Singh 2006; Maiti 2006, 2007)

			No. of replicates			
Name of plants	Location	Year	R1	R2	R3	Average ± SD (CV)
D. sissoo	JCF	1985	1,460	2,260	1,420	1,713 ± 473 (27%)
	KDH	1994	1,740	1,360	1,000	1,225 ± 414 (33.8%)
C. siamea	JCF	1985	3,180	3,980	4,020	3,720 ± 473 (12.7%)
	KDH	1994	4,320	4,160	4,320	4,075 ± 390 (9.6%)
	Piparwar	1998	480	380	440	433 ± 50 (11.5%)
	KDH	2000	340	300	620	420 ± 174 (41%)
A. auriculiformis	Piparwar	1998	520	360	460	380 ± 131 (34.5%)

Table 11.9 Analysis of spoil materials and VAM spores in SECL (Date: 5-6-2003) (Maiti 2006)

Site description	Plant rhizosphere	Year of plantation	pH	EC (dS m^{-1})	No. of living VAM spores/10 g spoil
Gevra dump	*Cassia siamea*	1985	6.50	38.8	28
	Non-rhizosphere	1985	5.68	22.3	—
Lakshman project	*Acacia nilotica*	1989	5.23	165	87
	Non-rhizosphere	1989	3.94	90.9	—
Dipka dump	*Grevillea robusta*	1990	4.70	59.3	100
	Non-rhizosphere	1990	4.45	23.9	—
Kusmunda project	*Cassia siamea*	1990	6.22	56.5	92
	Non-rhizosphere	1990	4.40	61.0	—
Manikpur area	*Grevillea robusta*	2000	3.97	156	69
	Non-rhizosphere	2,000	3.57	675	—

11.15 Concluding Remarks

The OB dumps initially lack viable mycorrhizal fungal population, thus, the establishment of a vegetative community will be delayed. Mycorrhiza inoculation can enhance productivity of OB dumps by increasing drought tolerance of plants and phosphorus availability which are the main two limiting factors for plant establishment. A number of interaction factors affect the successful colonisation of VAM fungi, among them are pH, soil nutrients, organic matter, moisture, temperature and age of disturbance sites. As mine spoils are nutritionally and microbiological impoverished, it is recommended that plants should be inoculated in nursery. Studies reported that all five species of VAM fungi are generally present in revegetated coal mine overburden dumps, namely, *Glomus, Gigaspora, Acaulospora, Entrophospora* and *Sclerocystis. Glomus* is the most predominant genus found in the mining area.

Maximum infections are reported in *D. sissoo* (99%), followed by *Prosopis juliflora* (95%), *A. auriculiformis* and *Tectona grandis* (93%), *Melia azadarech* (83%), *Alstonia scholaris* (78%), *Polyalthia* (66%), *Cassia siamea* (44%), *Azadirachta indica* (35%) and *Eucalyptus* (0%).

In case of older dump, about 33% of the VAM spores are found alive and rest are dead, while in younger plantation, live spore counts increased up to 70-72% (Maiti 2006). Higher VAM spore density found in *C. siamea*, followed by *D. sissoo* and *A. auriculiformis.*

The VAM spores count in coal mine OB dumps is found to be double than non-mining areas due to higher disturbance. It has been observed that VAM spore density in the rhizosphere of *C. siamea* is found higher than *Dalbergia sissoo*. Similar type of plants those are growing on garden soil (ISM site), density of VAM spores are found lower, suggests that stress conditions favour higher colonisation VAM. As age of tree species increases, density of VAM also increases. Density of VAM spores reduces as profile depth increases from surface to 45-cm depth.

Therefore, survey of the status of VAM association in coal mine overburden dumps should be based on vegetation cover, year of plantation, climatic conditions and physico-chemical and biological properties. Percentage of root infection of plants growing in overburden dumps and density of VAM spores in the rhizosphere of host plants should be assessed during selection of tree species for ecorestoration. Selection of suitable host plant for different geo-climatic and dump characteristics is essential.

The adaptability of selected VAM fungi should be studied before application in the degraded site. The selected VAM fungi must be adapted to a wide range of environmental and edaphic factors. Measurement of mycorrhizal growth response should be conducted both in pot and field trial basis. The amount of inoculum to be applied and rate of root infection in the host plants should be correlated.

Study of the interaction between indigenous microbial population and applied VAM fungi should also be studied. The VAM association will be more successful where indigenous soil biota is low, like coal mine degraded land. Possibilities of using a combination of symbiotic/asymbiotic nitrogen fixers, phosphate solubilising microbes and VAM together will give maximum benefit to the plants growing under stress conditions.

References

Allen EB, Allen ME (1980) Natural re-establishment of VAM following strip-mine reclamation in Wyoming. J Appl Ecol 17:139–147

Duvert PR, Perrin R, Plenchette C (1990) Soil receptiveness to VA Mycorrhizal association: concept and method. Plant and Soil 124:1–6

Gaur A, Adholeya A (1994) Estimation of VAMF spores in soil: a modified method. Mycorrhiza News 6 (2):10–11

Habte M, Munjunath A (1991) Categories of vesicular-arbuscular dependency of host species. Mycorrhiza 1:1–12

Juwarkar AS, Malhotra AS (1994) Manganese mine spoil dump reclamation using pressmud and biofertilizer—a case study. In: Banerjee SP (ed) Minerals and ecology. Oxford/IBH, Calcutta, pp 95–102

Juwarkar AS et al (1994) Reclamation of coal mine spoil dump through integrated biotechnology approach. In: Shringarputale SB et al. (eds) Proceedings of the international symposium on environmental issues of mineral industries. VNIT, Nagpur, CSM, Oxford/IBH Pub Co Pvt Ltd., New Delhi, pp 121–136

Killham K (1994) Soil Ecology. CUP, Cambridge. p 242

Maiti SK (1997) Importance of VAM fungi in coalmine overburden reclamation & factors effecting the establishment of VAM Fungi on overburden dumps. Environ Ecol 15(3):602–608

Maiti SK (2006) An assessment of overburden dump rehabilitation technologies adopted in CCL, NCL, MCL and SECL Mines. Report sub MOEF (India) no. J-15012/38/98-IA II (M)

Maiti SK (2007) Minesoil properties of different aged reclaimed coalmine overburden dumps of Korba, Gevra and Kusmunda area of SECL, India. Minetech 28(2 &3):93–98

Maiti SK, Shee C (2003) Status of VAM infections and spores in an afforested coalmine overburden dumps—a case study from Jharia coalfield. In: Srivastava BK et al (eds) Proceedings of the environmental management in mines. Mining Engineering Department, BHU, Varanasi, pp 257–262

Maiti SK, Singh G (2006) Ecorestoration status of coalmine overburden dumps in Korba, Gevra and Kusmunda area of SECL, India. In: Shringarputale SB et al (eds) Proceedings of the international symposium on environmental issues of mineral industries. VNIT, Nagpur and CSM, pp 217–224

Maiti SK, Shee C, Jha PC (2003) Status of VAMF- infections and spores in an afforested coalmine overburden dump. Minetech 24(4):48–53

Mosse B et al. (1981) Ecology of mycorrhiza and mycorrhiza fungi. In: Alexander M (ed) Advances in Microbial Ecology, vol. 5. Plenum Press, pp 137–210

Mukhopadhyay S, Maiti SK (2009) Reclamation of mine spoils with Vesicular Arbuscular Mycorrhiza (VAM) fungi—a review. Environ Ecol 27(2):642–649

Mukhopadhyay S, Maiti SK (2010) Natural mycorrhizal colonization in tree species growing on the reclaimed coalmine overburden dumps: case study from Jharia Coalfields, India. Bioscan 3:761–770

Norland MR (1993) Soil factors affecting mycorrhizal use in surface mine reclamation. US Department of the Interior, Bureau of mines, Information Circular/9345. p 21

Phillips JM, Hayman DS (1970) Improved procedures for clearing and staining parasitic and vesicular arbuscular mycorrhizal fungi for rapid assessment of infection. Trans Br Mycol Soc 55:158–161

Powell CL (1984) Field inoculation with VA mycorrhizal fungi. In: Powell CL, Bagyaraj DJ (eds) VA mycorrhiza. CRC Press, Boca Raton, pp 205–222

Schenk NC, Perez Y (1988) Manual for the identification of VA mycorrhizal fungi. International culture collection of VA mycorrhiza fungi, Florida, p 241

Smith SE, Pearson VG (1988) Physiological interactions between symbionts in vesicular-arbuscular mycorrhizal plants. Ann Rev Plant Physiol Plant Mol Biol 39:221–244

Tisdall JM (1991) Fungal hyphae and structural stability of soil. Aus J Soil Res 29(6):729–743

Biodiversity Erosion and Conservation in Ecorestored Site   12

Contents

12.1   Introduction

In 1992, India was one of 188 countries that ratified the Convention on Biological Diversity (CBD) at the Rio Earth Summit, which entered into force on 29 December 1993. Through this, the global community acknowledged that biodiversity is '*a common concern of humankind, and an integral part of the development process*'. It recognised that whilst biodiversity conservation can require substantial investments, it brings significant environmental, economic and social benefits in return. The convention recognises that ecosystems, species and genes are used for the benefit of humans.

The Convention has three main goals:

1. conservation of biological diversity (or biodiversity);
2. sustainable use of its components; and
3. fair and equitable sharing of benefits arising from genetic resources

In accordance with Article 26 of the Convention, Parties prepare national reports on the status of implementation of the Convention. The articles 5 to 21 deals with different aspects of biodiversity; like

- *Article 6*: Integrates, as far as possible and an appropriate, the conservation and sustainable use of biological diversity
- *Article 14*: 'Introduces appropriate procedures for EIA of its proposed projects that are likely to have significant adverse effects on biological diversity with a view to avoiding or minimizing such effect'

S.K. Maiti, *Ecorestoration of the Coalmine Degraded Lands*,
DOI 10.1007/978-81-322-0851-8_12, © Springer India 2013

Accordingly, MoEF (1998) published 1st report on Implementation of article 6 of the Convention of Biological Diversity in India; subsequently released 2nd National Report to the CBD (MoEF 2001), 3rd National Report to the CBD (MoEF 2006), 4th National Report to the CBD (MoEF 2009) and 5th National reports are due by 31 march 2014 and that reports should focus on implementation of the 2011-2020 Strategic Plan for Biodiversity and progress towards the *Aichi Biodiversity Targets* (CBD 2011). Immediate effect was to incorporate biodiversity concerns in the existing EIA procedure.

12.1.1 Conference of Parties (COP)

The CBD governing body is the Conference of the Parties (COP), consisting of all governments (and regional economic integration organizations) that have ratified the treaty. The CBD Secretariat, based in Montreal, it operates under the UNEP (United Nations Environment Programme). The year 2010 was declared as the International Year of Biodiversity. The 10th Conference of Parties (COP) to the CBD was held in 2010 in Nagoya (Japan). The Nagoya Protocol was adopted on 22 December 2010 and the UN declared the period from 2011 to 2020 as the UN-Decade on Biodiversity (CBD 2010).

12.1.2 Conference of Parties (COP) - 11

The 11th COP meet was held in Hyderabad, India (1-19 October, 2012), where Honorable Environmental Minister Smt. Jayanthi Natarajan, and president of the COP 11 said: "The present economic crisis should not deter us, but on the contrary encourage us to invest more towards amelioration of the natural capital for ensuring uninterrupted ecosystem services, on which all life on earth depends." The COP 11 released the National Biodiversity Plans, which are as follows (CDB 2012):

1. Developed countries agreed to double funding to support efforts in developing states towards meeting the internationally-agreed Biodiversity Targets, and the main goals of the Strategic Plan for Biodiversity 2011-2020.
2. Revolved practical and financial support for countries in implementing national biodiversity plans to meet the Strategic Plan for Biodiversity and the 2020 Aichi Biodiversity Targets.
3. New measures to factor biodiversity into environmental impact assessments linked to infrastructure and other development projects in marine and coastal areas.
4. Agreed to a number of measures to engage the main economic sectors, such as business and development organizations, to integrate biodiversity objectives in their plans and programmes.
5. Develop a new work in support of achieving Aichi Target 15 which calls for the restoration of 15% of degraded lands.
6. A decision on climate change and biodiversity called for enhanced collaboration between the CBD and UN climate change initiatives including Reducing Emissions from Deforestation and Forest Degradation (REDD+). Given that forests are home to more than half of all terrestrial species, initiatives such as gas Reducing Emissions from Deforestation and Forest Degradation (REDD+), where developing countries can receive payments for carbon offsets for their standing forests, can potentially help achieve international biodiversity targets, as well as those concerned with cutting carbon emissions. The decision covers technical advice on the conservation of forests, sustainable management of forests, and enhancement of forest carbon stocks.
7. UNEP's The Economics of Ecosystems and Biodiversity (TEEB) Initiative also presented a series of practical guides for governments at COP 11 for integrating the economic, social and cultural value of ecosystems into national biodiversity plans.

The 12th COP (COP 12) meet will be held in South Korea in 2014.

12.2   What Is Biodiversity?

Biological diversity or Biodiversity is all life on earth—plants, animals, fungi and microorganisms—as well as the variety of genetic variation they contain and the diversity of ecological systems in which they occur. It includes the relative abundance and genetic diversity of organisms from all habitats including terrestrial, marine and other aquatic systems.

Biodiversity, thus usually considered at three different levels: *genetic diversity, species diversity* and *ecosystem diversity.*

- *Genetic diversity* refers to the variety of genetic information contained in all living things. Genetic diversity occurs within and between populations of species as well as between species.
- *Species diversity* refers to the variety of living species.
- *Ecosystem diversity* relates to the variety of habitats, biotic communities and ecological processes, as well as the diversity present within ecosystems in terms of habitat differences and the variety of ecological processes.

Evolutionary change results in an ongoing process of diversification within living things. Biodiversity increases when new genetic variation is produced, a new species evolves or a novel ecosystem forms; it decreases when the genetic variation within a species decreases, a species becomes extinct or an ecosystem is lost or degraded. This concept emphasises the interrelated nature of the living world and its processes.

12.3 Biodiversity Conservation – a Stairway to Ecorestoration

The UNEP declared 2010 as International Year of Biodiversity- to celebrate the life on earth and of the value of biodiversity for our lives. The world has been invited to take action in 2010 to "safeguard the variety of earth-biodiversity" and appropriately declared 2011-2020 as decade on biodiversity. On this occasion, The UNEP published a report on *"Dead Planet, Living Planet - Biodiversity and Ecosystem Restoration for Sustainable Development- A Rapid Response Assessment"* under the editorial ship of Nellemann and Corcoran (2010). The repot assessed that in 2010, nearly two-thirds of the globe's ecosystems are considered degraded as a result of damage, mismanagement and a failure to invest and reinvest in their productivity, health and sustainability. An analysis of 89 major ecosystem restoration projects worldwide concluded that ecological restoration increased provision of biodiversity and ecosystem services increased by 44% and 25%, respectively (Benayas et al., 2009). Increases in ecosystem services and biodiversity were positively related.

An Ecosystem is the dynamic complex of plant, animal and micro-organism communities and the nonliving environment interacting as a functional unit. Ecosystem Services are the benefits that people obtain from ecosystems. They can be described as supporting services, provisioning services, regulating services and cultural services (MEA 2005).

- Supporting services (e.g., soil formation, photosynthesis, nutrient cycling),
- Provisioning services (e.g., timber, fish, food crops),
- Regulating services (e.g., regulation of climate, floods, water quality, soil characteristics), and
- Cultural services (e.g., aesthetic value, recreational and spiritual).

Forest restoration can restore many ecosystem functions and recover many components of the original biodiversity. Approaches to restoring functionality in forest ecosystems depend strongly on the initial state of forest or land degradation and the desired outcome, time frame, and financial constraints (Fig. 12.1). In many deforested, degraded and fragmented forest habitats investments in restoration and rehabilitation forests can yield high biodiversity conservation and livelihood benefits. Some of the benefits of forest restoration are:

- Increased and higher quality habitats for animals and plants;
- Prevention and reduction of land degradation;
- A secure source of biomass and biofuel energy;
- Environmentally sound and socially acceptable carbon sequestration;
- Adequate and sustainable income and employment opportunities for rural communities;
- Sound return on investment for forestry investors;
- Increased resilience and resistance to climate change;

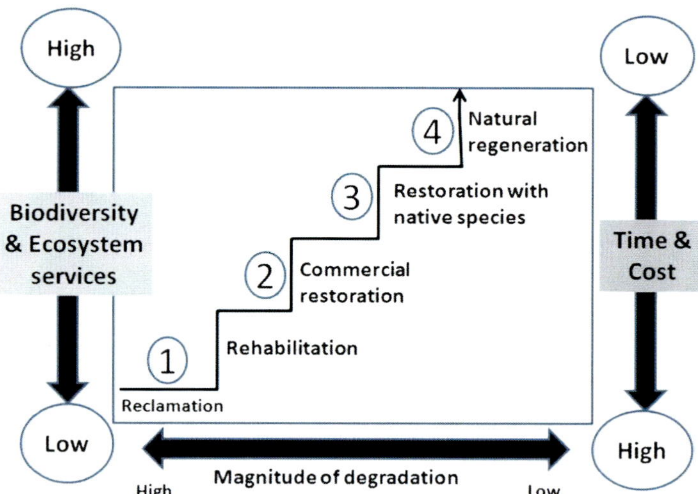

Fig. 12.1 The restoration stairway. Depending on the state of degradation of an ecosystem, a range of management approaches can at least partially restore levels of biodiversity and ecosystem services given adequate time (years) and financial investment (capital, infrastructure and labor). Outcomes of particular restoration approaches are (Chazdon 2008). (1) restoration of soil fertility for supporting ecosystems; (2) production of some products such as timber; or (3) Plantation of native tree species like, MPTs, fruit trees etc. (4) Recovery of biodiversity and ecosystem services

- Additional sources of non-timber forest products;
- Recreation and tourism opportunities;
- Increased property values near restored areas;

12.4 Ecorestoration and Biodiversity Conservation

Biodiversity is among the most commonly assessed outcomes of restoration efforts, but what factors dictate these biodiversity outcomes? Post-restoration biodiversity is a result of site-level factors, such as abiotic and biotic filters, landscape-level factors (e.g. connectivity between restoration sites and relict source populations) and various historical contingencies (e.g. species arrival order). Brudvig (2011) reviewed 'the restoration of biodiversity research' and proposed a 'conceptual model for the restoration of biodiversity' (Fig. 12.2). He stated that restoration efforts may seek to modify factors related to sites, landscapes or historical contingency. Whilst the biodiversity outcomes of restoration, in turn, are dictated by site, landscape and historical factors (non-bold arrows in Fig. 12.2), each of which may have some components that are directly influenced by restoration and others that are

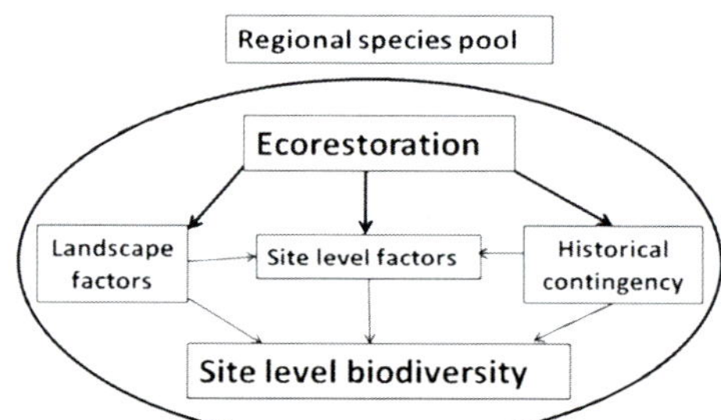

Fig. 12.2 Conceptual model of biodiversity restoration (After Brudvig 2011). Biodiversity development in a restored site is a function of site-level, landscape and historical factors imposed filters on the regional species pool, which are the potential natural sources of seed bank in a given site. During restoration activity, landscape, site and/or historical factors (*bold arrows*) are manipulated. Biodiversity, in turn, may be affected by local, landscape or historical factors that may or may not be directly influenced by restoration (*non-bold arrows*)

independent of restoration. The influence of three factors on site-level biodiversity in mine restored area is described described below:

12.4.1 Site-Level Factors

Site-level conditions create a series of filters that facilitate or hinder membership of plants and animals during restoration. Not surprisingly, manipulation of a site to make it suitable for a target community, is key in ecological restoration.

1. At highly modified sites, restoration may begin by reinstating basic abiotic and structural conditions. Past work reported a wide variety of approaches, ranging from replacing topsoil on former mine sites (Bradshaw 1997). Once abiotic and structural conditions have been restored, it may be important to restore a disturbance regime.

2. Simply creating a suitable site will not, however, guarantee restoration success, and many restoration efforts that adopted an '*if you build it, they will come*' approach have ultimately failed to support desired community members. Thus, biotic conditions may need to be actively reinstated, and, in many instances, the simple act of reintroducing individuals can dramatically influence post-restoration biodiversity levels.

3. Planting seedlings or introducing seeds is the most common approach with plants; however, a passive restoration approach, in which species are assumed to disperse to a site without human assistance, may be appropriate in some situations and is more common in the restoration of animal communities.

4. Reintroduction of focal species may not succeed without consideration of species interactions.

5. With the example of plants, successful restoration may be contingent upon not only reintroducing plant propagules but also promoting or controlling a suite of interacting species, including mycorrhizae, pollinators, seed dispersers, consumers and competing plant species. The effects of (native or exotic) invasive species on native biodiversity during restoration may be particularly important.

6. Considerations range from the properties of ecosystems that confer invasion resistance, to direct and indirect effects (mediated by changes in ecosystem properties) of invasive species on native biodiversity, to the promotion of exotic species by disturbances related to restoration and to the role of invasive species in forming alternative stable states that may be resistant to restoration efforts. Restoration approaches may range from eradication of invasive species, to utilisation of invasive species as tools during restoration and to management of highly invaded ecosystems as novel entities, with goals related to ecosystem functions and services rather than native species composition.

7. Finally, it is important to recognise that site-level factors not directly manipulated by restoration may also have impacts on the biodiversity outcomes of restoration (Fig. 12.2).

12.4.2 Landscape Factors

Landscape-scale factors can influence the site-level biodiversity outcomes of restoration efforts in a number of ways. One of the important facts is that that restored habitat patches are often too small to provide for self-sustaining populations.

1. In practice, the site-level biodiversity outcomes of restoration may be influenced by the composition of the surrounding landscape, by connectivity among patches of restored habitat and remnants or other restored habitats, or through the influence of patch geometry (the size or shape of patches undergoing restoration). In turn, restoration may seek to modify landscape-scale effects.

2. Landscape restoration strategies include construction of new habitat patches in specific locations that maximise biodiversity benefits, construction of landscape elements, such as corridors, to connect restored patches of habitat with each other or with remnants.

3. It is the blending of restoration sites with the elements of landscape ecology, where focus of restoration might move from individual patches to landscape-scale restoration.

12.4.3 Strong and Weak Linkages in Restoration of Biodiversity

12.4.3.1 Strong Restoration Linkages

- The major focus of restoration ecology during the past decade has been the restoration of site conditions and the subsequent impacts on biodiversity. The vast majority of studies (97%) investigated restoration of site-level factors. Out of these, most of them are about restoration of biotic conditions; however, restoration of abiotic and structural conditions

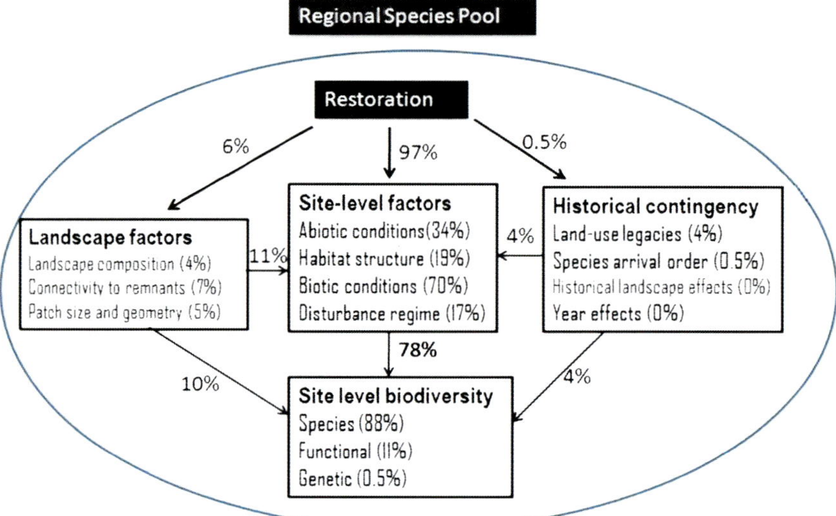

Fig. 12.3 Results of a literature review (percentage of 190 articles evaluated) that assessed how past research has addressed the conceptual model of biodiversity restoration. Past work has been overwhelmingly focused on site-level restoration, with assessment at the species-level of biodiversity. Relatively little effort has been directed toward understanding links between restoration and landscape processes or factors that determine historical contingency, nor has biodiversity been frequently assessed at the functional or genetic biodiversity levels. (Note: tallies may exceed 100% because of studies that investigated multiple links.) (Brudvig 2011)

and disturbance regime were reasonably well investigated.

- Biodiversity was a frequently assessed restoration outcome, and, perhaps not surprisingly given this focus on site-level restoration, about 78% of studies found biodiversity to be a function of site conditions (Brudvig 2011).
- Biodiversity is a major focus of restoration studies: About 88% of the research papers in this review assessed biodiversity in some way, and all but one of these assessed species-level biodiversity (Fig. 12.3). In addition to species diversity, 11% of papers assessed some facet of functional diversity—generally based on plant life-form group (e.g. Gramineae, forb and shrub).

12.4.3.2 The Less-Studied Facets of Biodiversity

Restoration has focused almost exclusively on the species level of biodiversity, with emphasis on only a few taxonomic groups—primarily plants, but also, to lesser extent, arthropods and vertebrates. Most research on biodiversity relies on most easily and inexpensively measured, requiring only botanical knowledge (for plants, at least) and survey time.

The controlled perturbations that accompany restoration may prove particularly useful for understanding linkages between species, functional and genetic levels of biodiversity.

Concepts from biodiversity–ecosystem functioning research are of fundamental interest to restoration ecology, and the linkages between these two disciplines have been explored in detail. At present, restoration research is highly focused on promoting species diversity primarily of plants within individual restoration patches.

12.5 Biodiversity Wealth of India

12.5.1 India as a Megadiversity Centre

India is recognised as a country uniquely rich in all aspects of biodiversity (ecosystem, species and genetic) by virtue of its tropical location, varied physical features and climatic conditions. In fact, it is one of the 17 'megadiversity' nations in the world. The other 'megadiversity' nations are Mexico, Columbia, Madagascar, Ecuador, Australia, Brazil, China, Columbia, Democratic Republic of the Congo, Ecuador, Indonesia, Madagascar, Malaysia, Mexico, Papua New

Guinea, Peru, Philippines, South Africa, United States and Venezuela.

12.5.2 India as a Vavilov Centre

India is also a 'Vavilov' centre of high crop genetic 'diversity', so named after the Russian agrobotanist N.I Vavilov who identified eight such centres around the world in the 1950s. Indian region alone has given nearly 167 economic plants whose centre of origin/diversity lies in India along with 320 species, their wild races and within the

---

One-Third of India Should Be Under Forest Cover?

Questioning the '*theology*' since 1952 that one-third of India should be under forest cover, then Honourable Minister of Env and Forest Mr. J Ramesh said there is a need to change the debate from the quantity of forests to the quality of forests.

He also said 40% of the country's total 70 million hectares of forest is open degraded forest. 'India today has about 70 million hectares under forest cover. Roughly about 21% of India's geographical area is under forest cover. The theology since 1952 has been one-third of India should be under forest cover.'

- *Definition of Biodiversity*
  1. '*Great variety that exists in living organism (plants, animals and microbes) on earth reflects Biodiversity*' (Jain 1994)
  2. *The variety and variability of all living organism (IUCN General Assembly, Costa Rica, Feb, 1988)*
  3. *The totality of genes, species and ecosystem* (Global Biodiversity Strategy, WRI, IUCN, UNEP, 1992)
  4. *The variability among living organism which includes diversity within species, between species and of their ecosystem* (Biodiversity Convention. Rio de Janeiro, June 1992)

---

species the variety is enormous. There were 50,000–60,000 varieties of rice grown in India.

Unlike much of the world, a large range of endemic species and ecosystems are supported in many parts of the India. Indian biodiversity comprises flora with numbers of about 45,000 species (12% global) and fauna with 75,000 species, which constituted 6.5% of world population. The numbers of floral species endemic in nature is 7,000 (15.5%), out of which 10% of flowering plants, that is, about 1,500 sp, are threatened.

12.6 Biodiversity, Society and Mining

Humanity is dependent on biological systems and processes for its sustenance, health, well-being and enjoyment of life. Biodiversity is the basis of numerous ecosystem services that keep the natural environment alive, ranging from maintaining watersheds that provide fresh water, to pollination and nutrient cycles and to the maintenance of clean air and atmospheric gases. We derive all of our food and many medicines and other products from wild and domesticated components of biodiversity.

- Biodiversity is also important for aesthetic, spiritual, cultural, recreational and scientific purposes.
- The interdependency between humanity and biodiversity is critical for all people, because all communities ultimately depend on biodiversity services and resources.
- In recent decades, ecosystems have degraded more rapidly and extensively due to human encroachment than at any time in history. This has placed serious threats on basic ecosystem services we all depend upon.
- *Through land disturbance, mining can have significant local and direct impacts on biodiversity.*
- Broad scale and indirect impacts may also result from associated land use changes.

At the same time, the mining industry has contributed considerable knowledge and expertise to the understanding of biodiversity management and rehabilitation. It is important that the industry recognises that it not only has a responsibility to manage its impacts on biodiversity but also has the opportunity to make a significant contribution to biodiversity conservation through the generation of knowledge and the implementation of initiatives in partnership with others.

> **'Go' and 'no go' areas in iron ore mining soon** [*Ramesh New Delhi, May 26, 2011, Decan Herald, PTI*]
>
> *Like coal, the iron ore mining sector in the country too will soon have 'go' and 'no go' areas where green clearance will be given depending on their **ecological sensitivity**. H*e underlined the need to ensure developmental activities whilst *protecting ecological concerns.*
>
> Present *'go'* areas would be those areas where, prima facie, the statutory Forest Advisory Committee (FAC) in the MoEF would consider proposals for diversion of forests land for coal mining purposes.
>
> Present *'no go'* areas would be those areas with rich forest cover and biodiversity where applications would not be entertained for forest land diversion.
>
> A 'no go' zone is a densely forested area where mining will not be allowed at any cost. '35% of coal mining areas fall in 'no go areas' where mining will not be allowed.
>
> But even in the *'go' areas*, projects will have to go through the due environmental and forest clearance process before being approved. 'Go' does not mean green signal. 'Go' area prima facie means that the ministry will only consider the proposal for approval or rejection.

12.6.1 Social Licences to Operate

Mining activities often run in remote environments where local communities engage in subsistent agricultural practices or sustainable livelihoods based on surrounding natural resources. In these circumstances, the human (*social and economic*) dimensions of biodiversity play a crucial role. This is particularly true in the rural areas of developing countries, where entire communities are directly dependent on biodiversity and ecosystem services and therefore more vulnerable to their degradation.

Public concern over biodiversity loss and ecosystem damage is reflected in a growing number of initiatives. These range from civil society and local community action to international, national and local laws, policies and regulations *aimed at protecting, conserving or restoring ecosystems*. To maintain their social licence to operate, mining companies are responding to expectations and pressures for stricter measures to conserve and manage remaining biodiversity. They are increasingly being called upon to the following:

- Make *'no go'* decisions on the basis of biodiversity values, which may include pristine, sensitive or scientifically important areas
- The presence of rare or threatened species or where activities pose unacceptable risks to ecological services relied upon by surrounding populations.
- A precautionary approach in relation to mitigating or avoiding impacts on biodiversity.
- Where practicable, mitigate impacts and positively enhance biodiversity.
- Responsible management of biodiversity, in conjunction with key stakeholder groups such as regulators and indigenous peoples, is a key element of leading practice sustainable development in the mining industry.

Broad range of leading practice approaches for sustainable development in the mining industry include

- The collection of seed from mature plants before disturbance
- Material segregation (topsoil) that contributes to the rehabilitation and the establishment of landforms

12.6.2 Biodiversity Assessment and Planning

1. Prior to undertaking any operations, mining companies need to identify the biodiversity values present in a particular area, determine key risks to biodiversity and enable the

design of the management programmes, rehabilitation and closure objectives.

2. Mining may be excluded from areas deemed to have significant biodiversity values through either regulation or the voluntary adoption of guidelines.

3. Landscape/catchment level planning enables mining companies to address both of the direct and indirect impacts of their activities.

4. Consideration should be given to cumulative impacts during planning.

5. To optimise biodiversity management, risk assessment procedures need to be closely linked to the assessment of impacts and to ensure relevant information is obtained and used in the decision-making process.

6. Biodiversity objectives should be developed in consultation with all stakeholders and linked to specific, measurable targets as part of the completion criteria developed for the mine closure plan.

7. Conservation and sustainable management of biodiversity values during mine closure planning is an ongoing process.

8. Leading practice requires i.e, starts from the earliest moments of project planning, development and continues throughout the life of the project.

12.6.3 General Overview of Baseline Monitoring

Prior to undertaking any operations, mining companies need to delineate the biodiversity values in a particular area. This is influenced by a range of social and economic factors, and the resulting information is essential for the identification of key risks to biodiversity and the effective design of management programmes, rehabilitation and closure objectives.

Baseline monitoring involves studying some element of biodiversity, which is not expected to change without being disturbed. In determining what baseline monitoring is required, it is crucial to understand the range of influential factors within a specific environment. Surveys and monitoring programmes should differentiate between the direct and indirect impacts of the exploration and mining operations and any other factors that may threaten local and regional biodiversity values.

12.6.4 Regional Planning

Where mining operations have many mines operating in a particular region, state governments can play a significant role in biodiversity management by setting up natural resource management plans. One such plan operates in a cluster.

- The plan conceptually proposes opportunities for revegetation across the mining areas in an integrated approach that considers biodiversity
- Agroforestry for amenity and commercial return
- Catchment protection
- Remodelling of mined landforms

Another recent initiative increasingly used by managers to help focus on a regional landscape perspective involves the development of *biodiversity action plans* (*BAP*). These plans are usually based on the following hierarchical approach:

- Avoid irreversible losses of biodiversity.
- Seek alternative solutions that reduce biodiversity losses.
- Use mitigation and rehabilitation to restore biodiversity resources
- Compensate for unavoidable loss by providing substitutes of at least similar biodiversity value.
- Seek opportunities for enhancement.
- The BAP is a structured approach for identifying priorities and mapping significant areas for native biodiversity conservation at the landscape and bioregional or biogeographical scales.
- The BAP attempts to take a strategic approach to conservation of threatened and declining species and assemblages by looking for opportunities to conserve groups of species in appropriate ecosystems.

The development of a BAP by a mine operator depends on the location and type of operation.

It may be at a local site level, on a slightly larger surrounding area or catchment level or may incorporate plans developed by governments or other stakeholders on a bioregional level.

12.6.5 Assessing Impacts on Biodiversity

Environmental and social impact assessment (ESIA) should be an iterative process of assessing impacts, considering alternatives and comparing predicted impacts on the established baseline. At a minimum, the following assessments should be made in and around the proposed project area:

- An assessment of the impact level (ecosystem, species and/or genetic)
- An assessment of the nature of the impact (primary, secondary, long term, short term, cumulative)
- An assessment of whether the impact is positive, negative or has no effect
- An assessment of the magnitude of the impact in relation to species/habitat richness, population sizes, habitat sizes, sensitivity of the ecosystem and/or recurrent natural disturbances

Many existing mining projects have conducted an ESIA very recently or in some cases, not at all. For these projects, it is important that biodiversity assessment and management considerations are built into their EMS (Environmental Management System) and any other relevant internal and regulatory systems and procedures.

When assessing biodiversity impacts, it should be recognised that the intensity of impacts varies over the life of a project. Typically low at the starts, the intensity of the impact increases markedly through the construction and operation phases, and diminishes as planned closure occurs.

A proposed activity can directly or indirectly impact biodiversity. Both types of impacts need to be identified and managed. Other aspects or types of impacts also need to be considered, and they are:

- Loss of ecosystem or habitats
- Habitat fragmentation

- Alteration of ecological processes
- Pollution impacts: They can affect air, water and soil that includes
  - Airborne pollutants
  - Water pollution from spillages or discharges
  - Mobile sediments from soil erosion
  - Disturbance impacts (soil disturbance, noise, vibration, artificial lighting)
  - Microclimate change affecting the suitability of sites for particular species

12.6.6 Setting Biodiversity Objectives

As with land and water use objectives, biodiversity objectives should be developed in consultation with all stakeholders and linked to a specific measurable targets and standards. They should form a part of the completion criteria developed for the mine closure plan.

Leading practice requires that these objectives be driven in part by the physical and biological components within the landscape. They should also be driven by the social and economic factors that are operating in the environment.

The objectives will depend on the biodiversity aspects identified, its requirement and opportunities to mitigate impacts. They can focus on specific local issues, such as, a plant or animal species, or they may be aimed more generally at the ecosystem level.

Objectives should be *realistic and achievable* and be set in conjunction with the biodiversity values identified by the company and stakeholders. All participants should seek opportunities to reduce negative impacts and increase positive impacts on biodiversity. Examples of goals and objectives may include:

- Successful reintroduction of key flora or fauna species to mined areas,
- Non-disruption of migration/movement patterns of faunal populations,
- Protection (non-interference) of designated high conservation value sites,
- Control of weeds and other pest species.

Each mine should set specific, realistic targets that clearly describe what is to be achieved and by when and that are linked into the overall rehabilitation and mine closure strategy. Each target should take into account of availability of resources, any technical limitations, the expertise of personnel and contractors, views of landowners and the community, as well as long-term land management requirements.

12.6.7 Planning for Closure

Conservation and sustainable management of biodiversity values during mine closure planning is a continuous process. The planning should starts from the beginning of project planning and development, and continues throughout the life of an operation. Good practice also requires open and effective dialogue with regulators, the local community, indigenous groups and traditional owners, conservation NGOs and any other stakeholders. Mine closure and decommissioning plans are dynamic documents that may need to be adjusted and updated in response to general requirements. For developing a mine closure and decommissioning plans, biodiversity aspects should also be taken with regards to:

- Baseline conditions
- Predicted impacts of the operation
- The physical extent of impacts associated with operations
- The operational plan agreed end use for the various components
- Future ownership and maintenance

The baseline sets the *'terms of reference'* (TOR) for decommissioning planning. It should clearly define biodiversity values of the receiving environment and the project's potential impacts. It also defines the decommissioning requirements arising from legal or regulatory controls and the expectations of other stakeholders in regard to decommissioning outcomes.

Regular review of the closure and decommissioning plans should identify knowledge gaps relevant to biodiversity management and conservation. These may include information gaps, potential issues or risks and ongoing monitoring, investigation and research needs. All operations require some flexibility in the planning phases as priorities and operational activities evolve. There are associated potential direct and indirect impacts associated with these changes.

12.7 Integrated Biodiversity Management

Management of biodiversity impacts due to mining should involve, in order of priority, the following:

- Avoid the impacts as far as possible,
- Reduce or minimize the impacts as far as practicable,
- Undertake remedial measures that includes mitigation, restoration and revegetation of degraded sites;
- Compensate the loss of biodiversity by taking appropriate site specific measures (offsets).

It is important that a holistic view is taken when managing biodiversity. It is important to minimise impacts on the floral and faunal communities of surrounding areas in order to achieve mine rehabilitation objectives.

Offset schemes are gradually being incorporated into the ESIA process and should be considered where appropriate.

- Community partnerships are effective means of achieving mutually beneficial conservation outcomes.
- Leading practice management of water quality goes beyond compliance and focuses on understanding and managing biodiversity values of the receiving environment.
- Fauna species, weeds and plant pathogens need to be monitored, and their impacts are to be understood and managed. They can significantly reduce an area's biodiversity values and retard development of the post-mining rehabilitated ecosystem.
- Where re-establishment of biodiversity is a priority, this should be taken into account during all stages of the operation, including topsoil management, seeding, planting and, wherever required, establishment of recalcitrant and rare species and habitat transfer.
- Restoring fauna habitat may require the use of specialised techniques for particular species.

12.8 Conclusive Remarks

1. The importance of biodiversity values is recognised. Society and the mining industry now recognise that biological diversity possesses intrinsic value.
2. Through surveys and research, we get valuable information on an area's biodiversity values, ecological processes and services and on the effectiveness of management and rehabilitation practices.
3. Rehabilitation of *nearby unmined but degraded areas* and the linkage of these to rehabilitated sites and remnant vegetation can significantly reduce overall impacts and help restore an area's local flora, fauna and associated values.
4. Much can be achieved through applying general biodiversity management procedures, even though each mine and its environment are unique.
5. Companies that achieve the best results are those that adopt a *'learn as you go'* approach and implement sound monitoring and research programmes.
6. Liaison with government, the community, including indigenous people, researchers, NGOs and others, is critical when developing *biodiversity management programmes* that achieve the best outcomes.
7. Once mining impacts is evaluated for a specific location, short-term measures can be considered that include avoid, minimise and mitigate the impacts (e.g. by rehabilitation). In addition, the long-term management solutions need to be put in place to ensure that the resources, funding and expertise necessary for ongoing biodiversity conservation are available.

Areas where there are frequent opportunities for improvement include:

- Recognition of whole-of-lease issues
- Improved establishment of floristic diversity through better topsoil handling
- Seeding methods
- Better liaison with stakeholder groups, particularly NGOs
- The importance of assessing cumulative impacts and integrating mining proposals into bioregional contexts and land use planning processes.

Greater recognition of the importance of monitoring and research programmes that will-enable continuous improvement of biodiversity and its rehabilitation is not simply a case of *'do and forget'* and requires management solutions which ensure that the values existing at mine closure are sustained or enhanced. Integration of the precautionary principle in a consistent manner in relation to biodiversity management is also an opportunity for improvement. Lastly, we should agree that "Biodiversity as an opportunity to be realized more than a problem to be solved" (CBD 2012).

References

Benayas JMR et al. (2009) Enhancement of biodiversity and ecosystem services by ecological restoration: a meta-analysis. Science 325:121–124

Bradshaw A (1997) Restoration of mined lands – using natural processes. Ecol Eng 8:255–269

Brudvig LA (2011) The restoration of biodiversity: where has research been and where does it need to go? Am J Bot 98(3):549–558. http://www.amjbot.org/content/98/3/549.full.pdf+html

CBD (2010) COP- 10 decisions. http://www.cbd.int/decisions/cop/?m=cop-10

CBD (2011) The Convention on Biological Diversity Year in Review 2011. http://www.cbd.int/doc/reports/cbd-report-2011-en.pdf or http://www.cbd.int/nr5/

CBD (2012) Press release. http://www.cbd.int/doc/press/2012/pr-2012-10-20-cop-11-en.pdf or http://www.cbd.int/cop11

Chazdon RL (2008) Beyond deforestation: Restoring forests and ecosystem services on degraded lands. Science 320: 1458–1460 http://www.sciencemag.org/content/320/5882/1458.figures-only

Jain SK (1994) Biodiversity; some perspective in study and conservation. Reg conv Min Env Forest, GOI, Lucknow, India

MoEF (1998) Implementation of article 6 of the Convention of Biological diversity in India - National Report. Ministry of Environment and Forests, Govt of India. http://nbaindia.org/uploaded/Biodiversityindia/1st_report.pdf

MoEF (2001) India's Second National Report to the CBD. MoEF, GOI. http://www.cbd.int/doc/world/in/in-nr-02-en.doc

MoEF (2006) India's Third National Report to the CBD, http://www.cbd.int/doc/world/in/in-nr-03-en.doc

MoEF (2009) India's Fourth National Report to the CBD, http://envfor.nic.in/downloads/public-information/in-nr-04.pdf

Millennium Ecosystem Assessment (2005) Ecosystems and Human Well being: Biodiversity Synthesis. World Resources Institute, Washington DC. http://www.maweb.org/en/index.aspx

Nellemann C, Corcoran E (eds) (2010) Dead Planet, Living Planet - Biodiversity and Ecosystem Restoration for Sustainable Development. A Rapid Response Assessment. UNEP. http://www.unep.org/pdf/RRA ecosystems_screen.pdf

Monitoring and Aftercare of Ecorestored Site

13

Contents

13.1 Introduction

The revegetation of plants on a site either by planting, habitat transplanting, development by seeding or combination of all, is only the beginning of the establishment process. *Aftercare* describes the crucial process of managing the soils and the vegetation systems after the initial revegetation in order to ensure that the desired land use is attained within a reasonable time period. The process would involve soil amelioration and vegetation management that is more intensive than normally associated with land in that particular use. The objectives of aftercare, which can last for 2–5 years, are:

1. To ensure that the plants become established and overcome the initial constraints on growth on derelict site.
2. To establish a viable soil–plant system, with sufficient nutrient *'capital'* and turnover to support the vegetation.

A fundamental aspect of aftercare is monitoring, that is, keeping tract of what is happening to the soil and the plant community, and it involves the following:

- Monitoring and assessment of vegetative cover, that is, species present, their relative abundance, the overall density and distribution of plant cover and arrival of new species, if any.
- Observation of soil development, root system, moisture improvement, periodic soil analysis and nutrient content.

S.K. Maiti, *Ecorestoration of the Coalmine Degraded Lands*,
DOI 10.1007/978-81-322-0851-8_13, © Springer India 2013

Table 13.1 Monitoring and maintenance of reclaimed site (aftercare) Maiti (2010)

1. Survival rate	Dead plants are to be replaced
2. Provision of watering in the drier area and application of slow-releasing fertiliser	Continuous surveillance
3. Physical monitoring	Stability, erosion control, re-establishment of drainage, damaged to fencing
4. Monitoring of nutrient accumulation and cycling	pH, organic matter and organic carbon, nitrogen, phosphorus and potassium
5. Biological monitoring	Plant growth (height, girth, crown cover), establishment of natural flora and fauna, ecological parameters—litter fall and litter decomposition, microbial community development
6. Anthropogenic disturbance	Continuous surveillance
7. Overgrazing of the area (if any)	Control overgrazing, check fencing
8. Illegal timber failing/cutting	Continuous surveillance

Aftercare involves addition of fertiliser, seeds, plants, harvesting, mowing, etc. The aim of aftercare is to build up the fertility until a natural cycle develops. If a reclaimed area is to be developed successfully, it is essential to ensure good soil profile and vegetation cover development. This is an integral and vital part of the whole ecorestoration scheme and should be considered right from the planning stage. Aftercare should begin as soon as the plantation is completed. For example, monitoring of soil pH, nutrient status round the year is important because nutrient availability depends on pH. It also helps to determine the amount of lime/fertiliser to be applied.

1. The management components are monitored to assess the progress and development of the soil profile and vegetation growth *vis-a-vis* addition of fertiliser.
2. The forestry management should aim to maintain the optimum stand of trees for the purpose. Important components are replacing lost trees or thinning as necessary, weeding to reduce competition around young trees and maintaining fertility. In plantation, there are many steps that can be taken to enhance the wildlife interest, even in those managed purely for timber production.
3. The build-up of soil nutrients, particularly nitrogen, can be monitored to assess the progress of the reclamation. Fertilising and vegetation management should aim to promote a steady increase in soil nitrogen to around 700 kg/ha (Coppin and Bradshaw 1982).

Once the soil has been *ameliorated and the plants established*, the system will eventually develop on its own. However, the process of development and *building of a complex ecosystem*, whether natural or man-made, may take so long as to be unacceptable. *Some degree of management and aftercare will always be necessary, related to the desired after use.*

The period of fairly intensive aftercare would eventually be replaced by normal management practices, when the soils and plants are sufficiently well established. The length of this period can vary from *2* to *10 years*, depending on both the starting point and the level to which the productivity is targeted to be achieved. Details of monitoring and maintenance at site are shown in Table 13.1.

Revegetation may be established and maintained in initial stages but may fail if nutrients are not supplied adequately. Organic matter in soil is important because humus increases microbiological activities and fungus growth in the top soils and improves soil structure and cation exchange capacity of the soil, thereby helping in *recycling of nutrients*. Humus is major source of nitrogen and sulphur. It increases moisture-retaining capacity and reduces toxicity levels of trace metals. The sources and distribution of SOC in restored sites are illustrated in Box 13.1.

> **Box 13.1** Soil Organic Carbon (SOC) in restored sites (Tibbett 2008)
>
> The primary source of SOC is from plant litter and this is related to the productivity of the vegetation biomass. In developing reforested systems, litterfall tends to be higher than in native stable systems. While belowground deposition is relatively difficult to estimate, aboveground litter stocks can be easily measured as a litter layer above the mineral soil; although the litter layer can sometimes be discriminated into upper and lower layers, with the lower layer having undergone greater comminution. Stocks of litter on the forest floor are typically greater in restored sites than native, unmined forests' sites. Litter accumulation has been reported as being almost four times greater than the surrounding Jarrah forest.

13.2 Nutrient Accumulation and Cycling

In a self-sustaining ecosystem, the nutrient demands of the plants must be met principally by nutrient cycling, supplemented by inputs in rainfall, dry deposition, rock weathering and, in the case of nitrogen, by fixation of atmospheric nitrogen. When the objective of restoration is agriculture or other more intensive uses, these processes can be augmented by fertiliser addition. Maintaining or increasing the ability of the soil to supply nutrients, to store and supply water and support root growth should be a major concern in developing a sustainable ecosystem. Improving water infiltration, relieving soil compaction and increasing the volume of soil accessible to plant roots are major aims of the deep ripping normally carried out as part of the erosion control measures in restoration. The re-establishment of nutrient cycles is essential to the sustainability of restoration.

Mining removes the vegetation and inevitably leads to the loss of some plant nutrients from the site. This is particularly important where the proportion of the total nutrients in the ecosystem, which are contained within the vegetation, and the plant litter on the soil surface is high, as is the case for many Indian ecosystems. In these cases, there must be an input of nutrients to the systems if they are going to reach a productivity level equivalent to the pre-mining ecosystem and be self-sustaining in the long term (Lyle 1987).

The nutrients which are usually most limiting to plant growth in mine-degraded derelict sites are nitrogen and phosphorus. Most of the nitrogen held in the soil is in the organic matter and is released for uptake by plants through mineralisation. As mineralisation rates are generally low, a large amount of nitrogen must be accumulated. The most practical way to increase the nitrogen capital of ecosystems is to establish nitrogen-fixing plants, usually legumes, which can quickly increase the nitrogen levels in the system. The accompanying increase in soil organic matter is important in improving soil structure and the ability of the soil to hold and supply water (Maiti and Saxena 1998). Most soils, except sandy soils, have a significant ability to retain phosphorus fertiliser. Some of this phosphorus may be available to plants, and an initial application of phosphorus fertiliser may supply sufficient phosphorus for plant growth until nutrient cycles are established. Symbiotic association with mycorrhizal fungi can increase the growth and survival of tree species (Maiti 1997).

13.3 Restoration of Faunal Population

Encouraging the native fauna to return to areas cleared for mining is a fundamental part of any restoration programme that aims at restoring a natural ecosystem. Some invertebrate species may be introduced if fresh topsoil is placed on the area, but most fauna species will need to recolonise from surrounding areas. Numerous factors, including the size of the restored area, the fauna populations in surrounding areas and the success of the revegetation programme, influence the rate of recolonisation by fauna. Many faunal groups will quickly colonise any areas which contain the resources that they require, such as food, shelter and breeding sites. In many cases, the main aim in a fauna return strategy should be the re-establishment of the native vegetation. If this is successful, then the fauna should colonise from surrounding areas.

Fauna may be returned slowly if their requirements are not available in young restored site. For example, some faunal species lives in Jarrah forest (Australia) require logs or stumps on the forest floor, and others require tree hollows to breed. These species may not be available for a long period of time after restoration. The return of these species can be expedited by creating fauna habitats and corridors during restoration using logs, stumps and other natural materials. Fauna corridors running from the surrounding areas to the centre of the restored areas encourage smaller species of mammals and reptiles, which are reluctant to traverse large distances of open ground, to colonise.

Tree hollows can be substituted by providing nest boxes of appropriate size for the target species in developing restored areas. Animals, particularly invertebrate species, are important in many ecological processes, such as nutrient cycling, litter decomposition, soil aeration, seed dispersal, seed predation and plants' ability to survive fire or set seed, so that they can become re-established after fire. Fire control strategies can include free breaks, hazard reduction bums in adjacent areas, prescribed cool bums in restored areas and weed control.

13.4 Management of Wildlife Conservation

The following measures can be taken to improve the quality of wildlife habitat in the restored sites. Initially emphasis should be given to enhance the plant diversity of the site by

- Increasing the canopy cover of the areas so that during day time, it creates shade and ameliorates temperature
- Planting understory and shrub layers below the tree canopy
- Leaving some dead and dying wood standing and some fallen logs and stumps for fungi and insects
- Assuring some scattered large old tree in the area

- Encouraging some species with flowers and fruits
- Ensuring paths are twisted (zigzag) and not straight
- Maintaining streams and ponds in the area (water body) where hiding place to be provided by planting evergreen trees with dense stratified canopy

13.5 Success Criteria and Monitoring of Ecorestored Site

It is essential to monitor the success of any restoration programme and to be prepared to rework any areas of restoration not developing adequately. Success criteria should be defined and agreed to all interested parties. In India there are no recognised criteria for determining when restoration is complete. The debate on whether an area has been completely restored cannot be resolved until there is widespread agreement on issues such as the following: what constitutes a viable ecosystem, what is an acceptable level of species diversity and how indistinguishable should a restored area be from neighbouring untouched areas. These are issues of continuing debate and research. However, the industry and regulatory agencies have had to address this issue on numerous occasions. Handing back responsibility for managing the land to the relevant landowner or government agency will often be possible when some mutually satisfactory end point has been reached. If the aim of restoration is to develop a new higher-value land use for the area, then the land manager may accept responsibility for the increased level of management that will be required before the mining company's responsibility is relinquished. Restoration can be considered to be successful when the site can be managed for its designated land use without any greater management inputs than other land in the area being used for a similar purpose. Restored native ecosystems may be different in structure to the surrounding native ecosystems, but there should be confidence that they will change with time along with or towards the make-up of the surrounding

Table 13.2 Emergency plan for monitoring of tree growth in the ecorestored site Maiti (2006)

Symptoms	Probable causes	Diagnosis	Correction measures to be taken
Trees fail to establish	Poor quality of sapling or bad handling and plantation	Health of sapling and planting condition	Carefully replant again or replant those species most suited
	Draught	Low rainfall, chosen wrong time for plantation, moisture retention capacity is very low. Termites attack on root	Add additional topsoil and organic matter in pit Put insecticides (BHC) on plantation pit
Tree growth is not proper (in terms of height or girth, crown cover)	Lack of watering	Measure both N and P in soil (0–15 depth) and leaves	Fertilise and water
	Nutrient deficiency	Take out few plants and observe the root growth (limited root extension)	Add additional topsoil on plantation pit
	Compaction	Low rainfall	Replace with more suitable species
	Diseases	Coarse soil, stoniness >80 % Check for symptoms of fungi (root-wilting fungi, Alternaria sp.)	Apply herbicides

area. The restored land should be capable of withstanding normal disturbances such as fire.

The success criteria should include

- Physical (stability, resistance to erosion, re-establishment of drainage)
- Biological (species richness, plant density, canopy cover, seed production, fauna return, weed control, productivity, establishment of nutrient cycles)
- Water quality standards for drainage water
- Public safety issues

13.5.1 Soil Structure and Development

Several important things occur during the development of soil, most being attributed to the influence of root growth, the presence of decomposed organic matter and activities of soil micro- and macrofauna (Coppin and Bradshaw 1982).

- There is a general reduction trend in reduction in bulk density, giving improved water retention and soil erosion.
- Increase in cation exchange capacity giving improved nutrient retention.
- Aggregation of fine particles into distinctive crumbs, improving the texture of the soils and imparting structure.

- Gradual improvement of infiltration and drainage due to increase organic matter, litter fall and improvements in soil structure.

Soil development can be monitored over a period of time by examining the physical changes that are occurring. This, together with measurement of the nutrients accumulating in the soil and plants, is the best way of maintaining check on the management strategy and residence of the newly developed soil.

Monitoring techniques must be designed to provide statistically valid results with the desired order of accuracy. The sampling intensity will usually have to be a compromise between the level of precision of the data collected and the cost of collecting the data. There will always be a compromise between the number of samples collected and the number of decimal points that their analysis is expressed.

13.6 Development of Emergency Plan for Monitoring

The development of emergency plan for subordinates (or field peoples) is essential for monitoring of ecorestored sites. An example of emergency plan is given below (Table 13.2), which should be updated by field observations.

References

Coppin NJ, Bradshaw AD (1982) Quarry reclamation. Mining Journal Books, London

Lyle ES Jr (1987) Surface mining reclamation manual. Elsevier, New York

Maiti SK (1997) Importance of VAM fungi in coalmine overburden reclamation & factors effecting the establishment of VAM Fungi on overburden dumps. Environ Ecol 15(3):602–608

Maiti SK (2006) Ecorestoration of coalmine OB dumps – with special emphasis on tree species and improvements of dump physico-chemical, nutritional & biological characteristics. MGMI Trans 102(1&2):21–36

Maiti SK (2010) Revegetation planning for the degraded soil and site aggregates in Dump sites. In: Bhattacharya J (ed) Project environmental clearance. Wide Publishing, Kolkata, pp 189–228

Maiti SK, Saxena NC (1998) Biological reclamation of coalmine spoils without topsoil: an amendment study with domestic raw sewage and grass-legumes mixture. Int J Surf Min Reclam Environ 12:87–90

Tibbett M (2008) Carbon Accumulation in Soils During Reforestation - The Australian Experience After Bauxite Mining. In: Fourie et al (eds) Mine Closure 2008. Johannesburg, South Africa. http://www.acg.uwa.edu.au/_data/page/5316/sample_chapter.pdf

Contents

14.1   Introduction

The first scientific approach on evaluation of reclamation successes in coal mine derelict sites has been systematically documented in the 'Annual meeting of the American Society of Surface Mining and Reclamation' held on 1990 under the leadership of Chambers and Wade (USDA Forest Services) in a symposium on 'Evaluating Reclamation Success: The Ecological Consideration'.

The reclamation of drastically disturbed lands results in construction of ecosystem literally from bedrock. Construction of new ecosystem begins as soon as soil and geological materials are placed in the disturbed site and that time itself basic characteristics of new ecosystem decided. The reclamationist by using energy, machinery, geological and biological materials decide the characteristics of new ecosystem at on-site. They emphasised that ecosystem attributes can be used for evaluation of reclamation success (Table 14.1).

Following indicator parameters which are easy to monitor and interprets the status of ecosystem recovery in the restored sites are discussed in this chapter:

- Soil quality (physico-chemical, biological)
- Microbial Biomass Carbon (MBC)
- Enzyme activities (i.e. Dehydrogenase)
- Litter accumulation and decomposition

S.K. Maiti, *Ecorestoration of the Coalmine Degraded Lands*,
DOI 10.1007/978-81-322-0851-8_14, © Springer India 2013

Table 14.1 Ecosystem attributes used for evaluation of reclamation success in coal mine derelict sites (Chambers and Wade 1990)

Ecosystem attributes	Description
1. Soil biological property and nutrient cycling	Faster nutrient cycling leads to improvement of good vegetation cover and erosion control and helps to develop stability of ecosystem. Capital of nutrient is required for an ecosystem to be self-sustaining
2. Soil microbial process	Soil microbial activity regulates the nutrient availability to plants. After disturbance, recovery of soil decomposer system is essential; otherwise reclamation will not lead to the effective development of belowground microbial process and interaction
3. Abundance of Mycorrhiza fungi	Mycorrhiza fungi form mutualistic association with desirable plant species. They regulate nutrient and water uptake. Many mycorrhiza fungi immigrate rapidly by animal and wind onto disturbed sites
4. Highly diverse and self-perpetuating faunal community	In view of importance of invertebrate to ecosystem response (e.g. pollination, herbivory, seed predation and dispersal, soil aeration, litter decomposition), analysis of faunal community development (tropic level, guild apportionments, population demographics and species turn over rate) can be used as an evaluation method for successful reclamation
5. Trajectory of succession process	Numerous biotic and abiotic factors affect succession. During restoration process, some are under control of humans and many are not. The key is to know what factors reclamationist induces that consequently affect rate, pattern and trajectory of succession. This information gives reclamationist an ability to select most appropriate land use and direct succession process towards that land use
6. Community-level process	Evaluating community-level process to determine reclamation success has been advocated by Allen (1990). The 1977 Surface Mining Control and Reclamation Act (SMCRA) refers to the reclamation of community-level structure and process in Sec. 515 (19), when it requires the operator to 'establish.... a diverse (e.g., *species diversity, structural diversity, species composition, dominance, and rare and uncommon species*) effective, and permanent vegetative cover of the same seasonal variety (*penology, life form*)... and capable regeneration (*productivity, competition, reproduction, stability, resilience*) and plant succession at least equal in extent of cover to the natural vegetation of the native area'
7. Landscape ecology	Landscape patterns are the consequences of numerous disturbances or perturbations creating patches and its works on both vertical and horizontal dimension. Horizontal dimension is a polygon that is described as map. The vertical dimensions are geology, soils, soil chemistry, water system, depth of soil cover, depth of water table, etc. The structure of vegetation and stratification is also a vertical dimension
8. Post-reclamation land use	Determination of post-reclamation land use or uses can be used as evaluation of reclamation success. An effort should be made to work with and not against ecological process that occurs during liability period which can enhance final reclamation process

14.2   Soil Quality Indicators

The definition of soil quality encompasses physical, chemical and biological characteristics, and it is related to fertility and soil health. According to USDA, soil quality indicators are classified into four categories that include visual, physical, chemical and biological indicators (Fig. 14.1):

1. *Soil physico-chemical indicators* are related to soil texture and development of soil structure, bulk density and pore space that reflecting effects on root growth and compaction, water-holding capacity, infiltration and rooting depth. Chemical indicators include pH and salinity, organic matter content, nutrients availability (NPK), cation exchange capacity and base saturation and nutrient cycling. Soil properties

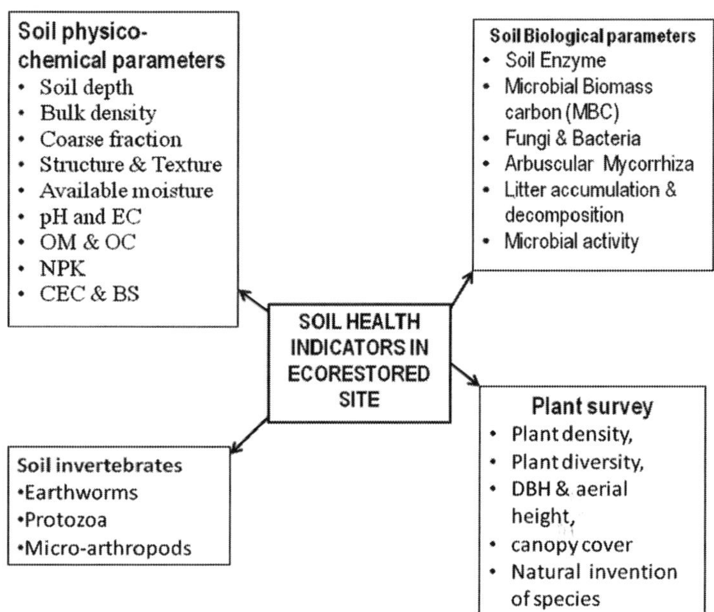

Fig. 14.1 Soil health indicator parameters that can be used for evaluation of reclamation success

associated with soil organic matter (SOM) have been recognised as key indicators and to have an effect on other properties. Soil organic carbon (SOC) is considered one of the most important indicators of soil quality; it has positive effects on soil physical properties like soil structure, promotes water infiltration and water storage and provides habitat for different groups of microorganisms that mineralize organic matter and enhance nutrient availability (Martinez-Salgado et al. 2010). The forms of soil carbon and its role for the improvement of degraded sites are highlighted in Box 14.1.

2. *Biological indicators* include measurements of micro- and macroorganisms and their activities or functions. Concentration or population of earthworms, protozoa, micro-arthropods, termites, ants, as well as microbial biomass and fungi can be used as indicators for their role in soil development, decomposition and mineralization process and fostering nutrient cycling. Biological indicators also include metabolic processes such as respiration, used to measure microbial activity related to decomposition of organic matter in soil, enzyme activity, mycorrhiza colonisation, litter fall and decomposition, microbial biomass carbon (MBC) and SOM to MBC ratio, which are associated to

Box 14.1 The carbon in soil (Tibbett 2008)

Soil carbon is the largest terrestrial pool of carbon on the planet. Increasing the size of this pool using land management techniques by just 5% has the potential to decrease the amount of atmospheric carbon by up to 16% (Baldock 2007).

Soil carbon can exist in numerous forms and these can be divided into organic and inorganic. The inorganic forms of carbon in soils are primarily carbonates ($CaCO_3$) but can also be found as calcite, dolomite and serite. In global estimates of soil profiles to 1-m in depth, soil may contain a ratio of inorganic carbon to organic carbon of 1:2 (695–748 Pg of inorganic compared to 1462–1548 Pg of organic carbon; Pg = petagram = 10^{15} g = 1 billion tons) (Batjes 1996).

Soil organic carbon (SOC) is derived primarily from the necromass of plant tissues and, to a lesser extent, animal corpses and microbes. All organic matter is thought to be processed eventually by the microbial biomass, and converted from litter into soil organic matter. The form, and hence the

(continued)

Box 14.1 (continued)

stability, of the organic matter, and thus the organic carbon therein can be grouped into four components. These are:

- *Inert organic matter*: This form of organic matter is usually black and may contain charcoal, charred material (that comes from burning) or coal. This form of carbon is very resistant to chemical and biological reactions.
- *Particulate organic matter*: This form of organic matter represents the material that has undergone comminution by soil invertebrates and has been exposed to the early stages of microbial attack. It is composed of organic fragments that retain some recognizable structure; it includes organic fragments >53 µm in diameter and the light fraction of organic matter that has a specific gravity of less than 1 (i.e. it floats on water). This form of organic matter is susceptible to microbial mineralization to CO_2.
- *Humus*: This form of organic matter is fully decomposed and finely divided organic matter. No recognizable structures can be identified from the original biological material from which it was derived. It typically has a dark color and its chemical structure provides a large area for chemical reactions and water holding. It is often considered the solid fraction <53 µm. This form of organic matter is resistant to microbial mineralization to CO_2.
- *Dissolved organic matter*: This form of organic matter is made up of very fine materials <45 µm diameter, found in solution. It represents the major loss of carbon via water movement down the soil profile.

Soil organic matter (SOM) is an essential component of soils necessary for the physical, chemical and biological functioning of the soil. SOM provides structural stability to soils and can strongly affect the capacity of soils to hold water. These features are of particular importance to retaining top soil integrity after mine site rehabilitation. The carbon in SOM provides a source of energy for biological processes and a nitrogen, phosphorus and sulphur reservoir of nutrients. Chemically, the structure of SOM can provide exchange sites for nutrient retention, buffer pH change, and can reduce the concentration of potentially toxic elements.

mineralization of organic substrate per unit of microbial biomass (Bastida et al. 2008).

(a) *Enzyme activities* have been associated with indicators of biogeochemical cycles, degradation of organic matter and soil remediation processes, so they can determine, together with other physical or chemical properties, the quality of a soil. Many research reports emphasised enzymes as good indicators because (a) they are closely related to organic matter, physical characteristics, microbial activity and biomass in the soil; (b) provide early information about changes in quality; and are more rapidly assessed. Soil microorganisms play a key role on phosphate solubilization with the release of low molecular weight organic acids and production of extracellular enzymes as phosphatases.

- Dehydrogenase enzyme activity has been considered as soil health indicator, because microbes used this enzyme for oxidation of organic matter, and involved in electron transport systems of oxygen metabolism and requires an intracellular environment (viable cells) to express its activity.
- The enzyme β—Glucosidase is widely distributed in the environment, and its activity has been detected in soil, fungi and plants. It has been used as a key soil quality indicator due to its importance in catalytic reactions on cellulose

(continued)

degradation, releasing glucose as a source of energy to maintain metabolically active microbial biomass in soil.

(b) *Presence of earthworms* in ecorestored site is taken as key soil quality indicator. They produce aggregates and pores (i.e. biostructures) help in aggregate stability, to improve water-holding capacity, pore size and infiltration rate, thus affecting its physical properties, nutrient cycling and plant growth. In the Netherlands, they are used as a biological indicator of soil quality and in Germany as a factor for soil biological site classification, a practical form to define the use of soil, based on the structure of the earthworm community, their abundance and biomass (Rombke et al. 2005).

(c) The *health of existing vegetation* can be used as indirect measurement of soil indicators, because the healthier the plant community, the better the soil health. Plant parameters such as development of stratified canopy, canopy cover, increase in plant density and diversity, increase in DBH (diameter at breast height) and aerial height, and natural colonisation of species can be monitored.

Three most important biological indictor parameters, (a) microbial biomass carbon (MBC), (b) dehydrogenase activity and (c) litter accumulation and decomposition, are discussed in the following sections.

14.3 Microbial Biomass Carbon: As an Indicator Parameter

The microbial biomass carbon (MBC) acts as the transformation agent of the organic matter in soil and also acts as the centre of all biological activity. It acts as source and sink of the nutrients C, N, P and S contained in the organic matter. To properly understand biological activity in soil, one must therefore have knowledge of the microbial biomass. The MBC consists of living and active part of soil organic matter, which is reestablished after disturbance of the land.

Study of recovery time of microbial biomass is important for the development of self-sustaining ecosystem in mine degraded land. Increase in MBC and organic carbon contents increases the functional diversity and stability of ecosystem. It varies with depth, season and altitude; type of tree species; vegetation cover; soil quality; and management practices.

The MBC increases with accumulation of organic matter during soil development process. Although the MBC constitutes only 1–3% of total soil carbon, the microbial biomass N (MBN) is made up to 5% of total soil N, and they are the most labile C and N pools in soils (Jenkinson and Powlson 1976). Therefore, nutrient availability and productivity mainly depend on the size and activity of the microbial biomass. Microbial biomass is a more dynamic soil quality parameter than those based on physical and chemical properties and, therefore, has the advantage of serving as early signal of soil degradation or soil improvement. The MBC has been correlated with enzyme activities, decomposition of organic matter, nutrient cycling, availability of plant nutrients, soil structuring, organic matter storage, biological control and suppression of plant pathogens, and as an index of success of restoration of soil microbial populations. It is also frequently used as an early indicator of changes in soil chemical and physical properties resulting from soil management and environmental stresses.

Several researchers emphasised that ratio of MBC to SOC should be used as indicator of reclamation successes, and the higher the ratio, the better is the quality of restored ecosystem. This ratio has been proposed as a measure of success of reclamation efforts (Insam and Domsch 1988). In mine reclaimed sites, continuous increase in MBC to SOC ratio indicates steady state is yet to be achieved (Sinha et al. 2009). Increasing contents of soil organic carbon (SOC) and MBC may result in increased functional diversity of soil microbial communities

Table 14.2 Status of microbial biomass carbon (MBC), microbial biomass nitrogen (MBN), microbial biomass phosphorus (MBP) and MBC/SOC ratio in different mine restored sites (After Mukhopadhyay and Maiti 2011)

Land use	MBC (mg/kg)	MBN (mg/kg)	MBP (mg/kg)	MBC/ SOC	References
Coal mine degraded land (Dhanbad, India)	Rhizospheric soil: *M. oleifera*—600 *A. marmelos*—590	–	–	0.58–8.03	Sinha et al. (2009)
Coal mine restored land (Jayant project, Singrauli coalfield, MP, India)	Increase from 4 to 6 year: 129.34–363.17	20.03–43.09	10.11–17.32	3.92–6.88	Singh et al. (2004)
Chronosequence study (0–2, 5–7, 18–20 and 38–42-year-old coal mine reclaimed sites) (Southwestern Virginia, USA)	Recently reclaimed sites: 131–138; 16–20-year-old sites: 280 38–42-year-old sites:244	–	–	–	Clayton et al. (2009)
Agricultural and forest chronosequence of open-pit mine reclamation soils	Agricultural land: 570; forest: 430 (after 15 year)	–	–	–	Insam and Domsch (1988)
Surface coal mine sites (semiarid regions of Wyoming, USA)	Reclaimed soils: 45–510. Undisturbed soils: 79–158	–	–	–	Anderson et al. (2008)
Reclaimed coal mine soil (Czech Republic)	372–499 (0–20 cm), 52–1,544 (reclaimed soil)	–	–	0.54–4.50	Ruzek et al. (2003)

and thus increased functionality and stability of soil ecosystems.

A general trend of increasing MBC from a low level in the more recently reclaimed sites to a higher level of MBC in the older sites was reported, which indicated a recovery of soil microbial communities with time. Determining recovery time of microbial biomass in reclaimed systems is important because microbial associations with organic matter and nutrient cycling are critical for ecosystem function and availability of essential plant nutrients, particularly nitrogen and phosphorous (Clayton et al. 2009).

Several studies reported that microbial diversity and MBC increase over time in the restored mine degraded land and used as one of the criteria parameter for the assessment of reclamation successes. Similarly, chronosequence studies have been carried out in the coal mine reclaimed areas to study the rate and manner of accumulation of MBC and compared with natural forest MBC level, accumulation of MBC at different depth, seasonal variation of MBC, microbial quotient (MBC/SOC) and influence of different plant species on MBC in rhizospheric soil. Status of MBC, MBN and MBP and MBC to SOC ratio in different mine restored sites are given Table 14.2.

14.3.1 Measurement of Microbial Biomass Carbon (MBC)

Several methods have been proposed for the estimation of MBC; however, chloroform fumigation and extraction with 0.5 M potassium sulphate is still most popular. The soil MBC is often measured by the fumigation–extraction method (FE) proposed by Vance et al. (1987). The principle of the method is that killing and lysing of soil microbial cells by chloroform fumigation make intracellular organic matter extractable with 0.5 M K_2SO_4. In this method two types of extracts are produced: (1) extract of non-fumigated soil containing extracellular organic matter only and (2) extract of fumigated soil containing both intracellular (i.e. biomass) and extracellular organic matter. Soil samples are sieved (2 mm) and stored at 4°C at its field moisture content. Microbial analysis is carried out within 2 weeks after sampling. It is advisable to determine microbial biomass on a fresh sample or within 3 days of sampling. The MBC is calculated as extractable carbon in fumigated samples minus extractable carbon from non-fumigated samples. Soil organic matter extracted with 0.5 M K_2SO_4 from non-fumigated soil samples can be used as a measure of the SOC, which is actually available for microorganisms. It represents the light fraction that is well degradable by microorganisms and is closely related to microbial activities and growth.

Different values of coefficient of conversion or extraction efficiency factor (Kc) were suggested for MBC calculation, because chloroform does not apparently render all cell components soluble, leaving some microbial components unextracted; hence, Kc factor is required to compensate for the unextracted C of microbial origin. The Kc values were summarised for different studies as 0.43 ± 0.10 ($n = 74$); 66% of the Kc values was found to lie within a range of 0.33–0.53 and 90% lie within the range of 0.23–0.63 (Mukhopadhyay and Maiti 2011).

14.4 Dehydrogenase Enzyme Activity in Ecorestored Site

The dehydrogenase activity (DHA) has been proposed as a measure of overall microbial activity and used as index of the soil microbial biomass (Mukhopadhyay and Maiti 2010a). The microbiological and biochemical status of the soil has often been proposed as an early and sensitive indicator of soil ecological stress or restoration processes in different ecosystems. Enzymes play a key role in soil nutrient cycling; its activity is essential in both the mineralization and transformation of organic carbon and plant nutrients in soil ecosystem. Study of soil enzyme gives information about the release of nutrients in soil by means of organic matter degradation and microbial activity as well as indicators of ecological change. The success of ecorestoration of mine degraded land is usually assessed based on measurement of plant cover and primary productivity, but it is equally important to measure the soil enzyme activities to evaluate the success of reclamation. In the reclaimed land, it has been shown that soil enzyme activities could also be used as an index of evaluating soil fertility and soil health (Mukhopadhyay and Maiti 2010a).

DHA is measured by two methods using TTC and INT as substrate; however, various authors reported poor results when TTC is used as substrate. Most workers have used incubation periods of 24 h or longer at 37°C to study DHA (Casida 1977). Some researchers modified to use a 6-h, 37°C incubation with either glucose or yeast extract as the electron donating substrate. Soil aeration parameters, water content and bulk density have significant effects on DHA in soil. Highest DHA was reported in rainy season and positively correlated with soil pH, Ca, Mg, K and water content.

Soil aeration parameters, water content and bulk density have significant effects on DHA in soil. In soils having low moisture level, DHA was reported close to zero. The highest DHA was reported in rainy season and the lowest in winter or summer seasons. DHA also found less in heavy metals contaminated soils. DHA is positively correlated with soil pH, Ca, Mg, K and water content.

In opencast coal mine restored sites, dehydrogenase activities are generally found low due to damage of soil microflora and lack of organic matter. Addition of organic amendments increases enzyme activity in mine soil. The incorporation of organic amendments to soil stimulated DHA because the added material contain intra- and extracellular enzymes and may also stimulate microbial activity in the soil. However, enzyme activity of polluted soils was restored; after 15–20 years, it was quite similar to the enzyme activity of natural soils. It has been reported that enzyme activity on 15-year-old dumps was 3- to 14-fold higher than on younger (3-year-old dumps). It has been observed that most enzymes, including DHA, peaked in the next 1 or 2 years after

Table 14.3 Dehydrogenase activity (DHA) in different types of ecorestored mine soils

Site description	Types of soil	DHA (μg TPF g^{-1} dry soil 24 h)$^{-1}$	References
Road side soil	Disturbed soil	18.2–56.75	Joshi et al. (2010)
Forest soil	Undisturbed soil	28–78.3	
Urban soil	Upper 20 cm	10.7–258.4	Kizilkaya and Askin (2007)
Opencast coal mine restored areas	Disturbed soil	10–220	Harris and Birch (2007)
	Undisturbed soil	140–580	
Vegetated coal discard sites, S. Africa	Rehabilitated sites	67.53–578	Claassens et al. (2005)
	Reference sites	302.6–482.4	
Coal mine degraded sites, Dhanbad, India	Rhizosphere samples	92.3	Sinha et al. (2009)
Old sulphur mines, Ukraine	Forest topsoil	289	Levyk et al. (2007)
	Disturbed soil (open-pit mining)	123	

reclamation with topsoiling and declined thereafter. A 4-year-old topsoiled site was statistically similar to the undisturbed soil (Clayton et al. 2009). A comparative assessment of DHA in natural and mine soil are given in Table 14.3.

14.5 Litter Accumulation and Decomposition: An Indicator

The accumulation and decomposition of plant litter is one of the most important processes in the initiation of nutrient cycling in the establishment of self-sustaining ecosystem in the reclaimed sites. The accumulation of plant litter depends on type of tree species, age of plantation, density and seasons, whilst decomposition depends on climatic conditions, soil moisture, microbial activity and physico-chemical properties of the mine soil (Mukhopadhyay and Maiti 2010b). Litter fall transfers organic matter, nutrients and energy from vegetation to soil and is a dominant link in the biogeochemical cycling of matter. The analysis of litter quality and quantity and its rate of decomposition is highly important for the understanding of energy flow, primary productivity and nutrient cycling in forest ecosystems.

Litter decomposition is one of the key biogeochemical processes in forest ecosystems, and it is estimated that the nutrients released during litter decomposition can account for 69–87% of the total annual requirement of essential elements for forest plants (Waring and Schlesinger 1985). The study of litter decomposition is also an important part of the most intensively studied nutrient cycling processes in forest ecosystems. The rate of litter decomposition is largely a determining factor for productivity or biomass of every terrestrial ecosystem in general and of forest ecosystems in particular.

The average litter accumulation of five tree species growing in a reclaimed coal mine overburden dumps was studied by Mukhopadhyay and Maiti (2010b). Maximum litter accumulation under tree with broad leaves (*G. arborea*), Whilst analysing the composition of litter, only 35% consists of fresh leaves of *G. arborea*, 25% woody fractions, and 40% belongs to other category. Higher litter accumulation also observed under *C. seamea*, which consists of fresh leaves of *C. seamea* (60%), *A. auriculiformis* (15%), leaves woody parts (15%) and other (10%). (Mukhopadhyay and Maiti 2010a, b). The litter accumulation and their decomposition rate under different tree canopies are given in Table 14.4.

In the studied tree species, it was found that broad leaves *(G. arborea)* decomposed rapidly compared to smaller leaves *(P. juliflora)*. Highest decomposition of litter was found under

Table 14.4 Litter accumulation and decomposition rate in a reclaimed coal mine overburden dumps (all weights in g dry wt/m^2) (Mukhopadhyay and Maiti 2010b)

Sl no	Tree cover	Fresh litter	Partially decomposed litter (>2 mm) (A)	Decomposed litter (<2 mm) (B)	Total decomposed (A + B)	% decomposition
1.	*C. seamea*	510.03 (429–578)	45.09 (31–66)	64.45 (45–82)	109.54 (84–148)	17.57 (15–20)
2.	*D. strictus*	437.41 (418–453)	65.11 (61–76)	63.95 (61–69)	129.05 (121–138)	22.77 (121–138)
3.	*P. juliflora*	390.90 (372–410)	27.40 (25–31)	48.60 (43–56)	76.00 (67–88)	16.24 (15.3–17.6)
4.	*F. religiosa*	388.42 371–405	43.13 (32–59)	38.57 (33–46)	81.7 (70–105)	17.26 (15–20)
5.	*G. arborea*	652.08 (631–704)	99.76 (89–119)	82.44 (73–99)	182.19 (162–218)	21.77 (20–24)

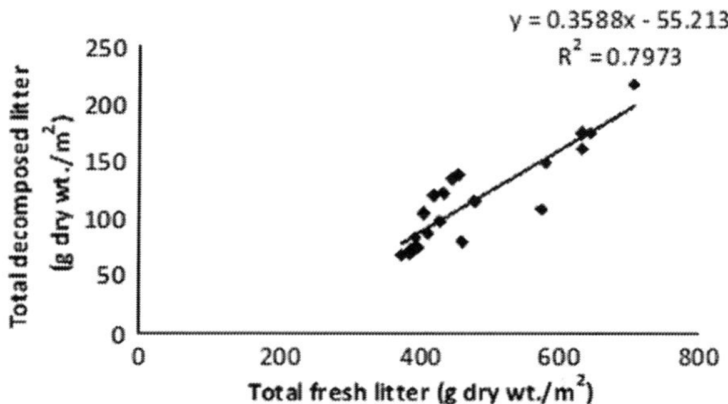

Fig. 14.2 Correlation between total litter fall and total decomposition rate in a reclaimed coal mine site (Mukhopadhyay and Maiti 2010b)

bamboo (*Dendrocalamus strictus*) plantation, which indicates higher release of nutrients compared to other trees. The average deposition rate in reclaimed coal mine dumps is approximately 19% to the total litter at a given time (Fig. 14.2).

References

Allen EB (1990) Evaluating community–level process to determine reclamation success. In: Chambers JC, Wade GL (eds) Evaluating reclamation success: the ecological consideration, organized USDA forest service. American Society of Surface Mining and Reclamation, Charleston

Anderson JD, Ingram LJ, Stahl PD (2008) Influence of reclamation management practices on microbial biomass carbon and soil organic carbon accumulation in semiarid mined lands of Wyoming. Appl Soil Ecol 40:387–397

Baldock JA (2007) Composition and cycling of organic carbon in soil. In: Marschner P and Rengel Z (eds) Nutrient Cycling in Terrestrial Ecosystems, Springer-Verlag, Berlin Heidelberg

Bastida FZA, Hernandez H, Garcia C (2008) Past, present and future of soil quality indices: a biological perspective. Geoderma 147:159–171

Batjes NH (1996) Total Carbon and Nitrogen in the Soils of the World. Eu J Soil Sci 47:151–163

Casida LE Jr (1977) Microbial metabolic activity in soil as measured by dehydrogenase determinations. Appl Environ Microbiol 34(6):630–636

Chambers JC, Wade GL (eds) (1990) In: Evaluating reclamation success: the ecological consideration organized USDA, USDA Forest Service. Am. Soc. of Surface Mining and Reclamation, Charleston, West Virginia, 23–26 Apr 1990

Claassens S et al (2005) Soil microbial properties in coal mine tailings under rehabilitation. Appl Ecol Environ Res 4(1):75–83

Clayton HG, Wick AF, Daniels WL (2009) Microbial biomass in reclaimed soils following coal mining in Virginia. Paper was presented at the 2009 national meeting of the American Society of Mining and Reclamation, Billings, MT, Revitalizing the environment: proven solutions and innovative approaches, May 30–June 5 2009 R.I

Harris JA, Birch P (2007) Soil microbial activity in opencast coal mine restorations. Soil Use Manag 5(4):155–160

Insam H, Domsch KH (1988) Relationship between soil organic carbon and microbial biomass on chronosequences of reclamation sites. Microb Ecol 15(2): 177–188

Jenkinson DS, Powlson DS (1976) The effects of biocidal treatments on metabolism in soil – V. A method for measuring soil biomass. Soil Biol Biochem 8: 209–213

Joshi SR, Kumar R, Saikia P, Bhagobaty RK, Thokchom S (2010) Impact of roadside pollution on microbial activities in sub-tropical forest soil of North East India. Res J Environ Sci 4:280–287

Kizilkaya R, Aşkin T (2007) The spatial variability of soil dehydrogenase activity: a survey in urban soils. Agric Conspec Sci 72(1):89–94

Levyk V, Maryskevych O, Brzeziñska M, Wlodarczyk T (2007) Dehydrogenase activity of technogenic soils of former sulphur mines (Yavoriv and Nemyriv, Ukraine). Int Agrophysics 21:255–260

Martinez-Salgado M et al. (2010) Biological soil quality indicators: a review, current research, technology and education topics. In: Mendez-Vilas A (ed) Applied microbiology and microbial biotechnology, www.formatex.info/microbiology2/319-328.pdf

Mukhopadhyay S, Maiti SK (2010a) Dehydrogenase activity in natural and mine soil – a review. Indian J Environ Prot 20(11):921–933

Mukhopadhyay S, Maiti SK (2010b) Ecorestoration of coalmine overburden dumps- with emphasis on minesoil properties, natural VAM colonization, litter accumulation and tree growth. Minetech 31(2): 16–26

Mukhopadhyay S, Maiti SK (2011) Status of microbial biomass in reclaimed mine degraded land and non-mining areas: a review. Indian J Environ Prot 31(8): 642–657

Rombke J, Jansch S, Didden W (2005) The use of earthworms in ecological soil classification and assessment concepts. Ecotoxicol Environ Safety 62(2): 249–265

Růžek L et al (2003) Chemical and biological characteristics of reclaimed soils in the most region (Czech Republic). Plant Soil Environ 49(8):346–351

Singh AN, Raghubanshi AS, Singh JS (2004) Impact of native tree plantations on mine spoil in a dry tropical environment. For Ecol Manag 187:49–60

Sinha S, Masto RE, Ram LC, Selvi VA, Srivastava NK, Tripathi RC, Joshy G (2009) Rhizosphere soil microbial index of tree species in a coal mining ecosystem. Soil Biol Biochem 41(9):1824–1832

Tibbett M (2008) Carbon Accumulation in Soils During Reforestation – The Australian Experience After Bauxite Mining. In: Fourie et al (eds) Mine Closure 2008. Johannesburg, South Africa. http://www.acg.uwa.edu.au/__data/page/5316/sample_chapter.pdf

Vance ED, Brookes PC, Jenkinson DS (1987) An extraction method for measuring soil microbial biomass C. Soil Biol Biochem 19:703–707

Waring RH, Schlesinger WH (1985) Forest ecosystems: concepts and management. Academic Press, Orlando, FL

Contents

15.1 Introduction

The environmental and forestry clearances given by the Ministry of Environment and Forests are based on the (i) Environment Protection Act, 1986, (ii) The Forest Conservation Act, 1980 and (iii) The Wildlife Protection Act, 1972. The very welcome intervention of the Supreme Court has made the forestry clearance process (and also that of wildlife clearance under the Wildlife Protection Act, 1972) very comprehensive and detailed: The detailed acts and rules are:

1. The Indian Forest Act, 1927
2. The Forest Conservation Act 1980 with Amendments Made in 1988
3. Forest (Conservation) Rules, 2003
4. Forest Rights Act (FRA), 2006
5. Procedure for Forest Clearance—Policy Initiatives of MoEF
6. FAC stipulation
7. CAMPA (Compensatory Afforestation Fund Management and Planning Authority)—2004, State CAMPA 2009
8. Net Present value (NPV)

S.K. Maiti, *Ecorestoration of the Coalmine Degraded Lands*,
DOI 10.1007/978-81-322-0851-8_15, © Springer India 2013

9. Wildlife (Protection) Act, 1927
10. Wildlife (Protection) Amendment Act, *2002*
11. Critical Wildlife Habitats Guidelines (04-05-2011)
12. Wildlife (Protection) Amendment Act, 2010

15.2   Forest Conservation Acts

(i) The ***Indian Forest Act, 1927***: An Act to consolidate the law relating to forests, the transit of forest-produce and the duty leviable on timber and other forest-produce [It has 13 chapters].

Salient Features of the Wildlife Protection Act, 1972

- Applicable all over India except Jammu and Kashmir which has its own Act.
- Hunting of any scheduled animal prohibited: exceptions—mice, rats, common crow and fruit bats.
- Hunting also includes capturing and trapping a wild animal.
- In Wildlife (Protection) Amendment Act, *2002 (17* January 2003), punishment and penalty for offences have been made more stringent.
- It has *six schedules* which give varying degrees of protection.
- Schedule I and part II of Schedule II provide absolute protection—offences under these are prescribed the highest penalties. for example. mammals [e.g. 42. free-tailed bat]
- Species listed in Schedule III and Schedule IV are also protected, but the penalties are much lower.
- *Schedule V* includes the animals which may be hunted.
- The *plants in Schedule VI* are prohibited from cultivation and planting.
- Up to April 2010 there have been 16 convictions under this act relating to the death of tigers.

---

*(ii) The **Forest (Conservation) Act (1980) – with Amendments Made in 1988***

Restriction on the de-reservation of forests or use of forest land for non-forest purpose: 'non-forest purpose' means the breaking up or clearing of any forest land or portion thereof.

- *FC Act, 1980—a regulatory Act, not prohibitory.*
- *The Act is an interface between conservation and development.* Enacted to help conserve the country's forests.
- It strictly restricts and regulates the de-reservation of forests or use of forest land for non-forest purposes without the prior approval of Central Government.
- *Permits judicious and regulated use of forest land for non-forestry purposes.*
- A commitment of the Central Government for the cause of environmental conservation and sustainable development.

During 1950–1980, the rate of diversion of forest land for non-forestry purposes was *1,50,000 ha per annum.* After enactment of the FC Act, 1980, *the rate of diversion of forest land for non-forestry purposes came down to about 35,000 ha per annum.*

(iii) **Forest Rights Act (FRA), 29-12-2006**

The *Scheduled Tribes and Other Traditional Forest Dwellers (Recognition of Forest Rights) Act* (2006) recognises the rights of forest-dwelling scheduled tribes and other traditional forest dwellers over the forest areas inhabited by them and provides a framework for according the same.

There are important policies and laws which influence the ecorestoration activities of mining areas. The few important and relevant ones are discussed below:

15.3   National Forest Policy (NFP), 1988

The basic objectives of the National Forest Policy (NFP) (1988) among other things emphasise on the following:

1. Maintenance of environmental stability through *preservation* and where necessary, *restoration of the ecological balance* that has been adversely disturbed by serious depletion of the forests of the country

2. Conserving the natural heritage of the country by preserving the remaining natural forests with the variety of flora and fauna, which represent the remarkable biological diversity and genetic resources of the country

It states that 'The principal aim of Forest Policy must be to ensure environmental stability and maintenance of ecological balance including atmospheric equilibrium which are vital for sustenance of all life forms, human, animal and plant. The derivation of direct economic benefit must be subordinated to this principal aim'.

Section 4.3.1 states that 'Schemes and projects which interfere with forests and clothe steep slopes, catchments of rivers, lakes, and reservoirs, geologically unstable terrain and such other ecologically sensitive areas should be severely restricted'.

Section 4.4.1 on 'Diversion of forest lands for non-forest purposes' states that 'forest land or land with tree cover should not be treated merely as a resource available to be utilised for various projects and programmes, but as a national asset which requires to be properly safeguarded for providing sustainable benefits to the entire community. Diversion of forestland for any non-forest purpose should be subject to the most careful examinations by specialists from the standpoint of social and environmental costs and benefits. Construction of dams and reservoirs, mining and industrial development and expansion of agriculture should be consistent with the needs for conservation of trees and forests'.

Section 4.4.2 emphasises the need for restoration of mined areas and states that no mining lease should be granted without a proper management plan appraised from the environment point of view and enforced by adequate machinery.

15.4 National Wildlife Action Plan (NWAP), (2002)

The NWAP 2002–2016 was released in January 2002 and in its preamble has stated some of the following:

Effective ecosystem conservation is the foundation of long-term ecological and economic stability. Natural processes. . ..

India ranks sixth among the 12 mega biodiversity countries of the world. Conservation of biodiversity is directly linked with conservation of ecosystems and thus with water and food security. These together constitute a major plank of the Indian economy.

. . .development priorities to take into account ecological imperatives including the protection of wild species, which sustain and enhance natural habitats, even as they depend on such areas for their survival. Habitat loss caused by developmental project such as dams, mines, etc., compounds the problems of wildlife conservation.

15.4.1 Wildlife Conservation Strategy (2002)

The National Wildlife Action Plan (NWAP) 2002–2016 was released by the prime minister during the twenty-first meeting of the Indian Board for Wildlife (IBWL) held on 21 January 2002. At the above stated meeting, the IBWL adopted the 'Wildlife Conservation Strategy 2002'. Among the 17 points articulated are the following:

No diversion of forest land for non-forest purposes from critical and ecologically fragile wildlife habitat shall be allowed.

Lands falling within 10 km. of the boundaries of national parks and sanctuaries should be notified as ecofragile zones under section 3(v) of the Environment (Protection) Act and Rule 5 Subrule 5(viii) and (x) of the Environment (Protection) Rules.

Subsequent to this meeting of the IBWL, the Addl. DGF (Wildlife), MoEF, has written to the Chief Wildlife Wardens of all States and Union Territories (D.O. No. 6-2/2002 WLI dated 5 February 2002) asking them to list out such areas and furnish detailed proposals for their notification as eco-sensitive under the EPA.

15.4.2 Forest (Conservation) Act (FCA), 1980; Forest (Conservation) Rules, 2003

The FCA has been one of the most crucial legislations to be enacted with respect to forests, and it in essence puts a restriction on the de-reservation of forests or use of forests for non-forest purpose, which includes mining.

15.5   Biodiversity Act (2002)

Certain sections of the Biodiversity Act, which have implications on mining and biodiversity, are stated below:

Section 36(2) states that 'The Central Government shall, as far as practicable wherever it deems appropriate, integrate the conservation and sustainable use of biological diversity into relevant sectoral or cross-sectoral plans, programmes and policies'.

Section 36(3) states that the Central Government shall undertake measures 'wherever necessary, for assessment of environmental impact of that project which is likely to have adverse effect on biological diversity, with a view to avoid or minimise such effects and where appropriate provide for public participation in such assessment'.

Comment: This could be used for projects affecting areas rich in biodiversity but not covered by the EIA notification or the FCA presently.

Section 37(1) states that 'without prejudice to any other law for the time being in force, the State Government may from time to time in consultation with the local bodies, notify in the Official Gazette, areas of biodiversity importance as biodiversity heritage sites under this Act'.

15.6   Procedure for Forest Clearance

- Proposals recommended by the State/UT Governments forwarded to the Central Government for approval under Section 2 of the Act
- Proposals examined by Forest Advisory Committee (FAC) constituted under Section 3 of the Act
- Decisions taken on the basis of the recommendations of the FAC
- FC Rules, 2003 prescribe specific time limits for processing the cases

15.6.1   Forest (Conservation) Rules (2003)

1. Submission of the proposals seeking approval of the Central Government under *Section 2 of the Act*
   (a) Every user agency, who wants to use any forest land for non-forest purposes, shall make his proposal in the appropriate form.
   (b) *Form A* for proposals seeking *first-time approval under the Act.*
   (c) *Form B* for proposals seeking *renewal of leases* where approval of the Central Government under the Act had already been obtained earlier.
2. Every State Government, after having received the proposal and after being satisfied that the proposal requires prior approval under *Section 2 of the Act, shall send the proposal to the Central Government* in the appropriate forms:
   - *Within 90 days* of the receipt of the proposal from the user agency for proposals seeking first-time approval under the Act
   - *Within 60 days* for proposals seeking renewal of leases where approval of the Central Government under the Act had already been obtained earlier
3. The proposal referred to in Sub-rule (2) above, *involving forest land of more than 40 ha*, shall be sent by the State Government to the Secretary to the Government of India, MOEF, Pryavaran Bhavan, CGO Complex, Lodhi Road, New Delhi-110 003, with a copy of the proposal (with complete enclosures) to the concerned regional office.
4. The proposal referred to in Sub-rule (2) above, involving *forest land up to 40 ha*, shall be sent to the Chief Conservator of Forests or Conservator of Forests of the concerned Regional Office of the Ministry of Environment and Forests.
5. The proposal referred to in Sub-rule (2) above, involving *clearing of naturally grown trees in forest land or portion thereof* for the purpose of using it for reafforestation, shall be sent to the Chief Conservator of Forests or Conservator of Forests of the concerned Regional Office of the MoEF.

15.6.2 Committee to Advise on Proposals Received by the Central Government

1. The Central Government shall refer every proposal, complete in all respects, received by it under *Sub-rule (3)* of Rule 6 including site inspection report, wherever required, to the Committee for its advice thereon.
2. The Committee shall have due regard to all or any of the following matters while tendering its advice on the proposals referred to it under *Sub-rule (1)*, namely:
   (a) Whether the forests land proposed to be used for non-forest purpose forms part of a *nature reserve, national park wildlife sanctuary and biosphere reserve* or forms part of the habitat or any endangered or threatened species of flora and fauna or of an area lying in severely eroded catchment
   (b) Whether the State Government or the other authority has certified that it has considered all other alternatives and that *no other alternatives* in the circumstances are feasible and that the required area is the minimum needed for the purpose
   (c) Whether the State Government or the other authority undertakes to provide at its cost for the acquisition of land of an equivalent area and afforestation thereof
3. While tendering the advice, the Committee may also suggest any conditions or restrictions on the use of any forest land for any non-forest purpose, which in its opinion would minimise adverse environmental impact.

15.6.3 Action of the Central Government on the Advice of the Committee

The Central Government shall, after considering the advice of the Committee tendered *under Rule 7* and after such further enquiry as it may consider necessary, grant approval to the proposal with or without conditions or reject the same within 60 days of its receipt.

15.6.4 Proposals for First-Time Approval Under the FC Act

<u>**FORM A**</u> (for proposals seeking first-time approval under the Act)

**PART I
(to be filled up by user agency)**

1. Project details:
   (i) Short narrative of the proposal and project/scheme for which the forest land is required
   (ii) Map showing the required forest land, boundary of adjoining forest on a 1:50,000 scale map
   (iii) Cost of the project
   (iv) Justification for locating the project in forest area
   (v) Cost–benefit analysis (to be enclosed)
   (vi) Employment likely to be generated
2. Purpose-wise break-up of the total land required
3. Details of displacement of people due to the project, if any
   (i) Number of families
   (ii) Number of scheduled castes/scheduled tribe families
   (iii) Rehabilitation plan (to be enclosed)
4. Whether clearance under Environment (Protection) Act, 1986 required? (Yes/No)
5. Undertaking to bear the cost of raising and maintenance of compensatory afforestation and/or penal compensatory afforestation as well as cost for protection and regeneration of safety zone, etc., as per the scheme prepared by the State Government (undertaking to be enclosed)
6. Details of certificates/documents enclosed as required under the instructions

**PART II
(to be filled by the concerned Deputy Conservator of Forests)**

State serial no. of proposal______________

7. Location of the project/scheme:
   (i) State/union territory
   (ii) District
   (iii) Forest division
   (iv) Area of forest land proposed for diversion (in ha.)

(v) Legal status of forest

(vi) Density of vegetation

(vii) Species-wise (scientific names) and diameter class-wise enumeration of trees (to be enclosed. In case of irrigation/hydel projects enumeration at FRL, FRL-2 and FRL-4 m also to be enclosed)

(viii) Brief note on vulnerability of the forest area to erosion

(ix) Approximate distance of proposed site for diversion from boundary of forest

(x) Whether forms part of national park, wildlife sanctuary, biosphere reserve, tiger reserve, elephant corridor, etc. (If so, the details of the area and comments of the Chief Wildlife Warden to be annexed)

(xi) Whether any rare/endangered/unique species of flora and fauna found in the area—if so, details thereof

(xii) Whether any protected archaeological/heritage site/defence establishment or any other important monument is located in the area. If so, the details thereof with NOC from competent authority, if required

8. Whether the requirement of forest land as proposed by the user agency in col. 2 of Part I is unavoidable and barest minimum for the project. If no, recommended area item-wise with details of alternatives examined

9. Whether any work in violation of the Act has been carried out (Yes/No). If yes, details of the same including period of work done, action taken on erring officials. Whether work in violation is still in progress

10. Details of compensatory afforestation scheme:

(i) Details of non-forest area/degraded forest area identified for compensatory afforestation, its distance from adjoining forest, number of patches and size of each patch

(ii) Map showing non-forest/degraded forest area identified for compensatory afforestation and adjoining forest boundaries

(iii) Detailed compensatory afforestation scheme including species to be planted, implementing agency, time schedule, cost structure, etc.

(iv) Total financial outlay for compensatory afforestation scheme

(v) Certificates from competent authority regarding suitability of area identified for compensatory afforestation and from management point of view (to be signed by the concerned Deputy Conservator of Forests)

11. Site inspection report of the DCF (to be enclosed) especially highlighting facts asked in col. 7 (xi, xii), 8 and 9 above

12. Division/district profile:

(i) Geographical area of the district

(ii) Forest area of the district

(iii) Total forest area diverted since 1980 with number of cases

(iv) Total compensatory afforestation stipulated in the district/division since 1980 on:
(a) Forest land including penal compensatory afforestation
(b) Non-forest land

(v) Progress of compensatory afforestation as on (date) _____________ on:
(a) Forest land
(b) Non-forest land

13. Specific recommendations of the DCF for acceptance or otherwise of the proposal with reasons

Signature
Name
Official Seal

Date:_____________

PART III
(to be filled by the concerned Conservator of Forests)

14. Whether site, where the forest land involved is located, has been inspected by concerned Conservator of Forests (yes/no). If yes, the date of inspection and observations made in form of inspection note to be enclosed

15. Whether the concerned Conservator of Forests agree with the information given in Part B and the recommendations of Deputy Conservator of Forests

16. Specific recommendation of concerned Conservator of Forests for acceptance or otherwise of the proposal with detailed reasons

Signature
Name
Official Seal

Date:_____________
Place:_____________

PART IV
(to be filled in by the Nodal Officer or Principal Chief Conservator of Forests or Head of Forest Department)

17. <u>Detailed opinion and specific recommendation of the State Forest Department for acceptance of otherwise of the proposal with remarks</u>
   (While giving opinion, the adverse comments made by concerned Conservator of Forests or Deputy Conservator of Forests should be categorically reviewed and critically commented upon.)

Signature
Name and Designation
(Official Seal)

Date:______________
Place:______________

PART V
(to be filled in by the Secretary in charge of Forest Department or by any other authorised officer of the State Government not below the rank of an undersecretary)

18. <u>Recommendation of the State Government</u>
   (Adverse comments made by any officer or authority in Part B or Part C or Part D above should be specifically commented upon)

Signature
Name and Designation
(Official Seal)

Date:______________
Place:______________

Instructions (for Part I):

1. The project authorities may annex a copy of the approved project/plan in addition to filling col. 1 (i) for example, IBM-approved mining plan for major minerals/CMPDI plan with subsidence analysis reports, etc.
2. Map has to be in original duly authenticated jointly by project authorities and concerned DCF col. 1 (ii).

3. Complete details of alternative alignments examined especially in case of project like roads, transmission lines, railway lines, canals, etc., to be shown on map with details of area of forest land involved in each alternative to be given—col. 1 (iii).
4. For proposals relating to mining, certificate from competent authority like District Mining Officer about non-availability of the same mineral in surrounding/nearby non-forest areas
5. In case the same company/individual has taken forest land for similar project in the state, a brief detail of all such approvals/leases be given as an enclosure along with current status of the projects
6. The latest clarifications issued by the ministry under Forest (Conservation) Act, 1980, may be kept in mind. In case such information does not fit in the given columns, the same shall be annexed separately.

General Instructions:

1. On receipt of proposal, nodal officer shall issue a receipt to the user agency indicating therein the name of the proposal, user agency, area in hectare, serial number and date of receipt.
2. If the space provided above is not sufficient to specify any information, please attach separate details/documents.
3. While forwarding the proposal to the Central Government, complete details on all aspects of the case as per form prescribed above read with the clarifications issued by the Ministry of Environment and Forests, Government of India, New Delhi, should be given. Incomplete or deficient proposals shall not be considered and shall be returned to the State Government in original.
4. The State Government shall submit the proposal to the Central Government within stipulated time limits. In case of delay while forwarding, the reasons for the same are to be given in the forwarding/covering letter.

15.6.5 Proposals for Renewal of Leases (Forest Clearance Granted): Form B

FORM B

Form for seeking prior approval under <u>Section 2</u> of the proposals by the State Governments and other authorities in respect of renewal of leases, which have been earlier granted clearance under Forest (Conservation) Act, 1980

PART I
(to be filled up by user agency)

1. Letter no. and date vide which clearance under Forest (Conservation) Act, 1980, accorded by the Central Government (copy to be enclosed)
2. Project details:
   (i) Short narrative of the proposal and project/scheme for which the forest land is required
   (ii) Map showing the required forest land, boundary of adjoining forest on a 1:50,000 scale map
   (iii) Cost of the project
3. Purpose-wise break-up of the total land required (already broken and to be broken)
4. Details of certificates/documents enclosed as required under the instructions

Signature

Date:_______________   (Name in block letters)
Place:_______________   (Official Seal)
                        Address (of user agency)
   State serial no. of proposal_______________

(to be filled up by the Nodal Officer with date of receipt)

15.7 Forest Advisory Committee (FAC) (2002)

- FAC is a seven-member committee under the chairmanship of DGF & SS, MoEF.
- Three non-official members—eminent experts in forestry and allied disciplines—appointed for a period of 2 years.
- ADGF, MoEF; Additional commissioner (soil conservation), Ministry of Agriculture.
- IGF (FC)—Member secretary.
- FAC committee meets at least once every month discuss all proposals received by The Central Government under the Forest (Con-

servation) Act, 1980 involving forest area of more than 20 ha.
- Quorum is three.

Certain conditions are stipulated at the time of granting approval under FC Act:

- Compensatory afforestation
- Catchment area treatment
- Phased reclamation of mining area
- Safety zone area
- Rehabilitation of project affected families, if any
- Muck disposal plan
- Wildlife management plan, etc.

15.8 Compensatory Afforestation Fund Management and Planning Authority (CAMPA)

In exercise of the powers conferred by Subsection (3) of Section 3 of the Environment (Protection) Act, 1986 and the Hon'ble Supreme Court's order dated the 30 October 2002, the Central Government hereby constitutes an authority to be known as *Compensatory Afforestation Fund Management and Planning Authority (CAMPA)* with effect from *23 April 2004* for the purpose of management of money towards compensatory afforestation (CAMPA 2004). The State Compensatory Afforestation Fund Management and Planning Authority (State CAMPA 2009) received money under compensatory afforestation (CA), net present value (NPV), etc. It is an instrument to accelerate activities for preservation of natural forests, management of wildlife and infrastructure development in the sector. The monies received in CAMPA from a state or the union territory as per para 6.2 and the income thereon after deducting expenditure incurred by the CAMPA on its establishment cost, monitoring and evaluation on a prorata basis shall be used only in that particular state or the union territory.

15.8.1 Net Present Value (NPV)

- On the directions of the Apex court in 2002, a net present value (NPV) of the forest land being diverted is being charged from the user agencies. NPV—intrinsic cost of the land, tangible and intangible benefits of the forest area.

- NPV charged at Rs.5.8–9.2 lakh per ha depending on the type and density of the diverted forest land. At present, all the funds received from the state governments/user agencies are being deposited in ad hoc CAMPA.

- Used on 73 Cr and admitted fund utilisation is poor (As per SN Trivedi, PCCF and nodal officer, CAMPA).
- Jharkhand Government has accumulated Rs.1200 Cr of which centre will release 10% every year.
- Hurdles: manpower crunch, inexperience of massive afforestation activity.

15.8.2 Aims and Objectives of CAMPA

State CAMPA shall seek to promote:
(a) Conservation, protection, regeneration and management of existing natural forests
(b) Conservation, protection and management of wildlife and its habitat within and outside protected areas including the consolidation of the protected areas
(c) Compensatory afforestation
(d) Environmental services, which include:
  (i) Provision of goods such as wood, non-timber forest products, fuel, fodder and water and provision of services such as grazing, tourism, wildlife protection and life support
  (ii) Regulating services such as climate regulation, disease control, flood moderation, detoxification, carbon sequestration and health of soils, air and water regimes
  (iii) Nonmaterial benefits obtained from ecosystems, spiritual, recreational, aesthetic, inspirational, educational and symbolic
  (iv) Supporting such other services necessary for the production of ecosystem services, biodiversity, nutrient cycling and primary production
(e) Research, training and capacity building
  Current status of CAMPA in Jharkhand (Source: The Telegraph, Ranchi, 28 May 2011)
- Under CAMPA, Jharkhand Government received 198 cr in two phases: 93 cr in Feb 2010 and 105 Cr in Nov 2011 (under 2010–2011 fiscal).

15.9 Critical Wildlife Habitats (CHW) Guidelines

- Ministry had issued guidelines for determination and notification of CWH on 25 October 2007. The revised guidelines were issued on 7 February 2011 in supersession of the previous guidelines (CWH Guidelines 2011).
- This protocol has been framed to determine and notify CWH within national parks and wildlife sanctuaries to harmonise the provisions of the FRA, 2006.
- The Wildlife (Protection) Act, 1972, and to address concerns of conservation of wildlife and its habitat, while safeguarding the forest rights of the scheduled tribes and other forest dwellers.

Purpose: How to determine critical wildlife habitats within national parks and wildlife sanctuaries as required by the Scheduled Tribes and Other Forest Dwellers (Recognition of Forest Rights) Act, 2006?

Objectives: To ensure the conservation of, and the prevention of damage to, wildlife and its habitat within the determined area.
- The Wildlife (Protection) Act, 1972, provides for the conservation and management of national parks and wildlife sanctuaries.
- The FRA, 2006, applies to national parks and sanctuaries, where forest rights are being recognised and vested in scheduled tribes and other traditional forest dwellers in such areas.
- These rights can only be modified within or resettled outside of the CWH as per the provisions of the FRA, 2006.

15.9.1 What Are CWH?

- Each expert committee shall identify areas within national parks and sanctuaries required to be kept inviolate for the purpose of wildlife conservation.
- The identification of CWH shall be the joint responsibility of the qualified scientific institution, qualified ecologists, and other wildlife experts on the expert committee. These members should conduct necessary field visits and undertake identification of CWH based on scientific and objective criteria.
- A CWH may extend to the entire area comprising a national park or wildlife sanctuary, or only a part of it, as is scientifically and objectively determined by the expert committee.

15.10 Wildlife (Protection) Amendment Act, 2010

The Wildlife (Protection) Amendment Bill, (2010)—This Act may be called the Wildlife (Protection) Amendment Act, 2010:

- Section 2 (Subsection 16): *'Electrocuting' shall be inserted after the* word 'trapping'.
- ***Prohibition on leg hold traps****: (1) No person shall manufacture,* **sell,** *purchase, transport or use any leg hold trap.*
- *'**12A grant of permit for scientific research.'**
- *'Chapter vb: regulation of trade in endangered species of wild fauna and flora.'*
- *Section 49 (D): (e) '**Exotic species**' means species of animals and plants not found in the wild in India.*
- *(i)'**Plan**' means any member, alive or dead, of the plant kingdom, including seeds, roots and other parts thereof.*
- **25. Insertion of new Schedule VII.** The following schedule shall be inserted after Schedule VI of the principal Act:

Appendix I   Appendix II   Appendix III

15.11 Action Plan

- Develop stratified canopy, by planting large, medium and small tree species, shrubs and grasses (*Vetiver, lemon grass, pedicellatum,* etc.) to provide hiding space for animals.
- Develop water bodies.
- Some invertebrate species may be introduced if fresh topsoil is placed on the area. Animals, particularly invertebrate species, are important.
- A range of factors like (a) size of the restored area, (b) the faunal population in surrounding areas and (c) the successes of the revegetation programme influence the rate of re-colonisation by fauna.
- Provide food, shelter and breeding sites.
- The return of the species can be expedited by creating fauna habitats and corridors during restoration by using logs, stumps and other natural materials. Fauna corridors running from the surrounding areas to the centre of the restored areas encourage smaller species of mammals and reptiles, which are reluctant to travel large distance of open ground to colonise.
- Tree hollows can be substituted for by providing nest boxes of appropriate size for the target species in developing restored areas.

15.12 Schedule Animals The Wildlife (Protection) Act, 1972

Chapter VII: Miscellaneous
62. Declaration of certain wild animals to be vermin 2 [The Central Government] may, by notification, declare any wild animal other than those specified in Schedule I and Part II of Schedule II to be vermin for any area and for such period as may be specified therein, and so long as such notification is in force, such wild animal shall be deemed to have been included in Schedule V.

Schedule I

Part I Mammals (41 nos)
2. Black Buck
3. Brow-antlered Deer or Thamin
3[3-A. Himachal Brown bear (Ursus Arctos)]
1[3-B. Capped Langur
4. Caracal
5. Cheetah
1[5-B. Chinkara or Indian Gazelle]
6. Clouded Leopard
7. Dugong (Dugong dugon)
8. Fishing Cat
1[8-A. Four-horned antelope]
3[8-D. Gangetic dolphin]
3[8-E. Gaur or Indian bison]
9. Golden Cat
10. Golden Langur
3[10-A. Giant squirrel]
3[10-B. Himalayan Ibex]
1[10-C. Himalayan Tahr]
11. Hispid Hare
12. Hoolock
1[12-B. Indian Elephant]
13. Indian Lion
14. Indian Wild Ass
3[15. Indian Wolf]
16. Kashmir Stag
2[16-A. Leaf Monkey]
2[16-B. Leopard or Panther]
17. Leopard Cat
18. Lesser or Red Panda
19. Lion-tailed Macaque
20. Loris
21. Lynx
22. Malabar Civet
23. Marbled Cat
24. Markhor
4[24-A. Mouse Deer]
25. Musk Deer
26. Nyan or Great Tibetan Sheep
27. Palla's Cat
28. Pangolin
29. Pygmy Hog
30. Rhinoceros
31. Rusty spotted Cat (Felis rubiginosa)
4(31-A. Serow
1[31-B. Clawless Otter]
1[31-C. Sloth Bear]
32. Slow Loris
1[32-A. Small Travencore Flying Squirrel]
33. Snow Leopard
34. Spotted Linsang
35. Swamp Deer
36. Takin or Mishmi Takin
1[36-A. Tibetan Antelope or Chiru]
1[36-B. Tibetan Fox]
37. Tibetan Gazelle
38. Tibetan Wild Ass
39. Tiger
40. Urial or Shapu
41. Wild Buffalo
2[41-A. Wild Yak
1[41-B. Tibetan Wolf]

Part II: Amphibians and Reptiles (17 nos)
2[1. Agra monitor lizard]
2. Gharial
1[1-B. Audithia Turtle]
1[1-D. Crocodiles]
4. Green Sea Turtle
7. Indian Egg-eating Snake
10. Large Bengal Monitor Lizard
11. Leathery Turtle
12. Logger Head Turtle
13. Olive Back Logger Head Turtle
14. Peocock-marked Soft-shelled Turtle
1[14-A. Pythons]

Part IIA: Fishes
[1. Whole shark]
s[2. Shark & Ray]
3. Sea horse
4. Giant Grouper

Part- III: Birds (18 nos)
4[1. Andaman Teal]
1-A. Assam Bamboo Partidge
3[1-B. Bazas]
2[1-C. Bengal Florican]
1-D. Black-necked Crane
1-E. Blood Pheasants
2. Cheer Pheasant
2[2-A. Eastern White Stork]
3[2-B. Forest-spotted Owlet
3[2-C. Frogmouth
3. Great Indian Bustard
4. Great Indian Hornbill
2[4-A. Hawks]
3[4-B. Hooded Crane]
3[4-C. Hornbills
1[4-D. Houbara Bustard]
1[4-F. Indian Pied Hornbill]
5. Jerdon's Courser
6. Lammergier
7. Large Falcons
1[7-A. Large Whistling Teal]
2[7-B. Lesser Florican]
2[7-C. Monal Pheasants]
8. Mountain Quail
9. Narcondam Hornbill
10. Nicobar Megapode
1[10-A. Nicobar Pigeon]
1[10-B. Osprey or Fish-eating Eagle]
1[10-C. Peocock Pheasants]
11. Peafowl
12. Pink-headed Duck
13. Scatter's Monal
14. Siberian White Crane
1[14-B. Tibetan Snow-Cock]
15. Tragopan Pheasants
16. White-bellied Sea Eagle
17. White-eared Pheasant
2[17-A. White Spoonbill]
18. White-winged Wood Duck

PART IV- Crustacea and Insects
2[1.] Butterflies and Moths
1[1-A. Coconut or Robber Crabe]
1[2. Dragon Fly]

Schedule II

Part I (mammals)
3[1-A. Assamese macaque]
4[2. Bengal porcupine]
3[3-A. Bonnet macaque]
3[4-A. Common langur]
7. Ferret badgers
11. Himalayan crestless porcupine
3[11-A. Himalayan newt or salamander]
16. Pig-tailed macaque
2[17-A. Rhesus macaque]
19. Stump-tailed macaque
22. Wild dog or dhole
2[24. Chameleon]
25. Spiny-tailed lizard

Schedule III
2. Barking deer
5. Chital (Axis axis)
11. Hog deer (Axis porcinus)
12. Hyaena
14. Nilgai
16. Sambar
19. Wild pig
20. Sponges

Schedule V [Vermin]
1. Common crow
3. Fruit bats
5. Mice
6. Rats

PART II: [1.] Beetles
1[1-B. Common fox]
1[1-C. Flying squirrels]
1[1-D. Giant squirrels]
1[2. Himalayan brown bear]
2[2-A. Himalayan black bear]
1[2-B. Jackal]
4-B. Red fox
3[5. Sloth bear]
11. Indian cobras
12. King cobra
14. Russel's viper

Schedule IV
3[3-A. Five-striped palm squirrel]
4. Hares (Black-naped, Common Indian, Desert, Himalayan mouse hare)
2[4-E. Indian porcupine]
4[6-A. Mongooses]
4[7-A. Pole cats]
6[11. **Birds 4**]- Falcons, kingfishers, pigeons (Columbidae) except the blue rock pigeon (Columba livia),
1[12. **Snakes**] 2[other than those species listed in Schedule I, Part II and Schedule II, Part II]

Schedule VI [plants]
1. Beddomes cycad (Cycas beddomei)
2. Blue vanda (Vanda soerulec)
3. Kuth (Saussurea lappa)
4. Ladies slipper orchids (Paphiopedilum spp.)
5. Pitcher plant (Nepenthes khasiana)
6. [Red Vanda (Renanthera imschootiana)]

References

Forest (Conservation) Act (1980) http://forest.and.nic.in/fca1980.pdf

The Scheduled Tribes and Other Traditional Forest Dwellers (Recognition of Forest Rights) Act (2006) The Gazette of India, Extraordinary, Part-II-Section-I. http://www.stscodisha.gov.in/pdf/FRA_English.pdf

National Forest Policy (NFP) (1988) Ministry of Environment and Forests, GOI, New Delhi. http://www.moef.nic.in/downloads/about-the-ministry/introduction-nfp.pdf

National Wildlife Action Plan (NWAP) (2002) http://awsassets.wwfindia.org/downloads/national_wildlife_action_plan_2002_2016_2.pdf

Wildlife Conservation Strategy (2002) http://envfor.nic.in/pt/str2002.html

Biodiversity Act (2002) http://nbaindia.org/content/25/19//act.html

Forest (Conservation) Rules (2003) http://ebookbrowse.com/rules-forest-conservation-rules-2003-pdf-d26188394

FAC (Forest Advisory Committee) (2002) http://moef.nic.in/divisions/forcon/groups.html#fac

CAMPA (2004) Compensatory Afforestation Fund Management and Planning Authority (CAMPA). http://envfor.nic.in/modules/recent-initiatives/campa/

State CAMPA (2009) The Guidelines on State Compensatory Afforestation Fund Management and Planning Authority (state CAMPA). http://envfor.nic.in/downloads/public-information/CAMPA-guidelines.pdf

CHW Guidelines (2011) Critical Wildlife Habitats (CHW) guidelines. http://moef.nic.in/downloads/public-information/rev-gdlns-CWH.pdf

The Wildlife (Protection) Amendment Bill (2010) http://envfor.nic.in/downloads/public-information/final-draft-bill.pdf

Ecological Impact Assessment of Surface Mining Project

16

Contents

16.1 Introduction

Indian EIA in general and those of mining projects in particular lack focus and direction. Born out of executive directions and legal mandates, they often end up as documents with loads of information of which much are of little use. This has lead to a situation where, notwithstanding the notable expectations, EIA documents are not yet to appropriate use in environmental management.

Ecological Impact Assessment (EcoIA), an important component of any EIA, is often carried out mechanically. Mining industry is only an intermediate user of land, but land degradation caused by it is so devastating that an altogether new ecosystem is developed in the derelict site. The main thrust should be on habitat quality assessment and how it will help to ecosystem regeneration.

This chapter stresses the need to carry out focused EcoIA. This is of paramount importance in mining, particularly surface mining, which invariably inflicts significant damages on the receiving ecosystem. This chapter, apart from highlighting the principal ecological impacts caused by surface mining, attempts to chalk out the steps that would, in the authors' opinion, overcome the shortcomings of existing processes and practices. Highlighting cases from Indian coal sector, the authors encapsulate a holistic view of the problems and propose pragmatic solution.

S.K. Maiti, *Ecorestoration of the Coalmine Degraded Lands*,
DOI 10.1007/978-81-322-0851-8_16, © Springer India 2013

It has been stressed that in order to make EcoIA effective, an appropriate procedure needs to be adopted. While fulfilment of legislative requirement cannot be overlooked, ensuring data adequacy within the allocated finance poses a major challenge. The author stress that unless a freehand is given to the assessor in planning and executing impact assessment studies, there is little chance of improvement in the EcoIA study effectiveness. Only with effective coordination with appropriate feedforward and feedback mechanisms can an EcoIA be turned into an effective tool for ensuring environmental sustainability (Maiti and Sinha 2004).

16.2 Ecological Impact Assessment (EcoIA)

Ecological Impact Assessment (EcoIA) '*is a formal process of defining, quantifying and evaluating the potential impacts of defined actions on ecosystem*'. The basic components EcoIA are *baseline studies, impact assessment, impact prediction and evaluation, and mitigation measures.* Some researchers also incorporated the *monitoring* components after suggested mitigation measures as feedback to EIA process. As a formal discipline, EcoIA had its origin in the National Environmental Policy Assessment (NEPA) which become law in the USA in 1969 and established a legislative requirement for proponent of an action to assess potential environmental impacts.

Significant irreversible ecological damage is often associated with surface mining. A part and parcel of surface mining activities is stripping of overburden. Such activities lead to near-total destruction of ecological habitat forcing total destruction of two of the most basic ecosystem components, namely, plants and decomposers. At times even some consumers are badly affected. This calls for development of a new land habitat for development of a fresh ecosystem, which, for habitat quality variation, can sustain an ecosystem that is at significant variance with the pre-mining ecosystem. A unique feature of EcoIA of mining projects is that such assessment laid stress not only on identifying the key parameters of ecosystem regeneration but

also on assessing the impacts on these parameters during ecorestoration process.

16.3 Approach of EcoIA

Under the EIA notification (2006), ecological components include only floral and faunal aspects and that too are project-specific (Box 16.1). While conducting an EcoIA, floral and faunal components are studied in both the core and buffer zones. The study is usually conducted by taking representative samples from the core zone (within the project area) where detailed

Box 16.1 Check List of Environmental Impacts – Vegetation and Fauna (EIA Notifications 2006)

1. Vegetation

1.1. Is there any threat of the project to the biodiversity? (Give a description of the local ecosystem with it's unique features, if any)

1.2. Will the construction involve extensive clearing or modification of vegetation? (Provide a detailed account of the trees & vegetation affected by the project)

1.3. What are the measures proposed to be taken to minimize the likely impacts on important site features (Give details of proposal for tree plantation, landscaping, creation of water bodies etc along with a layout plan to an appropriate scale)

2. Fauna

2.1. Is there likely to be any displacement of fauna- both terrestrial and aquatic or creation of barriers for their movement? Provide the details.

2.2. Any direct or indirect impacts on the avifauna of the area? Provide details.

2.3. Prescribe measures such as corridors, fish ladders etc to mitigate adverse impacts on fauna.

Table 16.1 Conceptual approach for study of biological environmental impacts

Step	Methodology
1. Identification of biological impacts of the proposed project	Interaction matrix, simple and descriptive checklist, network, etc.
2. Description of existing biological condition (*baseline studies*)	Field survey
3. Procurement of relevant laws, regulations, guidelines, etc.	–
4. Impact prediction	Land use or habitat change
5. Assessment of predicted significant impacts	Magnitude and sensitivity or value of ecological system
6. Mitigation measures	Reclamation/restoration

quantitative data are collected. Similar studies are also conducted for buffer zone. A six-step or six-activity model is suggested for planning and conduction of ecological studies (Canter 1996) as shown in Table 16.1

16.3.1 Identification of Biological Impacts of the Proposed Project (Step 1)

The first step is to quantitatively identify the potential impacts of the proposed project (or activity) on biological resources, including habitat and species. In Fig. 16.1, impacts of mining on biological resources are given as an example. Each activity has a negative direct and indirect affect on flora and fauna. There are different sources of information's available, which could be useful. How a particular activity is going to effect biological resources could be identified by impact identification methods such as *interaction matrix, simple and descriptive checklist and network.*

The typical impacts of different phases of mining on flora and fauna are given in Table 16.2, while Table 16.3 described at what extent these impacts could actually be assessed and incorporate in ecological impact assessment report.

16.3.2 Description of Existing Biological Condition (Baseline Studies) (Step 2)

The description of the environmental setting (also referred to as 'baseline', 'existing', 'background' or 'affected environment') is an integral part of an environmental impact study.

This involves three phases of study in terms of intensity of study:
- Phase I: Habitat survey
- Phase II: Species composition
- Phase III: Quantitative information

16.3.2.1 Phase I: Habitat Survey

It includes a general description of habitat or vegetation type within a study area and tries to fit this to a standard classification so that it can be easily understood and compare. This may be like woodland (or forest), shrub land, and grassland or pastureland. The phase-I survey, better to call as reconnaissance survey, is done easily by the help of land use map or remote sensing map of the area. The types of vegetative cover can be

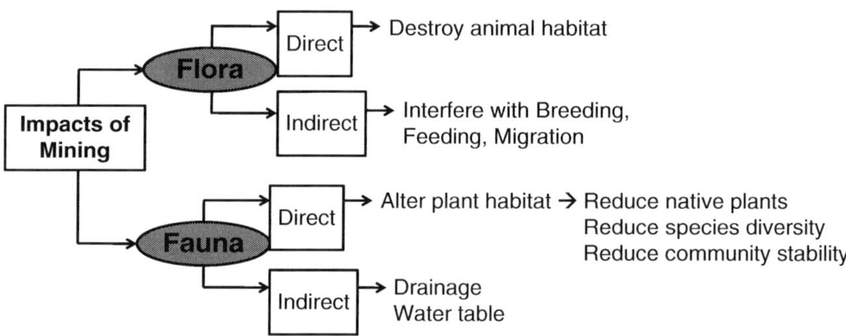

Fig. 16.1 Impacts of mining on flora and fauna (After Maiti 2002)

Table 16.2 Impacts of mining on flora and fauna (After Maiti 1997; Maiti and Pathak 1998)

Stage of mining operation	Types of impacts
Pre-production (development phase)—short duration	• Loss of terrestrial habitat.
Impact on vegetation	• Loss of standing crop, aesthetics and ameliorative properties of ecosystem.
Removal of vegetation, topsoil and subsoil	• Total loss of decomposer components including nutrient pool. • Fragmentation of ecosystem. • Invasion of weeds.
Production phase	• Deposition of settable dust and retardation of growth. • Rapid invasion of unwanted weeds (*Lantana, Eupatorium,* etc.) due to modification of habitat (rocky). • Scarcity of moisture limits normal succession process. • Alteration of landform and creation of rock habitat encourage vigorous growth of weeds that permanently occupy the site and never allow other late successional species to establish. • Increase in anthropogenic disturbance to the surrounding ecosystem components. • Increase in run-off and loss on fine-texture soil-forming materials. • Habitat transformed into stony surface, higher compaction, either very high or very low infiltration, increase in temperature and encouragement of shallow rooted plants. • Enhance the heavy metals recycling in ecosystem. • Lowering of water table.
Post-production (reclamation)	• Change in landform and topography (natural angle of repose in external dumps around 34–36°). • Reduction of local species and diversity in species, increases dissimilarity of ecosystem (reduces ecosystem stability). • Slow recovery of decomposers cycles. • Development of impoverished habitat. • Enhances recycling of heavy metals in ecosystem. • Regeneration of water table (?)
Impacts on wildlife	• Directly destroy animal habitat. • Indirect interfere with breeding, feeding and migration.

Table 16.3 Stages of impact assessment and availability of information (After Maiti 2007)

Stages of impact assessment	Types of information and data generation activity
1. Quantification of biological impacts	• Year-wise habitat degradation (excavation, dumping and ancillary development). • Loss of trees (timber). • Reduction in diversity of fauna (including avifauna, quantitative). • Invasion of unwanted weeds/shrubs (area wise, density). • Reduction in crop yield/productivity (effect of production). • Overall loss of habitat due to indirect activities.
2. Description of ecological environment setting	• Yes, all in quantitative terms (density, diversity, frequency, timber value, sensitive species, red data book category species, key stone species, etc.).
3. Procurement of laws and guidelines	• Yes
4. Impact prediction	• Macroscale: loss of habitat/chance in habitat. • Reduction in habitat quality and suitability for vegetation growth. • Area invaded by weeds/loss of habitat other than direct mining activities.
5. Assessment of predicted significant impacts	• Fragmentation of ecosystem (macroscale). • Microscale quantification of species lost.
6. Mitigation measures (most valuable)—repair the damage by reclaiming the area and assert the recovery by the continuous monitoring	• Successful recovery of ecosystem depends on surround seed banks, dumps morphology, nature of spoil materials, planting practices, amendments uses, and aftercare of reclaimed sites.

marked directly in the top sheets or sometimes the type of forest—tropical throne forest (6A), C1, C2, DS1 and DS2, or dry tropical forest (5A), C1 (1a, 1b), C2, C3, etc. (Champion and Seth 1968). Marsh (1991) adopted five levels of vegetation classification based on the following vegetative structure:

- *Forest*: Trees with average height greater than 15 ft with at least 60% canopy cover.
- *Woodland*: Trees with average height greater than 15 ft with 20–60% canopy cover.
- *Orchard or plantation*: Same as woodland or forest but with regular spacing.
- *Bush*: Trees and shrubs generally shorter height less than 5–6 m (15–20 ft) with multiple stem (high density of stem), but variable canopy. A large number of plants may become either shrubs or trees, depending on the growing conditions they experience.
- *Grassland*: Herbs with grasses dominant.
- *Field*: Tilled or recently tilled farmed.

16.3.2.2 Phase II: Plant and Animal Community Composition and Species List

Floral Components

The list of plants should have local name along with Hindi or English, botanical name and family. This includes

- Tree species
- Shrubs
- Herbs (legumenous herbs)
- Grasses
- Climbers
- Succulents (cactus)
- Others (moss, ferns, etc.)
  Faunal Components
- Amphibians
- Reptiles
- Fishes
- Birds
- Mammals

The list should be supplemented with qualitative description such as degree of occurrence (C, O, R) or by + notations:

- C = Common, occurring in many localities in large number (+++)
- O = Occasional, occurs in several localities in small number (++)

- R = Rare, highly localised and restricted by scarcity of habitat or low number (+)

Habitat: Open land, forest area, roadside, aquatic, wetland, etc.

Red Data Book Species List: Threatened, endangered or critically endangered species, if any

16.3.2.3 Phase III: Community Attributes

It involves more intensive sampling to provide detailed quantitative information on species and/ or community attributes:

- Vegetation structure in terms of phenology and phenograms, abundance, life form (growth form, phanerophyte, chamaephyte, hemicryptophyte, cryptophyte, geophyte and therophyte), vertical structure (stratification), vitality, disseminule type and age structure
- Individual tree: Aerial height, DBH, CC, etc.
- Biomass and net primary productivity
- Quantitative character: Density, dominance (crown cover and basal cover), frequency, importance value and species diversity

The study of community structure can be useful in woodland survey, but is less commonly applied to other communities.

Planning of Phase III

Five important decisions needed in planning a detailed field survey. These are

- Selection of sampling location and sample size
- Sampling pattern
- Species abundance measure
- Relevant environment factors
- Method of data analysis

Sampling Pattern Option

- Random sampling
- Systemic sampling

Quadrate Sampling

0.1 m^2 for mosses, lichens, and other small mat-like plant
1 m^2 for herbaceous vegetation
10–20 m^2 for shrubs and sapling up to 3-m tall
100 m^2 (10 × 10 m) for tree communities

Optimum size of the quadrate can be determined by species–area curve method:

Transect method: Line transects and belt transects.

Plotless sampling

16.3.2.4 Diversity and Similarity

Species Diversity (SD)

Species diversity is to count the number of species. In such count, only residence species are counted, and accidental or temporary immigrants should not be considered. Species richness is the oldest concept of species diversity. Consider this example:

There are three indices used for species richness calculation (d):

$$d_1 = \frac{S-1}{Log_e N}; \quad d_2 = \frac{S}{\sqrt{N}}; \quad d_3 = \frac{S}{1,000}.$$

Index (d_1) is also known as Margalef index.

Where S is the number of species and N is the total number of individual plants.

In both case the species diversity is same. The drawdack of the species richness index is that it never considers the evenness of distribution. Hence, two indices are used:

- Shanon index or Shannon–Weaver index
- Simpson index of diversity

Shannon Index or Shannon–Weaver Index

Any ecosystem comprises of a long number of plant and animal species that forms diversity. Diversity is measure by Shannon–Weaver index (H):

$$H = -\sum \frac{ni}{N} \ln \frac{ni}{N},$$

where H is species diversity, ni is the proportion of individual species and N is the total number of individuals of all species.

16.3.2.5 Simpson Index of Diversity (D)

The Simpson's index measures the probability that two specimens picked at random in a community belong to different species:

$$D = 1 - \sum_{i=1}^{n} (P_i)^2,$$

where $Pi = ni/N$. The Simpson index ranges between a value of 0 (low diversity) and a maximum of $1 - 1/S$.

Similarity Index

Similarity index is used to compare two ecosystems. It is also useful for comparison of species diversity of a polluted site and unpolluted sites. The expression of similarity index is

$$S = \frac{2C}{A+B} \times 100,$$

where

S = percentage similarity
A = no. of species in polluted site
B = no. of species in unpolluted sites
C = no. of species common to both sites

The tree species composition in the newly restored sites is quite different from the original composition, which is shown in Table 16.4.

Faunal Survey

In general, animals are more difficult to sample than plants. Most of the studies contain only a faunal list. It can be argued that the ecological value of a site can be determined largely from a vegetation study, because most resident animals depend on plants for food and shelter. *Census survey* method is a commonly used to study birds and animals.

Table 16.4 Comparison of tree composition (dominance %) between mining sites and nearby natural forests (ECL area) (After Maiti et al. 1998)

Plant species	Mining sites	Natural forest
Acacia auriculiformis	58%	–
Eucalyptus	24%	–
Alstonia scholaris	12%	–
Azadirachta indica	6%	–
Bassia latifolia	–	56%
Anacardium occidentalis	–	19%
Shorea robusta	–	12.5%
Terminalia arjuna	–	6.25
Albizia lebbeck	–	6.25%

Table 16.5 Specimen data sheet for census survey

Distance from starting point (km)	Count of species		
	Snake (A)	Rabbit (B)	Deer (c)
0.5	1	–	–
1.0	–	6	1
–	–	–	–
5.0	2	5	1
Total roadside count	3	11	2

$$Census\ Index = \frac{M}{N},$$

where

M = no. of individuals in each species noted during survey

N = area travelled

With the CI, the biologist established the composition and structure of grassland and key species upon which birds and wildlife dependent for food and shelters. An example of census survey is given in Table 16.5

$$Census\ Index\ for\ A = \frac{M}{N} = \frac{3}{5} = 0.6;$$

$$CI\ for\ B = \frac{11}{5} = 2.2\ and$$

$$CI\ for\ C = \frac{2}{5} = 0.4$$

Valued Ecosystem Components (VECs) and Key Biological Process

'Key species' have been selected on account of their economic importance, their protected status, their rarity, their sensitivity to specific impacts or their representativeness of other species, which use a common environmental resource (Guilds) (Westman 1995).

16.3.3 Procurement of Relevant Laws, Regulations, Guidelines, etc. (Step 3)

The primary sources of information on pertinent legislation, regulation, criteria, or guidelines related to the biological environment are available with state or central government agency.

- The Forest (Conservation) Act 1980 (amendments made in 1988)
- The Forest (Conservation) Rules 1981 (subsequent amendments)
- Wildlife (protection) Act 1972

Rule 4: The Forest (Conservation) Rules 1981. Details of forest land involved

(a) Legal status of the forest (namely reserve, protected/ unclassified, etc.).

(b) Details of flora and fauna existing in the area.

(c) Density of vegetation.

(d) Species-wise and diameter-wise abstract of trees.

(e) Vulnerability of the forest area to erosion, whether it forms a part of a seriously eroded area or not.

(f) Whether it forms a part of National Park, wildlife sanctuary, nature reserve, biosphere reserve, etc.; if so, details of the area are involved.

(g) Item-wise break-up of the forest land required for the project/scheme for different purposes.

(h) Rare/endangered species of flora nd fauna found in the area.

(i) Whether it is a habitat for migrating fauna or forms a breeding ground for them.

(j) Any other significance of the area relevant to the proposal.

16.3.4 Impact Prediction (Step 4)

The most technically demanding step in addressing the biological environment is the prediction of the impacts of the project activity. Impact prediction for the biological environment is generally focused on *land use or habitat changes*. Some important changes that affect faunal population are

- Loss of habitat
- Habitat fragmentation and isolation
- Reduction of habitat quality and suitability
- Pollution
- Disturbance

Some additional issue that needs to be addressed:

- Likelihood of impact in a relative scale: High, medium or low
- Duration of impact: Short term or long term (anticipated duration)

- Reversibility of the impacts
- Relative resiliency of individual plants or animals within the study area
- Potential mitigation measure for a given project type

The list of potential effect on the biological system could be used as information source.

16.3.5 Assessment of Predicted Significant Impacts (Step 5)

Predicted impacts usually be considered a function of magnitude and sensitivity or value of ecological system affected. Interpretation of the anticipated impacts of a proposed project (or activity) should consider in terms of

Individual plant species	Role of the individual species food web relationship Resiliency of the plant
Effects on general characteristics of the affected habitat	Resiliency of the habitat Reduction in species diversity Economic importance
Effects on overall ecosystem	Fragility of the environmental setting Interruption in natural succession process

In UK, most widely used criteria for evaluation of ecosystem based on criteria proposed by Morris and Therivel (1995). These consist of primary criteria and secondary criteria.

Primary criteria	Secondary criteria
Diversity: change in diversity	Position in an ecological/geographic unit
Naturalness: seminatural, improved or artificial	Potential value
Rarity of species Fragility/sensitivity Uniqueness	Intrinsic appeal

One of the important aspects is identification of suitable indicator species, which can be based on

- *Functional aspects*: Change in biomass, productivity, etc.
- *Traditional approach*: Just to quantify impacts (e.g. may be in terms of reduction in density of plants species) before (baseline or control area) and after impacts occur (project site or impact zone) due to activity or proposed action

Drawback
- Lack of replication
- Identification of suitable site (baseline or control site)

Use of GIS-based techniques: Overlaying the maps of before and after projects, the degree of fragmentation of habitat can be identify.

16.3.5.1 Case Study: Quantification of Impacts of a Mining Project on Tree Density

This is a very simple case study, by which impact of mining in a teak (*Tectona grandis*) forest was assessed by measuring DBH and density of trees in the project area and compared with control area (which was 5–6 km of mining sites) assuming there are no impacts of mining.

1. Count the numbers of teak and non-teak species and their DBH as follows (Table 16.6):
2. The number of standing trees (5–6 km away from project site) which was used as baseline or control site was totally counted and shown in Table 16.7. The quadrate size was 100×100 m, and 20 numbers of quadrate were used.
3. *Observations*
   (a) In the control area (5–6 km away from project site), tree density is about four times (2,702 trees/ha) that of project site (718 trees/ha).
   (b) This is predominantly teak area; therefore, teak poles and trees are 9–10 times more in control site than project site.
   (c) Even in respect of non-teak species, it is about 3 times more in control site than project site.
   (d) Quantification of loss. Area damaged due to project 215.77 ha.

Density of species (trees/ha) in control area	Total no. of tree species would have been present, without project (A)	Existing number of plant in project site (B)
Teak 1,054	Teak $1,054 \times 215.77 = 2,27,421$	
Non-teak 1,640	Non-teak $1,640 \times 215.77 = 3,53,862$	
Total	5,81,283	1,55,236

Table 16.6 Standing trees (leased out area) in project site (215.77 ha)

Tree size (DBH)	Teak (*Tectona grandis*)	Non-teak species	Total
Less than 20 cm	12,830 (59)	61,772 (286)	74,602 (345)
21–30 cm	15,374 (72)	52,788 (244)	68,162 (316)
>31 cm	3,063 (14)	9,220 (43)	12,283 (57)
Total	31,267 (145)	1,23,780 (573)	1,55,047 (718)

Methods: total count
Figure in parenthesis shows no. of trees/ha

Table 16.7 Standing trees in the control site

Tree size (DBH)	No of teak/ha	No of non-teak/ha	Total (plants/ha)
<20 cm	876	1,155	2,031
21–30 cm	93	372	405
>30 cm	85	121	206
Total	*1,054*	*1,640*	*2,702*

(e) Thus, number of tree lost due to direct/indirect activity of the project is (A–B) = 5,81,283–1,55,047 = *4,26,236* nos.

Drawback: Identification of suitable control sites.

16.3.6 Mitigation Measures (Step 6)

Mitigation measures for biological impacts can include:

- Avoidance of Impacts
- Minimise the impacts
- Ameliorate the impacts
- Habitat replacement/restoration

Surface mining control and Reclamation Act, 1977 (SMCRA), delineate the mitigation requirements directed towards restoring habitat disturbed by surface mining operation. A summary of some mitigation measures, suggested under Section 501(b) by the Surface Mining Control and Reclamation Act of 1977 (SMCRA) which relate to biological system or biological impacts, pertinent to Indian conditions is presented in Table 16.8.

16.3.6.1 Rule 4 of the Forest (Conservation) Rules 1981

In India, under Rule 4 of The Forest (Conservation) Rules 1981, detailed guidelines for the compensatory afforestation scheme are discussed herein:

(a) Details of non-forestry area/degraded forest area identified for compensatory afforestation, its distance from adjoining forest, number of patches and size of each patch

(b) Map showing non-forest/degraded forest area identified for compensatory afforestation and adjoining forest boundaries

(c) Detailed compensatory afforestation scheme including species to be planted, implementing agency, time schedule, cost structure, etc.

(d) Total financial outlay for compensatory afforestation scheme

(e) Certificates from competent authority regarding suitability of area identified for compensatory afforestation for afforestation and from management point of view

(f) Certificate from the Chief Secretary regarding non-availability of the non-forest land for compensatory afforestation (if applicable).

16.3.6.2 Limits of Acceptable Change (LCAs)

In some biological impact assessment, LACs have been defined for specific environmental components, with agreement on subsequent mitigation measures. For example, LACs were defined for

- Percentage vegetation cover
- The average size of bare patches
- Width of footpath, corridors, etc.

Table 16.8 Summary of biological mitigation measures (After Maiti 2002)

Biological impacts	Possible mitigation measures
Erosion and sedimentation	• Surface run-off must be collected in sediment pond
	• Disturbed soil must be revegetated
Destruction of vegetation	• Affected land must be restored to pre-mining capacity
	• Topsoil must be removed, segregated, stored and redistributed with minimum loss or contamination
	• Topsoil and subsoil may be removed separately and replace in sequence
	• Native vegetation or appropriate substitutes after mining must be established
	• Agriculture lands must be returned to the same or greater productive capacity obtained under pre-mining conditions
Disturbance of aquatic organisms and aquatic habitats	• A regulatory programme designed for restoration, protection, enhancement and maintenance of aquatic life must be implemented
	• Surface and underground mine opening must be cased and sealed to prevent escape of acid and toxic discharge
	• Buffer strips must be left between mining operations and waterways
Loss of wildlife and wildlife habitat	• A wildlife protection plan is required as part of any mining permit application
	• Wildlife agencies must be consulted
	• Timing, shaping and sizing operations must be conducted to avoid breeding or nesting season and trees, protecting key food, cover and water resources
	• Fencing will keep large mammals from direct contact with toxic chemicals in sedimentation ponds and from roadways to reduce the number of roadkills
	• Revegetation must use species with high nutritional or cover value
	• Topsoil handling and replacement prior to revegetaion must be conducive to wildlife
	• Topsoil storage must be covered with vegetation, thus providing cover for wildlife
	• A 30-m buffer zone on each side of streams must be undisturbed

Others
- Creation of buffer zones around the project
- Transplanting of habitat, seeds and ecological value added trees

- Photographs and permanent quadrate sites can be useful to record changes.
- Compare the monitoring data with control/reference sites.

16.3.7 Monitoring (Last Step)

Some basic questions to be asked before any monitoring programme. These are
- Who will monitor?
- What parameters to be monitored? What are the methods of monitoring?
- When to monitor?
- Duration of monitoring.
- How this monitored data will be used? And who will use?
  The simple way these can be performed:
- Monitoring should be done by implementing authority or project authority.
- Monitoring activities will provide the degree of mitigation measures.
- The monitoring database will be use for future EIAs.
  Methods: Those methods used for baseline studies.

16.3.8 Drawbacks of Existing Ecological Impact Assessment Methods

Although there is no dearth of theoretically appealing EcoIA methods in available EIA literature, application of such methods in Indian mining projects leads to few successes. Such limitations owe their origin to the fact that ecosystem degradation due to mining activities are caused by large-scale land degradation causing loss of habitat. Thus, when the project activities ceases on a land area, the reclamatory task focuses on generation of a new habitat, which, because of moisture, nutrient and other ecological stresses, are bound to be different from the pre-mining habitat. Conventional methods of EcoIA fail to address this unique feature of mining activities and are, therefore, of little practical relevance.

The focus of EcoIA under such situations must be on avoidance and minimization of impacts. That is, the EcoIA for mining projects must accept the hard reality of habitat loss being inseparably connected with mining activities, especially the surface mining activities. Once this reality is understood and accepted, the EcoIA logically focuses on creation of new habitat with the objective of accelerating the process of establishing a self-sustaining ecosystem in the regenerated habitat. In other words the effectiveness of the new habitat in developing a productive ecosystem should be the central theme of EcoIA for surface mining projects. In order to be effective for changed paradigm of EcoIA as proposed above, the habitat-based method of ecological impact assessment, as suggested by many EIA scholars including Canter (1996), is more applicable than the conventional methods.

16.4 Habitat-Based Method for Biological Impact Prediction

16.4.1 Habitat Evaluation System (HES)

HES was developed in 1976 by US Army Corps Engineers (USA). The fundamental assumption underlying HES is that the presence or absence, abundance and diversity of animal population in a habitat or community are determined by basic biotic and abiotic factors that can be readily quantified. HES does not treat individual species, although techniques can be modified to evaluate habitats for specific species (Canter 1996).

The HES procedure involves six steps for evaluating impacts of a development project (Canter 1996). These steps are presented in Table 16.9.

Table 16.9 Steps involved in evaluating ecological impacts of surface mining projects

Steps of HES	Methodology
1. Obtaining habitat type or land use (area in ha)	• Land use for both project area (core zone) and buffer zone is obtained from land use map. Subsequent changes during the project (involvement of forest land, area of compensatory afforestation, external dumping area, internal dumping area) after project could be accurately predicted
2. Deriving habitat quality index (HQI) scores	• Crucial step for the successfulness of HES. Determination of key variables that affect ecosystem health and functioning is important which may be of project specific, location specific or both. Some of the *HQ indicators* for coal mine overburden dumps that influence ecosystem development are listed below:
	• Dump heights and slope
	• Source of natural seed banks
	• Matching landscape
	• Stone content in the rooting depth (up to 100 cm)
	• Bulk density (g/cc)
	• Field moisture content (%)
	• Infiltration rate (cm/h)
	• pH
	• Organic carbon content (%)
	• Available N & P
	• CEC
	• Microbial activity
	• Density of VAM spores
	• Types of amendments envisaged
	• Problems of toxic metals
3. Deriving habitat unit values (HUVs)	• $HUV = HQI \times habitat~size$
4. Anticipated HUVs during the project and without project	• Changes in land use during project life, 5-year reclamation plan preparation
5. Using HUVs to assess impacts of project alternatives (even in terms of reclamation alternatives)	• Estimate impacts in terms of change in habitat value (HV)
6. Determining mitigation requirements.	• Reclamation/restoration practices

Step 1: Obtaining Habitat Type or Land Use

Delineate the study area in terms of habitat type. Study the land use habitat pattern for existing condition and project for future with project and without project:

- Define the study area.
- Delineating cover type.
- Selecting evaluation species—sensitive species, key species and species with high public interest, economic values or both and ecological interest (food web). A typical HES study will incorporate 4–6 species.

Terrestrial: Natural Forest, plantation area (afforested area), grassland (grazing land) and shrub land

Aquatic: Stream, lakes, wetland (marshy land), etc.

Step 2: Deriving Habitat Quality Index (HQI) Scores

Derive the HQI scores for each land use category or habitat type. Data are obtained on several key viable for each habitat type from field measurement, literature, and historical information. Assign weightage for key variable.

Important Stages

- Identification of key ecological attributes for particular habitat.
- Assignment of relative weights to each variable. Table 16.10 displays the variable and their relative weights for a forest ecosystem. In other types of ecosystem like grassland, productivity may be important variable. The assignment of relative weight will be site specific.

In the next step, value function curve is developed for each key variable for the said *habitat.* Thus, in an area, if there are other kinds of habitats

Table 16.10 Land use type: forest land

Key variable	Relative weights
Tree species association	20
Fruit trees (fauna attracting)	20
Canopy cover	30
Large tree (e.g. >30 cm DBH)	30
Percentage ground cover	10
Total	100

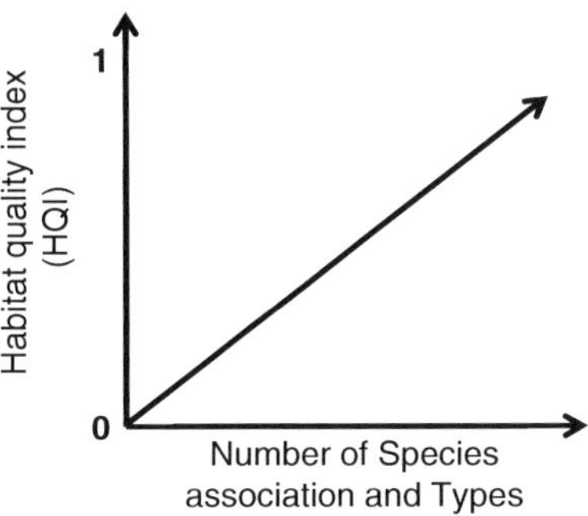

Fig. 16.2 Conversion of key variables to HQI

like plantation area and wetlands, then the key variables are to be identified and value function for each variable should be developed for the particular habitat. The values may be convert into HQI score (Fig 16.2).

HQI score is based on scale 0–1 (maximum value or height rating). HQI scores for a given habitat assigned a weight between 0 and 100. Conversion of HQI score is given in Table 16.11.

Step 3: Deriving Habitat Unit Values (HUVs)

The size of the given habitat is multiplied by aggregated HQI score to produce a habitat unit value (HUVs) for the habitat.

Thus, $HUV = habitat\ quality\ index\ (HQI) \times habitat\ size\ (acre\ or\ ha)$.

Step 4: Projecting HUVs for with Project and Without Project

Project the habitat unit value (HUV) over the project life and for various alternatives (reclamation/restoration scenario). HUV must be derived for each time increment, usually 10-year interval, over the project life against without project and each alternative plan (e.g. even for alternative reclamation plan).

Step 5: Using HUVs to Assess Impacts of Project Alternatives

$$Impact = With\text{-}project\ HUV - Without\ Project\ HUV$$

Table 16.11 Land use type: forest land (sampling plot 100 × 100 m; 20 plots)

Key variable	Field data	HQI score (value function curve)	Weights	Wt. HQI score
Species association	*Teak-Sal-Mohua*	0.80	20	16.0
Fruit trees (Fauna attracting)	6 (2 good, 2 moderate, 2 not so important)	0.60	20	12.0
Canopy cover	80% (good is >60%)	1.00	30	30.0
Large tree (e.g. >30 cm DBH)	10 trees (5 is over 45 cm)	1.00	30	30.0
Percentage ground cover	60% (palatable and 3 species rich)	0.80	10	08.0

Total HQI score = 96.0
Aggregate HQI = 0.76 (before project) and 96.0

Step 6: Determining Mitigation Requirements
Envisage mitigation plan. Mitigation may be defined as 'any measure taken to return the "with project" environmental quality of the area to the same level as the "without project" condition' (Canter 1996).

16.5 Advantages and Disadvantages of HES

Advantages
1. Highly dissimilar environmental characteristics are quantifiable in standardised terms (habitat quality index). Adverse and beneficial effects are clearly identified in comparable terms.
2. The system provides an objective method for comparing environmental effects of project alternatives and various mitigation measures.
3. HES results are reproducible. Functional curves are based on quantitative measurements of key variables, which are converted to an HQI score. Thus, once key variable are identified, the HQI is fixed.
4. Use of the HES is rapid and efficient and requires a minimum of field and laboratory data on terrestrial habitat. Data for most aquatic functional curves can be obtained from historical data sources.
5. HES is a flexible method. If Ecologists in a specific geographical area feel that the functional curves do not correctly represent conditions of that area, curves can be altered and reweighted.

Disadvantages of HES
The curves and weights assigned to each variable are subjective to some extent.
HES describe habitat quality for a broad range of species, rather than attempting to predict the density of a particular species.

16.6 Conclusive Remarks

More than three decades of global experience with EIA has amply demonstrated the multiple weaknesses of conventional EIA methods in assessing ecological impacts. Such weaknesses are more conspicuous in surface mining projects where habitat destruction is inevitable and conscientious.

Over the years a general conscientious has emerged about the structure and sequence of EIA process (Sinha 2001). However, in India only the base line studies are conducted in a comprehensive manner. Other sequential steps, particularly the prediction and evaluation of impacts, are poorly understood and hence are presented vaguely.

Vast improvement in mitigation planning may be achieved by restoring to quantitative techniques where due care should be taken to allow for risks and uncertainties.

While habitat-based system of ecological impact assessment seems to be a promising technique for overcoming the limitations of the conventional techniques, care should be taken to document the thought process that goes into quantifying the ecological impacts. Otherwise, the inherent subjectiveness of the habitat evaluation system may be buried under the spurious objectiveness.

A disquieting feature of the Indian EIA/EMP reports is that the reclamation plans developed for mining projects are seldom based on the prediction and evaluation of ecological impacts. The reclamation plans so developed are thus poor in both form and content. Adopting an appropriate EcoIA method may go a long way in improving the reclamation plan towards regenerating both flora and fauna to match with the local ecological setting.

References

Canter LW (1996) Environmental impact assessment, 2nd edn. McGraw Hill Inc., Singapore

Champion HG, Seth SK (1968) A revised survey of forest types of India. The Manager of publications Delhi-6. http://www.icfre.org/UserFiles/File/Education/Report/Annexure-II.pdf

EIA Notifications (2006) http://envfor.nic.in/legis/eia/so1533.pdf

Maiti SK (1997) Economic valuation of environmental impacts of open-cast coal mining project- an appraisal in Indian context. In: Upadhaya VP (ed) Issues of environment and sustainability. Nandighosh, Bhubaneswar, pp 112–125

Maiti SK (2002) Ecological environment. In: Saxena NC et al (eds) Environmental management in mining areas. Scientific Publishers, Jodhpur, pp 110–141

Maiti SK (2007) Ecological impact assessment of surface mining project. In: Singh KK et al (eds) Environmental degradation and protection, vol II. MD Publication Pvt. Ltd., New Delhi, pp 1–31

Maiti SK, Pathak K (1998) Economic evaluation of environmental impacts of opencast mining project- an approach. In: Proceedings of the VII national symposium on environment. ISM, Dhanbad, pp 174–180, 5–7 Feb 1998

Maiti SK, Sinha IN (2004) Ecological impact assessment of mining project: a pragmatic approach. In: Proceedings of the technology and management of sustainable exploitation of minerals and natural resources (TAMSEM 2004). Department of Mining Engineering, IIT, Kharagpur (India), pp 335–351

Maiti SK et al (1998) Noise attenuation through green belt – a case discussion. In: Proceedings of the VII national symposium on environment. ISM, Dhanbad, pp 227–234

Marsh WM (1991) Landscape Planning: Environmental Applications, 2nd edn. John Wiley and Sons, New York

Morris P, Therivel R (1995) Methods of environmental impact assessment. UCL Press, London

Sinha IN (2001) A framework of EIA for environmental sustainability. ENVIS monograph no.8, ENVIS CME, ISM, Dhanbad

Westman WE (1995) Ecology, impact assessment & environmental planning. Wiley, New York

Mine Closure

17

Contents

17.1   Introduction

Mine closure—as a process—refers to the period of time when the operational stage of a mine is ending or has ended and the final decommissioning and mine rehabilitation is being undertaken. Closure may be only temporary in some cases or may lead into a programme of care and maintenance. In this sense, the term mine closure encompasses a wide range of drivers, processes and outcomes. The goal of mine closure is mine completion (Ghose 2011).

Mine closure has been defined under MCDR (1988) as 'Mine closure means steps taken for reclamation, rehabilitation measures taken in respect of a mine or part thereof commencing from cessation of mining or processing operation in a mine or part thereof'. Progressive mine closure and final mine closure plan have also been defined in MCDR (1988). Progressive mine closure plan 'means a progressive plan, for the purpose of providing protective, reclamation and rehabilitation measures in a mine or part thereof that has been prepared in the manner specified in the standard format and guidelines issued by Indian Bureau of Mines'. While 'final mine closure plan' means a plan for the purpose of decommissioning, reclamation and rehabilitation in the mine or part thereof after cessation of mining and mineral processing operations that has been prepared in the manner specified in the standard format and guidelines issued by the Indian Bureau of Mines (2003)'.

S.K. Maiti, *Ecorestoration of the Coalmine Degraded Lands*,
DOI 10.1007/978-81-322-0851-8_17, © Springer India 2013

Mine completion ultimately defines what is left behind as a benefit or legacy to future generations. If mine closure and completion are not undertaken in a planned or effective manner, a site may continue to be hazardous and a source of pollution for many years to come. The overall objective of mine completion is to prevent or minimise adverse long-term environmental, physical, social and economic impacts and to create a stable landform suitable for some agreed land use. The basic and ultimate objective is to ensure planned, structured and systematic mine closure and completion of mines in the context of sustainable development (Ghose 2011).

Mines only close when their mineral resources are exhausted and a mine closure plan is in place and progressively implemented. However, in the real world mines may close prematurely due to other reasons too which include (Ghose 2011):

- Economic, such as low commodity prices or high costs
- Geological surprises, such as grade or size of ore body
- Technical such as adverse geotechnical conditions or equipment failure
- Regulatory, due to safety or environmental breaches
- Social or community pressures
- Closure of downstream industry or markets
- Flooding or inrush

Usually final mine closure activities should start a few years before depletion of economically extractable reserves of a mine. However, in case of coal deposits having multiple seams, some of which occurring in the upper horizons are amenable to opencast mining, the matter of final mine closure may assume complexity. This is because there may be considerable time lag between completion of opencast mining in upper seams and completion of underground mining in the lower seams. Such geo-mining scenario is usually noticed in Raniganj, Jharia, East and West Bokaro, North and South Karanpura, Talcher and many other coalfields of India. Final mine closure plan under such geo-mining conditions should be prepared considering the deposits in totality (Chaudhuri 2008). Poorly closed and derelict mines surface a legacy issue and ultimately tarnish the mining industry image.

17.2   Objectives of Mine Closure

Effective planning and realisation of closure objectives and measures need to be determined on the basis of local circumstances at the individual mine site. Establishing appropriate closure objective requires local *environmental attributes* (e.g. soils, climate and biodiversity), the *location of the mine* (proximity to habitation, potential pressures on post-closure land use and water use issues), as well as *technical matters* (such as methods used in quarrying and implementing closure) (Heikkinen 2008). The mine closure process can only be considered complete when the closure objectives have been satisfactorily met.

In planning for closure, there are *four key objectives* that must be considered:

1. Achieve a productive use of the land or a return to its original condition or an acceptable alternative acceptable to the local bodies, mine authorities and regulatory bodies.
2. Protect public health and safety or make the mine site safe for the surrounding habitat
3. Physical and chemical stability of sites (repair environmental damage and ensure there is no future pollution) and improve aesthetics of area as far as possible.
4. To the extent achievable, provide for sustainability of social and economic benefits resulting from mine development and operations.

If all the above objectives are achieved, the mine closure operation will contribute towards sustainable development (Chaudhuri 2008).

17.3   Issues Related to the Mine Closure

The World Bank suggested the following issues to be addressed in a closure plan (Debnath et al. 2011):

1. *Environmental Protection and Reclamation*
   Good practice today requires removing unwanted plant and equipment, stabilising and

securing waste dumps and impoundments, detoxifying hazardous materials, protecting groundwater, addressing any acid rock drainage issues (which in the worst cases can be a severe problem) and reclaiming, rehabilitating, and revegetating land in a manner compatible with local vegetation.

2. *Disposal of Asset*

Most mine production assets have little value at closure. Those that can be sold should be sold or converted to another productive purpose. Otherwise, plant and equipment need to be demolished or dismantled and removed. Some assets such as open pit and underground mine workings and dumps cannot be removed, and these should be stabilised and made secure and safe.

3. *Legal Framework*

Without a good legal framework for mine closure, mining companies do not know their obligations and potential future liabilities, and mining communities do not know their rights or responsibilities. The absence of a comprehensive legal framework for mine closure can also lead to inefficiencies and confusion among different ministries and government units at the central, regional, and local levels. Unless they know the lines of authority and responsibility, they will not be able to ensure that mine closure takes place properly and that adequate monitoring occurs after closure.

17.4 Mine Closure Planning

Mine closure planning provides a conceptual framework that guides a mining project through the life cycle of the operation from pre-feasibility to site decommissioning and abandonment. Mine closure planning and implementation requires a multidisciplinary team with skills in impact assessment, contamination assessment, engineering problem solving and practical operational competencies. Details of mine closure plan will differ from mine to mine depending on a host of factors. However, broadly, the closure planning must deal with the following aspects (Chaudhuri 2008):

1. Technical aspects
2. Environmental aspects
3. Social aspects
4. Closure action plan
5. Financial aspects

Pathak (2011) recommended following guiding principles to be followed in developing a mine closure plan:

1. Improving environmental performance after mine closure.
2. Operating mine closure in an economic, safe and responsible manner.
3. Reclaiming the landscape on an ongoing basis (progressive reclamation).
4. Provide a landscape that will be physically and chemically sustainable for the long term.
5. Achieve closure and custodial transfer in an economical, timely and secure manner.

As per the author, following aspects are to be considered during mine closure planning:

1. Land use plan and site rehabilitation plan of different categories of mines are to be collected based on size (i.e. small, medium and large scale based on magnitude of land degraded or land required) and types of mineral mined.
2. In case of waste dump and tailings, critical attributes, like-detoxifications of toxic metals, acid–base accounting, cover development aspects, air pollution (due to wind-borne dust) and water pollution aspects are to be assessed.
3. Site water management like harvesting and reuse of water for various purposes.
4. Site safety needed to be assessed with respect to how to make mine site safe, physical and chemical stability and ensuring that there is no more future pollution.
5. Final landform will be assessed with the surrounding matching landscape.
6. Restoration of land surface of sufficient quality to support pre-mining land use potential.
7. Management plan for implementation of closure will be given in flow chart form for different categories of mines and minerals.

8. Decommissioning of infrastructure.
9. Under socio-economic programme, issues like prevention of job losses, management of job losses and regeneration of local economies.
10. Post-closure monitoring proposal (air, water, biodiversity, soil development, etc.)
11. Time and cost of closure.

17.5 Environmental Impacts of Mine Closure

Impacts that change conditions affecting these objectives are often broadly discussed as the 'impacts' or the environmental impacts of a site or a closure plan. It is convenient to consider potential impacts in four groupings:

- *Physical stability*—buildings, structures, workings, pit slopes, underground openings, etc., must be stable and not move so as to eliminate any hazard to the public health and safety or material erosion to the terrestrial or aquatic receiving environment at concentrations that are harmful. Engineered structures must not deteriorate and fail.
- *Geochemical stability*—minerals, metals and 'other' contaminants must be stable, that is, must not leach and/or migrate into the receiving environment at concentrations that are harmful. Weathering oxidation and leaching processes must not transport contaminants, in excessive concentrations, into the environment. Surface waters and groundwater must be protected against adverse environmental impacts resulting from mining and processing activities.
- *Land use*—the closed mine site should be rehabilitated to pre-mining conditions or conditions that are compatible with the surrounding lands or achieves an agreed alternative productive land use. Generally the former requires the land to be aesthetically similar to the surroundings and capable of supporting a self-sustaining ecosystem typical of the area.

- *Sustainable development*—elements of mine development that contribute to (impact) the sustainability of social and economic benefit, post-mining, should be maintained and transferred to succeeding custodians.

17.6 Steps in Closure Plan Development

The typical steps for closure planning are shown in Fig. 17.1. These steps also provide a logical order in which to develop and present the various sections of a *Closure Plan Report*. They provide the reader with a progressive description of the material required to understand the need for, nature of, effectiveness of and cost of the closure plan. Any closure plan must consider the long-term physical, chemical, biological and social/land use effects on the surrounding natural systems (aquatic, groundwater, surface water, etc.):

1. Understanding the pre-mining environment (*step 1*) and the effects of past and future mine development (*step 2*) on the pre-mining environment.
2. Operational control measures must be selected during mining operation and implement such that it will minimise the impact on the surrounding ecosystems (*step 3*).
3. Impact assessments must be done prior to measures selection as well as periodically during operations in order to determine the success of the measures implemented (*step 4*).
4. Alternative mine closure measures are developed and assessed during mine design to ensure that there are suitable closure measures available to remediate the impact of the selected mine development (*step 5*) and (*step 6*). If suitable remediation or closure measures cannot be identified or achieved, then it may be appropriate to revise the type of mine development proposed (*return to step 2*).
5. Once a technically acceptable mine development and closure plan has been developed, it is necessary to prepare a monitoring and maintenance plan that will monitor the system

Fig. 17.1 Typical steps in the closure plan development process (Robertson and Shaw 2004)

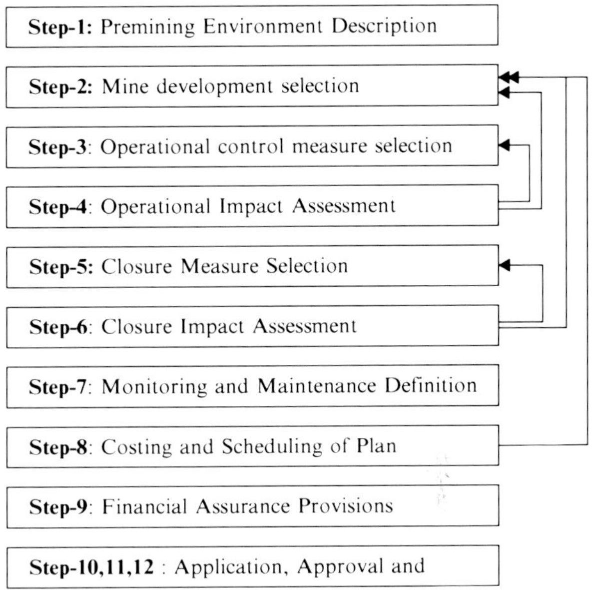

performance during operations and post-closure and provide for the maintenance necessary to ensure the long-term functionality of the system components (*step 7*).

6. Throughout this process, costing and scheduling evaluations (*step 8*) are completed, if the costs are too onerous, or if fatal flaws in the design are identified, the process returns to the design phase (*step 2*) and alternative measures are evaluated.

7. Once an acceptable plan is completed, an acceptable form of financial assurance is developed and provided (*step 9*) in order to cover the costs of plan implementation, long-term operations, monitoring and maintenance of the site post-closure.

8. The final stages of the closure plan process involve the application for closure, approval by the regulatory agencies of the closure plan and implementation at the end of mine life (*steps 10–12*).

17.7 Mine Closure Guidelines in India

The Indian Bureau of Mines (IBM) notified the mine closure plan [The Central Government vide Notification No. GSR 329 (E) dated 10.04.2003 and No. GSR 330 (E) dated 10.04.2003 amended the Mineral Concession Rules, 1960, and Mineral Conservation and Development Rules, 1988, respectively]. As per these amendments all the existing mining lessees are required to submit the '*progressive mine closure plan*' along with prescribed financial sureties within 180 days from date of notification. Further, the mining lessee is required to submit '*final mine closure plan*' 1 year prior to the proposed closure of the mine. In the notification it has been enumerated that the '*progressive closure plan*' and '*final closure plan*' should be in the format and as per the guidelines issued by the Indian Bureau of Mines. Judging the global need of post-closure mining operation, the Ministry of Coal and Ministry of Mines have formulated mine closure guidelines during 2009 and 2003, respectively. The responsibility of cleaning up of operation of mine closure lies with the operator on the *principle of polluter pays for degradation and pollution.*

Mine closure encompasses rehabilitation process as an ongoing programme designed to restore physical, chemical and biological quality disturbed by the mining to a level acceptable to all concerned. It must aim at leaving the area in such a way that rehabilitation does not become a burden to the society after mining operation is over. It must also aim to create as self-sustained ecosystem.

Mine closure operation is a continuous series of activities starting from day one of the initiation of mining project. Therefore, progressive mine closure plan will be an additional chapter in the present mining plan and will be reviewed every 5 years in the scheme of mining. As progressive mine closure is a continuous series of activities, it is obvious that the proposals of scientific mining have had included most of the activities to be included in the progressive mine closure plan. Therefore, reference to relevant paragraphs and a gist of the same in progressive mine closure plan will be sufficient.

'*Final mine closure plan*', as per statute, shall be considered to have its approval at least 9 months before the date of proposed closure of mine. This period of 9 months is reckoned as preparatory period for final mine closure operations. Therefore, all proposals for activities which have to be carried out after production of mineral from the mine or mining is ceased shall be included in the final mine closure plan. The final mine closure plan will thus be a separate document with detailed chapters as per guidelines given below.

The guidelines include the specific activities both in progressive mine closure plan and final mine closure plan, which is enclosed at the end of this chapter.

17.8 Key Elements of Mine Closure (Coal Mining)

All mine owners have to prepare mine closure plan during the mine planning stage as per the guidelines (Debnath et al. 2011). The mine closure plan will have two components, that is,

1. *Progressive or concurrent mine closure plan*
2. *Final mine closure plan*
   - The final mine closure plan activities will continue to its life cycle and till the area is restored to an acceptable level.
   - Statutory obligations have to be followed by mine operators.

- Funds which will be levied for mine closure activities are towards the security to cover the cost of closure in case the mine owner fails to complete the relevant closure activities.
- After the closure, the reclaimed leasehold area and any structure thereon, which is not to be utilised by the mine owner, shall be surrendered to the state government concerned.

17.9 Preparation of Mine Closure Plan Report (Coal)

The mine closure plan for the coal projects inter alia includes the technical, environmental and socio-economic details for calculating the mitigation cost of post-mine closure (Debnath et al. 2011). A typical detail of mine closure plan is given in Table 17.1.

However, possibility shall be explored for handing over the residential and non-residential buildings and other infrastructures including the reclaimed land to state government for the benefit of local villagers and strengthening the area infrastructures. The end use of these facilities shall be decided by state government with the help of local government and village panchayat.

17.10 Economic Repercussions of Mine Closure

The level of socio-economic implication of mine closure depends very much on the region where the mine is situated. If the community on the region is highly depended on the mining activity, the mine closure will certainly give negative impact on socio-economic conditions of the region. According to the regional condition, the mine closure plan will have different approaches (Pathak 2011). However, broad guidelines suggested by Debnath et al. (2011) are highlighted below:

- Manpower of the project
- Assessment of income scenario of local people

Table 17.1   Preparation of mine closure report for coal mine (After Debnath et al. 2011)

1. Closure planning details of mine	*Mined-out land and proposed final land use*
	(a) Total mined-out area (ha)
	(b) Concurrently/backfilled area (ha)
	(c) Balance left mined-out area (ha)
	(d) Land use of this balanced left mined-out area
	(e) Final land use to be developed including water body in the last cut
	(f) Details of past subsidence and reclamation of subsided land (location of past subsided area should be enclosed)
2. Water quality management	(a) Drainage pattern of the area (pre- and post-closure) as approved in the EMP. In case of projects, having no approved EMP, the existing drainage patterns have to be followed
	(b) Water quality status of surface and groundwater
	(c) Measures for control of pollution and water balance of the area (from EMP)
	(d) Acid mine drainage source (if any), the existing practice of control and future plan
	(e) Underground water/quarry water management after closure (specifying its usage like domestic water supply, irrigation and pisciculture or stabilising the groundwater regime)
	(f) Water quality monitoring for 3 years after closure (specify the monitoring sampling station and frequency)
	(g) The sampling stations shall be composed of (1) one mine water with quarterly frequency and (2) two groundwater samples in core and buffer zone with quarterly frequency
3. Air quality management	(a) Air quality (monitored data) monitoring for next 3 years (three samples should be taken at quarterly frequency for 3 years). One sample will be at core zone and one sample each in upwind and downwind directions of the project
	(b) The air quality scenario of the core and buffer zone area
	(c) Proposed air quality management
4. Waste disposal	(a) External OB dump and internal backfilling details (specify the reclaimed backfilled area, area of voids for water reservoir and also the OB dump area height and volume) prior to closure of mine or during progressive mine closure
	(b) Stabilisation of external OB dumps and backfilled area (technical reclamation)
	(c) Topsoil/soil amendment application
	(d) Plantation on external and backfilled area, avenue and block plantation with type of plantation, that is, local/native species. Name the local species for plantation
	(e) Disposal of coal beneficiation process reject
5. Details of surface structures proposed for dismantling (brief description)	(a) Industrial/mine structures (b) Residential buildings
	(c) Service buildings (d) Telephone cables
	(e) Substations (f) Transformers
	(g) Community services (h) Water line
	(i) Water treatment plants (j) Railway siding
	(k) Effluent treatment plant (ETP) (l) Power line
6. Disposal of plants and machineries	(a) Disposal or reuse of existing HEMM, CHP, workshop and railway siding
	(b) Disposal or reuse of haulage system, ventilation, CHP, workshop, and railway siding for UG
	(c) Disposal or reuse of transmission and substation.
7. Safety and security arrangement	(a) Details of fencing around abandoned quarry indicating the length of the fencing
	(b) Mine entry sealing arrangements and subsidence management for UG mines. Sealing details and dimensions shall also be provided. The thickness of mine sealing will be 100 cm RCC (M20) with nominal reinforcement. For incline entry, the mine entry path of 5 m will be filled with debris and clay before sealing the mine
	(c) Providing one-time lighting arrangement
	(d) Slope stability arrangement for high-wall and backfilled dumps

- Views of society and expectation on closure of mine
- Vocational training for local people for sustenance

Generally, reclaimed and afforested areas are to be handed over to State Forest Department for the better maintenance and preservation of ecosystem. The forest wealth can also be utilised by local people or tribal in the form of fruits and fodders. The water reservoir in the mine voids will be utilised for pisciculture, irrigation, domestic drinking water or stabilisation of the groundwater regime. Landscaping during closure of mine will make the spot for tourist attraction.

Post-mining regeneration priorities for developmental context include

- Restoration of land surface of sufficient quality to support pre-mining land use potential.
- Restoration of the ecological function of mined land, and in the case of previously degraded land, the ecological function must be improved.
- Efficient alternative use of mine infrastructure should be encouraged where this can be economically justified; where no economic alternative uses exist, mine infrastructure must be removed and the site rehabilitated to pre-mining condition.
- Job creation through education and stimulation of economic activity.
- Development projects to enable equitable participation in post-mining economies by all members of the community, especially marginalised groups.
- Enhancement of leadership capacity within the community and local government may be required to ensure that development continues post-closure.
- Skills and literacy training for community members.

17.11 Mine Closure Activity

Generally mine operator think that their job is over once the job of decommissioning and rehabilitation of mine is completed. The mine closure phase is supposed to be closed when mine is decommissioned, facilities at site are removed, management of waste dump is completed, and site is released in ecologically sustainable state. The time required to complete the job depends on the mine site situations (Pathak 2011). The final abandonment plan and schedule of operations are prepared when no further changes to mining operations are expected. Pathak (2011) suggested closure activity should be divided into four phases which are shown in Table 17.2.

Debnath et al. (2011) suggested very briefly the mine closure activities to be taken in Table 17.3.

17.12 Closure Costs (Abandonment Cost) and Financial Sureties

Closure costs are almost as site-specific as geology, but generalisations can be used to indicate the range of possible costs. Early cost estimates are critical: accurate, timeout estimates are necessary to ensure that sufficient funds are available towards the end of the mine's life. The estimates should be updated systematically every 5 years (for a 30-year life of mine, every 2 years for a 10-year life).

To take closure practice forwards, work needs to be done on

- How to finance closure
- How to address the social issues
- Developing a body of good and bad practice case studies
- Developing a closure scorecard

As a rule of thumb, at present, the mine closure cost will cover the following activities for which a corpus escrow account @ Rs. 6.0 lakhs per ha for OCP and @ Rs.1.0 lakh per ha for UG mine of the property leasehold shall be opened with the Coal Controller Organisation. In case of mines having acid mine drainage, post-closure acid mine drainage management cost shall also be included in the total closure cost (Debnath et al. 2011). The above cost expenditure will be met from the corpus escrow account deposited by the mine operator. However, the additional amount beyond the escrow account will be provided by the mine operator after estimating

Table 17.2 Stages of mine closure activity

Closure actions	Close activities
Active care actions	Site is re-profiled and rehabilitation work is undertaken for all the barren areas, waste dumps, etc.
	All sources of potential pollution and contaminants are identified, treated and made harmless
	At sites where progressive rehabilitation work has been carried out, these activities are also continued during the active care programme
	Continue the monitoring programme particularly aftercare and development of ecosystem (tree growth, cover species, soil fertility), erosion control, etc.
Passive care actions	Monitoring of post-mining situations and evaluating effectiveness of active care actions
	May be extended to 5-years period to cessation of mining, during which site investigation, review with regulatory authorities, estimates of final costs and contract preparation and tendering are also to be undertaken
	Explore and encourage other industries to take over the mine site for appropriate land and resource utilisation
Close-out actions	Close-out actions are taken after the mining company and the regulatory agencies agree that the close-out state has been reached
	Mining company prepares the close-out report and submits to the regulating authorities
Actions for disposal of site	After the regulators approve successful closure, finalise the close-out report, the mine abandonment and decommissioning
	The activities include mine decommissioning, demolition of structure, dismantling and removal of building, plant and other infrastructure elements and arrangement of selling of items with resale values

Table 17.3 Close activities for coal mines

Dismantling of structures	Service buildings
	Residential buildings
	Industrial structures like CHP, workshop and field substation
Fencing of the area	Permanent fencing of mine void and other dangerous area random rubble masonry/mud wall fencing/barbed fencing
Grading and reshaping	Grading of high-wall slopes/levelling and grading of high-wall slopes
Reclamation/ Restoration	Dump reclamation including handling/dozing of overburden dump into mine void and bioreclamation including soil spreading, plantation and maintenance
Landscaping	Landscaping of the cleared land for improving its aesthetic
Plantation	Plantation over area obtained after dismantling
	Plantation around the fencing
	Plantation over the cleared OB dump
Monitoring and aftercare	Monitoring/testing of parameters for post-closure 3 years
	Air quality
	Water quality
Entrepreneurship development	Vocational/skill development training for sustainable income of affected people
Others	Supervision and wages for 3 years (post-closure period)

the final mine closure cost 5 years prior to mine closure (as per the mine closure guideline).

The progressive mine closure will deal the land reclamation as per the calendar plan of project report as well as other environmental management measures. The cost of progressive mine closure is already part of the project cost. Therefore, the escrow account for mine closure will deal only the final mine closure. The activities engaged for mine closure will involve considerable expenditure of capital and revenue nature which is shown in Box 17.1.

In India, there is no provision in the grant mining lease, a financial surety to indemnify the authorities against the reclamation and rehabilitation costs at the time of abandonment of mines. As a result, the fate of existing abandoned mines, which was left by the earlier mine owners due to their reckless and unscientific activities, now the government has to spend public money to close these mines (Pathak 2011). Therefore, a financial surety should come into force, if the operators or owners fail to meet their full obligations at the planned time of abandonment or in the event of premature, unplanned closure. These financial sureties should cover both technical and financial failure. A number of financial options are available for the mining company to provide the necessary financial security like

- A letter of credit
- A performance or surety bond
- A parent company guarantee
- The pledging of assets
- A trust fund

For successful mine closure, transparent and full consultation between all stakeholders are essential during the closure planning and implementation processes. The closure planning and implementation are part of an integrated mine plan and are considered as critical variables when any decisions affecting mining operations are taken. Environmental and social costs of mining operations must be internalised during the operational life of the mine to avoid high-cost mining legacies on closure.

17.13 IBM Guidelines for Mine Closure Plan (http://ibm.nic.in/mineclosuregl.htm) (2003)

1. *Introduction*

The name of the lessee, the location and extent of lease area, the type of lease area (forest, non-forest, etc.), the present land use pattern and the method of mining and mineral processing operations should be given:

1.1 *Reasons for Closure:* The reasons for closure of mining operations in relation to exhaustion of mineral, lack of demand,

---

> **Box 17.1** Major Cost Involved in *Progressive Mine Closure Plan* are (After Chaudhuri 2008)
>
> - Resettlement and rehabilitation
> - Reshaping the surface hydrology and drainage pattern
> - Re-contouring the internal and external dumps
> - Excavation, storing and spreading of topsoil
> - Biological reclamation
> - Scientific studies for soil development, slope stability, hydrogeology, etc.
> - Water supply to local communities
>
> Cost involvement in final mine closure plan:
> - Re-contouring internal and external dumps to suit final land use plan
> - Revegetation/afforestation/landscaping, etc.
> - Water distribution from the closed underground mine or residual void
> - Establishment of training facilities for mine workers and local people (part to be borne by the company)
> - Maintaining civic amenities
> - Scientific studies
> - Cost of demolition of structure, filling of voids, construction of embankment, etc.

---

uneconomic operations, natural calamity, directives from statutory organisation or court, etc., should be specified.

1.2 *Statutory Obligations:* The legal obligations, if any which the lessee is bound to implement like special conditions imposed while execution of lease deed; approval of mining plan; directives issued by the Indian Bureau of Mines; and conditions imposed by the Ministry of Environment and Forests, by the State of Central Pollution Control Board or by any other organisation describing the nature of conditions and compliance position thereof should be indicated here (the copies of relevant documents may be attached as Annexure).

1.3 *Closure Plan Preparation:* The names and addresses of the applicant and recognised qualified person who prepared the mine closure plan and the name of the existing agency should be furnished. A copy of the resolution of the Board of Directors or any other appropriate administrative authority as the case may be on the decision of closure of mine should be submitted.

2. *Mine Description*

2.1 *Geology:* Briefly describe the topography and general geology indicating rock types available and the chemical constituents of the rocks/minerals including toxic elements if any at the mine site.

2.2 *Reserves:* Indicate the mineral reserves available category-wise in the lease area estimated in the last mining plan/mining scheme approved along with the balance mineral reserves at the proposed mine closure including its quality available (for final mine closure plan only).

2.3 *Mining Method:* Describe in brief the mining method followed to win the mineral, extent of mechanisation, mining machinery deployed, production level, etc.

2.4 *Mineral Beneficiation:* Describe in brief the mineral beneficiation practice if any indicating the process description in short. Indicate discharge details of any tailings/middlings and their disposal/utilisation practice followed.

3. *Review of Implementation of Mining Plan/ Scheme of Mining Including 5-Year Progressive Closure Plan up to Final Closure of Mine*

It indicates in detail the various proposals committed with special emphasis on the proposals for protection of environment in the approved mining plan/scheme of mining including 5-year progressive closure plan up to the closure of mine vis-a-vis their status of implementation. It highlights the areas, which might have been contaminated by mining activities, and type of contaminants that might be found there. The reasons for deviation from the proposals if any with corrective measures taken should also be given.

4. *Closure Plan*

4.1 *Mined-Out Land:* Describe the proposals to be implemented for reclamation and rehabilitation of mined-out land including the manner in which the actual site of the pit will be restored for future use. The proposals should be supported with relevant plans and sections depicting the method of land restoration/reclamation/rehabilitation.

4.2 *Water Quality Management:* Describe in detail the existing surface and ground water bodies available in the lease areas and the measures to be taken for protection of the same including control of erosion; sedimentation; station; water treatment; diversion of water courses, if any; and measures for protection of contamination of ground water from leaching. Quantity and quality of surface water bodies should also be indicated, and corrective measures proposed to meet the water quality conforming the permissible limits should also be described. Report of hydrological study carried out in the area may also be submitted. The water balance chart should be given. If there is potential of acid mine drainage the treatment method should be given.

4.3 *Air Quality Management:* Describe the existing air quality status. The corrective measures to be taken for prevention of pollution of air should be described.

4.4 *Waste Management:* Describe the type, quality and quantity of overburden, mineral reject, etc., available and their disposal practice. If no utilisation of waste material is proposed, the manner in which the waste material will be stabilised should be described. The protective measures to be taken for prevention of siltation, erosion and dust generation from these waste materials should also be described. If toxic and hazardous elements are present in the waste material, the protective measures to be taken for prevention of their dispersal in the air environment, leaching in the surface and ground water, etc., should be described.

4.5 *Topsoil Management:* The topsoil available at the site and its utilisation should be described.

4.6 *Tailing Dam Management:* The steps to be taken for protection and stability of tailing dam, stabilisation of tailing material and its utilisation, periodic desilting, measures to prevent water pollution from tailings, etc., arrangement for surplus water overflow along with detail design, structural stability studies, the embankment seepage loss into the receiving environment and ground water contaminant if any should be given.

4.7 *Infrastructure:* The existing infrastructural facilities available such as roads, aerial ropeways, conveyer belts, railways, power lines, buildings and structures, water treatment plant, transport, water supply sources in the area and their future utilisation should be evaluated on case to case basis. If retained, the measures to be taken for their physical stability and maintenance should be described. If decommissioning is proposed, dismantling and disposal of building structures support facilities and other infrastructure like electric transmission line; water line; gas pipeline; water works; sewer line; telephone cables; underground tanks; transportation infrastructure like roads, rails, bridges and culverts; electrical equipments and infrastructures like electric cables and transformers to be described

in connection with restoring land for further use.

4.8 *Disposal of Mining Machinery:* The decommissioning of mining machineries and their possible post-mining utilisation, if any, to be described.

4.9 *Safety and Security:* Explain the safety measures implemented to prevent access to surface openings, excavations, etc., and arrangements proposed during the mine abandonment plan and up to the site being opened for general public should be described.

4.10 *Disaster Management and Risk Assessment:* This should deal with action plan for high-risk accidents like landslides, subsidence flood, inundation in underground mines, fire, seismic activities and tailing dam failure and emergency plan proposed for quick evacuation, ameliorative measures to be taken, etc. The capability of lessee to meet such eventualities and the assistance to be required from the local authority should also be described.

4.11 *Care and Maintenance During Temporary Discontinuance:* For every five yearly review (as given in the mining scheme), an emergency plan for the situation of temporary discontinuance or incomplete programme due to court order or due to statutory requirements or any other unforeseen circumstances should include a plan indicating measures of care, maintenance and monitoring of status of unplanned discontinued mining operations expected to reopen in near future. This should detail item-wise status monitoring and maintenance with periodicity and objective.

5. *Economic Repercussions of Closure of Mine and Manpower Retrenchments*
Manpower retrenchment, compensation to be given, socio-economic repercussions and remedial measures consequent to the closure of mines should be described, specifically stating the following:

5.1 Number of local residents employed in the mine, status of the continuation of family occupation and scope of joining the occupation back

5.2 Compensation given or to be given to the employees connecting with sustenance of himself and their family members

5.3 Satellite occupations connected to the mining industry, number of persons engaged therein and continuance of such business after mine closes.

5.4 Continued engagement of employees in the rehabilitated status of mining lease area and any other remnant activities

5.5 Envisaged repercussions on the expectation of the society around due to closure of mine

6. *Time Scheduling for Abandonment:* The details of time schedule of all abandonment operations as proposed in para 4 should be described here. The manpower and other resources required for completion of proposed job should be described. The schedule of such operations should also be supplemented by PERT (Programme Evaluation and Review Technique), bar chart, etc.

7. *Abandonment Cost:* Cost to be estimated based on the activities required for implementing the protective and rehabilitation measures including their maintenance and monitoring programme.

8. *Financial Assurance:* The financial assurance can be submitted in different forms as stated in Rule 23(F)(2) of Mineral Conservation and Development (amendment) Rules, 2003. In the mine closure plan, the manner in which financial assurance has been submitted, and its particulars have to be indicated. FOI' Model Bank Guarantee Form, please click here.

9. *Certificate:* The above mentioned actions have been taken to be stated clearly in the mine closure plan. A certificate duly signed by the lessee to the effect that said closure plan complies all statutory rules, regulations, orders made by the Central or State Government, statutory organisations, court, etc., has been taken into consideration, and wherever any specific permission is required, the lessee will approach the concerned authorities. The lessee should also give an undertaking to the effect that all the measures proposed in this closure plan will be implemented in a time bound manner as proposed.

10. *Plans, Sections, etc.:* The Chaps. 1, 2, 3 and 4 should be supported with plans and sections. The closure plan may also be submitted depicting photographs, satellite images on compact disc, etc., wherever possible.

Note

1. The mine closure plan in progressive stage will be prepared by paragraphs where subparagraphs may be added for detailed items whereas the final mine closure plan will be prepared in chapters with subchapters as necessary with adequate details.

2. The guidelines for the both the documents will be same as above.

References

Chaudhuri S (2008) Planning for closure of coal mines. In: Chaudhuri S, Singh G (eds) Environmental management in coal mining areas, EDC programmes. ISM, Dhanbad

Debnath AK, Shekhar S, Ranjan R (2011) Mine closure-World Bank approach vis-à-vis Indian context. J Mines Met Fuels 59(9):274–278

Ghose AK (2011) Mine closure- its nitty-gritty. J Mines Met Fuels 59(9):279

Heikkinen PM (ed) (2008) Mine closure handbook—environmental techniques for the extractive industries. http://arkisto.gtk.fi/ej/ej74.pdf

IBM (2003) IBM guidelines for mine closure plan, http://ibm.nic.in/mineclosuregl.htm

MCDR (1988) The mineral conservation and development rules, 1988 (as ammended upto 18.01.2000) http://mines.nic.in/mcdr.html

Pathak K (2011) Mine closure planning: concepts and concerns. J Mines Met Fuels 59(9):261–274

Robertson A, Shaw S (2004) Mine closure—closure criteria and indicators. Robertson Geo Consultants Inc. http://technology.infomine.com/enviromine/issues/ cls_criteria.html. Accessed 15 Nov 2004

Part II

Analysis of Soil, Mine soil and Vegetation

Soil Sampling Techniques

18

Contents

18.1 Introduction

For biological reclamation of coal mine derelict lands, its physical, chemical and biological properties must be known in advance. The purpose is to identify the plant growth-limiting factors. The overburden (OB) dumps should be surveyed before and after biological reclamation. This will give information on the potential of OB material, its fertility, productivity and the level of amelioration and aftercare required.

There are many limitations in testing of OB materials. Hence, different methods are employed. The first and foremost important aspect is how to do OB sampling in dumps. The information from the site survey and soil analysis should be clearly recorded for ease of interpretation and assessment of changes in the OB dumps after biological reclamation.

18.2 Objectives of Soil Testing

1. To provide an index of nutrient availability or supply in a given soil
2. To provide a basis for recommendation of different soil ameliorative programme such as amount of fertiliser and liming materials to be added
3. To select suitable plants based on soil test data
4. To assess the health status of soils (i.e. pollution level)

S.K. Maiti, *Ecorestoration of the Coalmine Degraded Lands*,
DOI 10.1007/978-81-322-0851-8_18, © Springer India 2013

18.3   Soil/Overburden Sampling

Soil *sampling* is one of the most essential and crucial steps in the soil testing process which must be performed accurately as the reliability and applicability of subsequent chemical analysis and recommendations will depend on it.

The principal objective of this step (as is true of all sampling efforts) is to collect samples that are *representative of the entire area* which has to be revegetated. If samples are collected from spots that are distributed over the entire area, subsequent chemical analysis and fertiliser and lime recommendations will also be true for the entire area.

On the other hand if only a *few samples* are collected from unrepresentative sites, the recommendations will not hold true for the total area to be revegetated, and the whole process of soil testing will end up in an expensive *exercise without desired results.*

Therefore, a few points must be clearly understood and remembered before proceeding to take samples. These are:

1. An area can be considered as a single sampling unit, only if it is appreciably uniform in all respect. This means that areas on the reclamation site that differs considerably in *elevation, colour* or *texture* must be treated as different sampling units and sampled separately.
2. Spots located on or near ponds, channels and marshy tracts and spots near the trees or well compost piles or any other location that can be considered non-representative of the general area are to be avoided.
3. *Sampling depth* is an important consideration. It is usually done to a *depth of 15 cm* from the surface. However, for reclamation work, a sampling depth of about *50 cm* has been considered.

The overburden (OB) samples are collected from different horizon depth, that is, 0–10, 10–20, 20–30, 30–40 cm, etc. Sometimes samples are collected from deeper horizon depth up to *110 cm* when deep-rooted plants are to be considered for plantation (Maiti 1995)

18.4   Steps to Be Followed During Sampling

The *aim* is to collect sufficient representative samples from the area to give a clear picture of the average and the variation in properties over the site. The following steps are followed to collect samples:

18.4.1   Selection of Sampling Locations

The area is usually inspected, and if considerable differences in colour, texture, elevation, etc., existed, the area is divided into sampling units or grids (Fig. 18.1a). This will allow the parameters to be mapped and any pattern or problem areas over the site identified. The intensity of the grid will depend on the site variability, the resources available and the importance attached to the survey results.

Sometimes transects are laid down in the area as shown in Fig. 18.1b. Each 5- or 10-m interval soil sample is taken for analysis. If there is no considerable difference, composite sample may be taken. *Precautions: A uniform portion is to be taken from surface to the desire depth and secondly same volumes of soil to be obtain from each location of same depth.*

Sampling area may be divided in uniform subareas, based on colour, texture, topography, vegetation and drainage pattern, and from each, subarea composite sample is taken (Fig. 18.1c).

If no considerable difference is noted, the whole area could be treated as a single sampling unit. Soil sample may be collected randomly, and composite is prepared by taking five to six samples (Fig. 18.1d).

18.4.2   Optimal Numbers of Samples

Optimal numbers of samples are taken from spots distributed at random covering the entire area. At each spot, surface litter is scraped and vegetation cover is removed before the collection of samples.

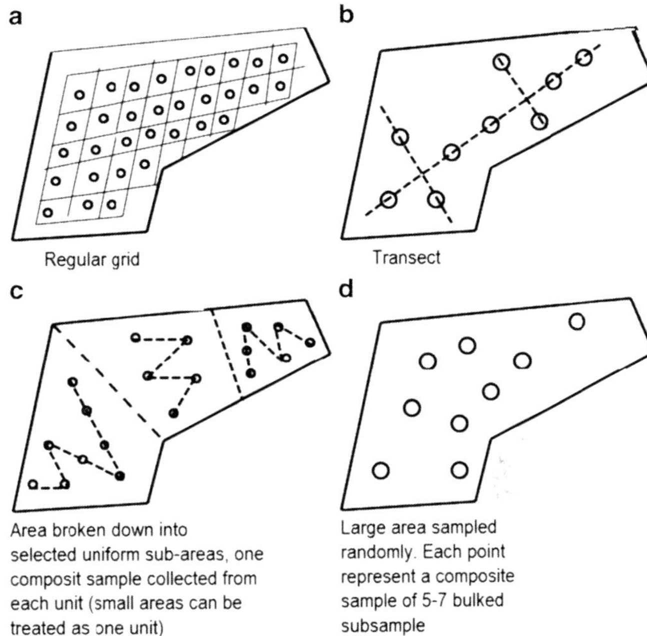

Fig. 18.1 (**a–d**) Soil sampling pattern for large area

18.4.3 Sampling Tools

Various tools are available, the simplest being a *spade* or a *pick axe* for hard stony material, which is always essential for OB sampling. Special *corers* and *augers* are available, which are suitable for less stony dumps and soil sampling (Fig. 18.2).

It is advisable to dig a pit on the dump by using pick axe and spade up to a depth of 50–100 cm (specify exactly) manually. For horizon sampling, sampling tube of different diameter can be use (steel pipe) and prominently marked in interval of 15, 30, 45 cm, etc.

18.4.4 Size of OB Samples Needed for Analysis

The size of sample needed will depend on the analysis performed and on the coarse fractions content of sample. Most analyses are performed on sieved samples (<2-mm size); large quantity of OB material (about 5 kg) should be collected.

If coarse fractions of the dump are more than 50–80%, it is advisable that a simple balance and few 2-mm sieve should be carried to the filled. In the field itself, measure the total OB material, that is, *soil fraction and coarse fraction, and note the weight.* Later, only soil fraction is transported to the laboratory.

The sample (<2-mm size) is mixed well by stirring with hands on a clean piece of cloth, paper or polythene sheet. *After coning and quartering, generally 500 g to 1 kg sample is taken for laboratory analysis.* A detail of coning and quartering method is shown in Fig. 18.3.

18.5 On-Site Tests and Description

On-site description in the site itself is very important, and proper care should be taken. The important informations to be collected in the site are

Qualitative parameters (*site description*)
1. Topography
2. Plant cover (grass, weeds, trees, etc.) appearance
3. Dump height and slope
4. Grazing land
5. Texture and stoniness (by horizontal map)

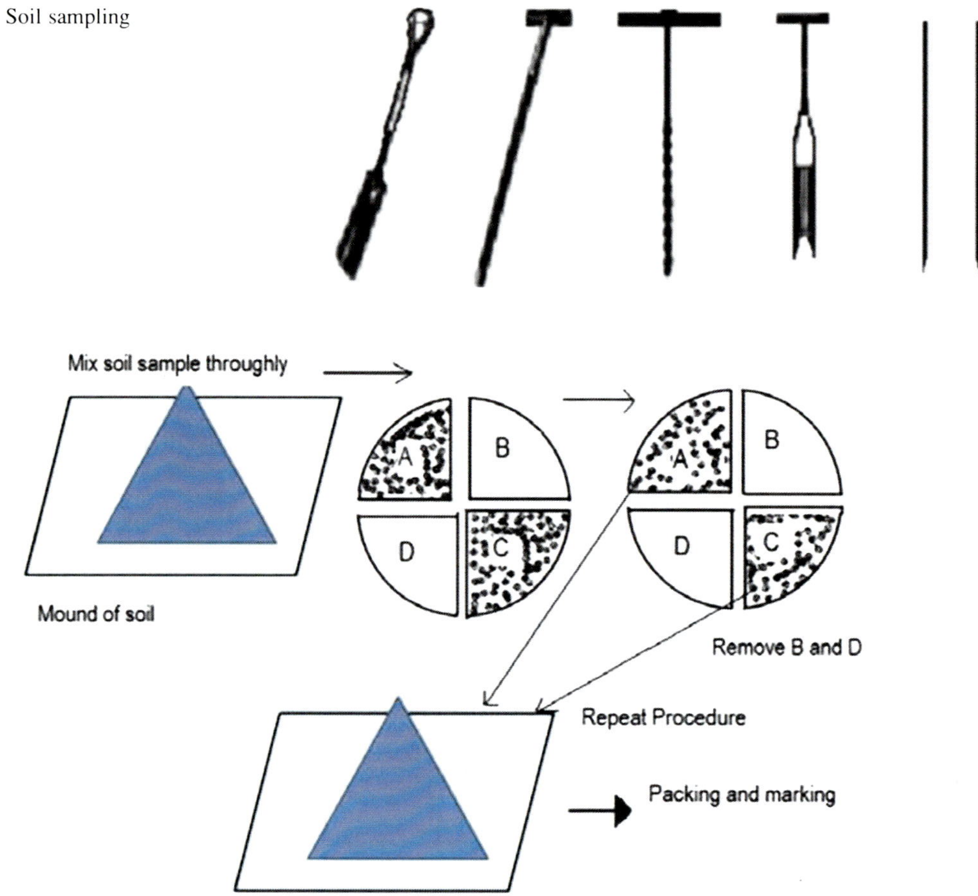

Fig. 18.2 Soil sampling tools

Fig. 18.3 Coning and quartering method (Maiti 2003)

Quantitative parameters

1. Bulk density
2. Field capacity
3. Initial moisture content should be done in the field itself, because during transportation fair amount of moisture is lost
4. Infiltration rate
5. Penetrability

18.6 Sample Preparation (For Sending Laboratory Analysis)

For biological reclamation work, it is very often required to send large number of samples in far-off laboratories for detailed analysis. The sample preparation devices mainly consists of pestle and motor, different sieves, brasses, spatula, plastic papers, etc., which are available in most of the standard soil laboratory.

The important steps to be taken for sending of the samples are as follows (Maiti 2003):

1. *Drying of samples*: As soon as possible air-dry fresh samples by laying them out in a shallow trays or plastic sheet or even newspaper or drawing sheet. Drying is done either by *sun-drying* method or *by room-drying* method.
   - Use 2–4 days (sun drying) to a maximum of 35°C
   - Up to 7 days for room-drying (care should be taken to put the sample number in the drying sheet).
   - Air-dried, sieved soil samples can be stored for subsequent analysis. For some parameters fresh samples are used, while for others, the soil sample is stored after

properly dried. The drying temperature for soil is critical, as the nature of many constituents is highly sensitive to temperature. High-temperature drying generally affects the exchangeable characteristics of clay and organic colloids.

2. *Sieving*: Weigh the air-dried soil, sieve through a 2 mm sieve, and reweigh. Record the proportion of soil <2-mm size. If some clods are formed due to clay particles, crush the clods and slowly grind the sample by mortar and pestle.

3. *Moisture content:* Weigh accurately about 10 g of the air-dried soil. *Oven-dry at 105°C for overnight, 12 h (minimum for 6 h)* and reweigh to determine the moisture content of the air-dried soil, given as

$$\text{Moisture}(\%) = \frac{\text{loss in weight (g)}}{\text{Weight of air - dried soil (g)}} \times 100$$

Any subsequent analysis results should be corrected for air-dry moisture and 2-mm gravel content.

4. *Labelling:* A durable label is placed inside and a tag outside the bag giving details of the location and depth of sampling. Soil sampling bags are numbered and tied or sealed properly.

18.7 Accuracy, Precision and Method Detection Limits (MDL)

The accuracy of an analytical measurement is how close a result comes to the true value. In other words, *Accuracy* is a measure of the agreement between an accepted true value and a measure results. For the determination of the measurement of accuracy, it usually requires calibration of the analytical method with a known standard.

Precision is the reproducibility of multiple measurements and is usually described by the *standard deviation, standard error*, or *confidence interval*.

It has nothing to do with accuracy. Precision describes the consistency of series of measurement. In other words, good precision tells how well a series of measurement cluster around the average results.

The EPA defines MDL as "the minimum concentration that can be determined with 99% confidence that the true concentration is greater than zero." This procedure is outlined in 40 CFR 136 (EPA 2003). Basically you make a solution of the analyte that is one to five times the estimated detection. Test this solution seven or more times, then determine the standard deviation of the data set. The method detection limit is calculated according to the formula:

$$\text{MDL} = \text{Student's t value} \times \text{the standard deviation} \tag{18.1}$$

Box 18.1 Method of determination of Minimum detection limit (MDL)

Example: Phosphate analysis by using spectrophotometer. As per instrument manual MDL was given as 0.045 mg/L.

(i) Prepare a solution of 0.175 mg/L of phosphate, which is approximately four times the estimated limit.

(ii) This solution is analyzed nine times over the course of several days.

(iii) Assume, the results were 0.194, 0.166, 0.174, 0.149, 0.183, 0.153, 0.144, 0.173, 0.190 mg/L.

(iv) Calculate the standard deviation of these results, which comes 0.018.

(v) Substituting the numbers into the equation (18.1)

$$\text{MDL} = \text{Student's t value} \times \text{std. deviation}$$
$$= 2.896 \times 0.018 = 0.052 \, \text{mg/L}$$

The student's t value changes with the number of data points.

18.8 Self Test

1. How to collect soil and overburden samples from the sampling site?
2. List the important test to be conducted in the field.
3. What are the prior important activities to be taken up before sending the sample for the laboratory analysis?

References

EPA (2003) 40 CFR. (7/1/2003 Ed). http://www.epa.gov/region9/qa/pdfs/40cfr136_03.pdf

Maiti SK (1995) Some experimental studies on ecological aspects of reclamation in Jharia coalfield. Ph.D. dissertation, Indian School of Mines, Dhanbad

Maiti SK (2003) Handbook of methods in environmental studies, vol 2. ABD Publications, Jaipur

Contents

S.K. Maiti, *Ecorestoration of the Coalmine Degraded Lands*,
DOI 10.1007/978-81-322-0851-8_19, © Springer India 2013

19.1 Introduction

The following parameters need to be analysed, and methodological details are as follows. Out of nine important physical parameters, four are determined in the field and rest five are in laboratory.

19.1.1 Parameters to Be Determined in the Field Itself

1. Field capacity
2. Bulk density and pore space
3. Infiltration rate
4. Compaction

19.1.2 Parameters to Be Determined in Laboratory

1. Coarse fragments
2. Texture analysis
3. Moisture content
4. Determination of wilting point
5. Water holding capacity

19.2 Coarse Fractions (>2-mm Size) (Sieving Method)

19.2.1 Introduction

The coarse fractions are measured by passing the air-dried sample through a 2-mm sieve or screen and reweighing the coarse materials. The >2-mm size stone again be divided into gravel (2–75-mm diameter), *Cobbles* (75–254-mm diameter) and *Stones* (greater than 254-mm diameter). The term 'boulders' is used where stones are larger than 600-mm diameter. The standard sieves as available in most of the soil laboratory are given in Table 19.1. Analysis of particle size distribution by dry sieving methods in an overburden dump is given in Table 19.2.

19.2.2 Calculation

% Coarse fractions

$$= \frac{\text{Weight of coarse fractions}(>2\text{mm size})}{\text{Weight of coarse fractions} + \text{dry weight of remaining soil}} \times 100$$

The name of coarse fractions is added as prefix during soil textural class reporting, like, gravely sandy loam (20–50% gravel) or very gravely loam (50–90% gravel).

Stoniness refers to the relative proportion of stones over 250-mm diameter in soil. Rockiness refers to the relative proportion of exposed bedrock over the area.

Table 19.1 Standards sieves set (mesh size and size opening in micron)

Sl. No.	Mesh size	Size of opening
—		12 in. (300 mm)
—		3 in. (75 mm)
—		2 in. (50 mm)
—		$1^{1}/_{2}$ in. (38.1 mm)
—		1 in. (25 mm)
—		$^{3}/_{4}$ in. (19 mm)
—		$^{3}/_{8}$ in. (9.5 mm)
1.	No. 3	5600 μm (5.6 mm)
2.	No.4	4750 μm (4.75 mm)
3.	No. 8	2,000 μm (2 mm)
4.	No. 16	1,000 μm (1 mm)
5.	No. 30	500 μm (0.50 mm)
6.	No. 40	425 μm (0.425 mm)
7.	No. 60	250 μm (0.250 mm)
8.	No. 72	212 μm (0.212 mm)
9.	No. 100	150 μm (0.150 mm)
10.	No. 150	100 μm (0.10 mm)
11.	No. 200	75 μm (0.075 mm)
12.	No. 240	62 μm (0.062 mm)
13.	No. 300	50 μm (0.050 mm)
14.	No. 325	45 μm (0.045 mm)
15.	No. 400	38 μm (0.038 mm)

Table 19.2 Particle size distribution (%) of levelled mined-out derelict sites (Maiti et al. 2002)

Profile depth (cm)	>5.6 mm	>2 mm	>1 mm	>0.5 mm	>0.212 mm	>0.15 mm	<0.15 mm
0–15 cm	51.23	15.33	3.71	5.77	16.16	5.62	1.98
15–30 cm	45.88	22.55	4.55	6.71	16.43	2.8	6.67
30–50 cm	40.56	17.4	4.2	7.42	23.33	2.4	4.69

ASTM (1998) Gives the Following Definitions for Various Types of Soil

For particles retained on a 3-in. (75-mm) US standard sieve, the following definitions are suggested:

- *Cobbles*: particles of rock that will pass a 12-in. (300-mm) square opening and retained on a 3-in. (75-mm) sieve.
- *Boulders*: particles of rock that will not pass a 12-in. (300-mm) square opening.

Gravel: particles of rock that will pass a 3-in. (75-mm) sieve and be retained on a No.4 (4.75-mm) sieve. *Coarse gravel*: passes a 3-in. (75-mm) sieve and is retained on a $\frac{3}{4}$-in. (19-mm) sieve. *Fine gravel*: passes a $\frac{3}{4}$-in. (19-mm) sieve and is retained on a No.4 (4.75-mm) sieve.

Sand: particles of rock that will pass a No.4 (4.75-mm) sieve and be retained on a No. 200 (75-µm) sieve with the following subdivisions:

- *Coarse sand*: passes a No. 4 (4.75-mm) sieve and is retained on a No. 10 (2-mm) sieve.
- *Medium sand*: passes a No. 10 (2-mm) sieve and is retained on a No. 40 (425-µm) sieve.
- *Fine sand:* passes a No. 40 (425-µm) sieve and is retained on a No. 200 (75-µm) sieve.

19.2.3 Comments

In coalmine overburden (OB) dumps, proportion of coarse fractions sometimes found to be in the order of 60–80% in Rajapur OB (Jharia coalfields); 49–67% in Block-II, of BCCL mines, etc. (Maiti 1995). In Ramagundam OB dumps of SCCL mines (Andra Pradesh), stone content was found to be low in the order of 20–40%. In a good soil, coarse fractions (>2-mm fraction) are found only 2–5% (garden soil, Maiti 1995). The coarse fractions content more than 50% is rated as poor quality of OB materials, which will limit the plant growth (Hu et al. 1992).

19.3 Texture Analysis (*USDA Method-International Pipette Method*)

The relative proportion of sand, silt and clay in a particular soil is determined by texture analysis. The sand fraction can also be broken into coarse, medium and fine sand. The sizes of particles of each fraction are given in Tables 19.3 and 19.4. The descriptive soil texture can be determined textural triangle (Fig. 19.1).

19.3.1 Principle

As the particles do not all exist as separate ones, the accuracy of estimation depends entirely on the complete dispersion of the sample prior to fractionation. For this purpose, all particles are first separated from each other by oxidising the organic matter and by using a dispersing agent, and the amount of each size group is determined by withdrawing an aliquot of the suspension after a pre-calculated time, based on Stokes' equation.

Table 19.3 The soil separates and their diameter range [USDA (United States Department of Agriculture) system]

Soil separate name	Diameter range (mm)	Visual size comparison of maximum size
1. Very coarse sand	2.0–1.0	House key thickness
2. Coarse sand	1.0–0.5	Small pin head
3. Medium sand	0.5–0.25	Sugar or salt crystals
4. Fine sand	0.25–0.10	Thickness of book page
5. Very fine sand	0.10–0.05	Invisible to the eye
6. Silt	0.05–0.002	Visible under microscope
7. Clay	Less than 0.002	Most are not visible even under a microscope

Table 19.4 Soil separates as per ISSS system (Intentional Society of Soil Science)

Soil separate name	Diameter range mm (µm)
1. Coarse sand	2.0–0.2 (2,000–200)
2. Fine sand	0.2–0.02 (200–20)
3. Silt	0.02–0.002 (20–2)
4. Clay	Less than 0.002 (<2)

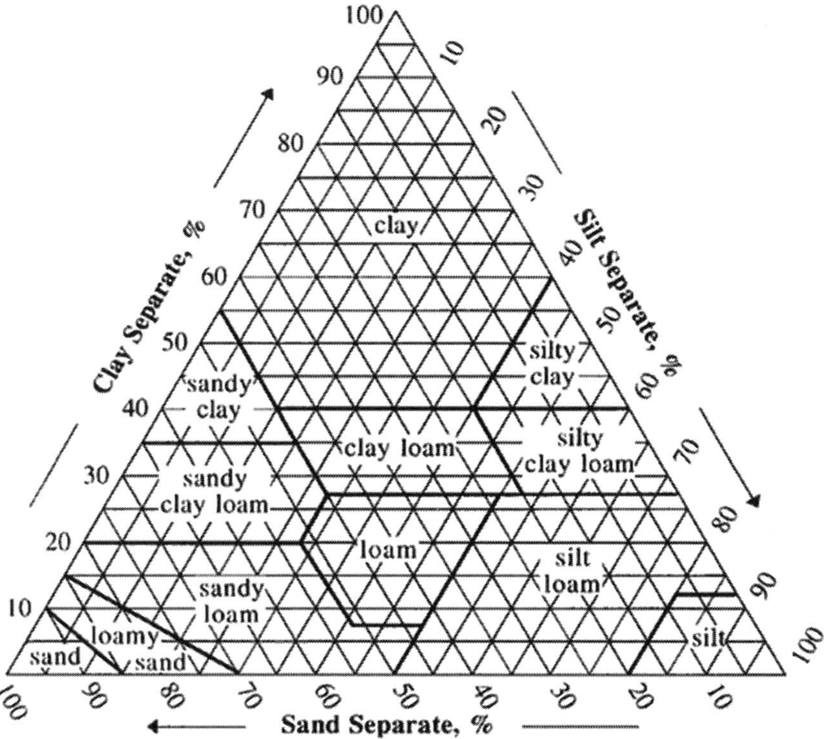

Fig. 19.1 Soil texture triangle based on the US particle size

The application of Stokes' law to the process of sedimentation is based on the following simplifying assumptions:

1. The condition of viscous flow in a still liquid is maintained.
2. There is no turbulence, that is, the concentration of particles is such that they do not interfere with one another.
3. The temperature of the liquid remains constant.
4. Particles are small spheres.
5. Their terminal velocity is small.
6. All particles have the same density.
7. A uniform distribution of particles of all sizes is formed within the liquid.

The following soil fraction can be determined by this method:

- Fine sand: 20–2,000 µm
- Silt: 20–2 µm
- Clay: less than 2 µm

Terminal velocities for approximate equivalent diameter particles are shown in Table 19.5.

Table 19.5 Terminal velocities of particles in suspension

Particles	Approximate diameter (µm)	Terminal velocity (mm/s)
Coarse silt	35	1
Medium silt	12	0.1
Fine silt	3.5	0.01
Clay	1.2	0.001

19.3.2 Materials and Equipments

1. Sieve—2.0 mm
2. Measuring cylinder (1,000 mL) with rubber stoppers and graduated length of 36 + 2 cm from the bottom on the inside
3. Buchner funnel, 7.5 cm in diameter
4. Centrifuge
5. Fine-grained filter paper (Whatman no. 1)
6. Suction pump
7. Beaker, 600 mL with covers
8. Weighing bottle or 100-mL beaker
9. Hot air oven and desiccator
10. Rubber policeman

11. Mechanical analysis pipette, 10 mL fitted on an adjustable stand
12. Hot plate
13. Stopwatch

Table 19.6 Quantity of materials for sedimentation tests

Materials	Pipette test (g)	Hydrometer test (g)
Sandy soils	30	100
Clayey soils	12	50

19.3.3 Reagents

1. Hydrogen peroxide, H_2O_2—30%, used for decomposing the soil organic matter
2. Glacial acetic acid
3. Hydrochloric acid, HCl, 0.1 N
4. Sodium hydroxide, NaOH, 0.1 N
5. Sodium hexametaphosphate $(NaPO_4)_6$ dispersing agent; 0.1 N or 5% solution. (Trade name is *Calgon*.) Used for dispersion of soil aggregates
6. Phenolphthalein indicator solution

19.3.4 Procedure

Pretreatment of Soil Sample

Pretreatment of soil sample is done to obtain and maintain maximum dispersion of soil particles during mechanical analysis. The following pretreatments are usually done:

Removal of organic matter: Organic matter is commonly decomposed by means of H_2O_2. Commercially 30% H_2O_2 is available in the market. Generally 6% H_2O_2 is used (Dilute 20 mL of 30% H_2O_2 to 100 mL gives 6% H_2O_2).

Removal of carbonates $(CaCO_3)$: $CaCO_3$ is decomposed by dilute HCl.

Removal of soluble salts and carbonates: Jackson (1973) recommended the use of sodium acetate.

Stage I: Selection of sample

The approximate size of sample required is indicated in Table 19.6.

Stage II: Removal of organic matter

1. Take 20 g of air-dried soil (<2-mm size) in a 600-mL beaker and add 5–6 mL (approx.) of 30% H_2O_2 and cover with watch glass (just to wet the sample).
2. After reaction is slowed down, more (4–5 mL) H_2O_2 is added and allowed sample to digest on hot plate.
3. If the organic matter has not yet been destroyed, further 20 mL of 30% H_2O_2 is added and the suspension gently boiled until reaction subsides. After all the organic matter is destroyed, the suspension is further heated for about 30 min to remove the remaining H_2O_2. If appropriate, HCl treatment is carried out next (stage III).

Stage III: Removal of Carbonates (Acid Treatment)

4. Next, add 100-mL *water, mix, and add* 2(N) *HCl* until the pH falls between 3.5 and 4.0, which can be tested by blue litmus paper. Allow the soil and acid to react for 10 min to destroy $CaCO_3$ present in soil. After acid treatment, the soil is extracted by (a) filtration with Buchner funnel or (b) by centrifuging.
5. The organic matter-free soil is filtered with Buchner funnel through a fine-grained filter paper. Wash the soil with successive *portions of distilled water* till the filtrate is neutral to litmus. Now this soil sample became hydrogen saturated and was reweighted (A).
6. *Or recover the soil by centrifuging.* Transfer the soil suspension to a centrifuge tube using a fine jet of distilled water from a washbottle, taking care not to lose any soil particles. Place the tube in a centrifuge tube. Run the centrifuge for 15 min at about 2,000 rpm. Remove from the tube and decant the clear supernatant liquid. Centrifuge 2–3 times until the supernatant becomes neutral to litmus. Carefully remove the soil and put in oven to dry overnight. Next morning, cool

Table 19.7 The times of sedimentation for the clay and silt fractions (depth of sedimentation = 10 cm)

Temperature, °C	Settling time, clay fraction (<2 μm)		Settling time, silt fraction (2–20 μm)	
	Hours	Minutes	Hours	Minutes
20	8	0	4	48
21	7	50	4	40
22	7	40	4	30
23	7	25	4	30
24	7	15	4	20
25	7	5	4	15
26	6	55	4	10
27	6	45	4	5
28	6	40	4	0
29	6	30	3	55
30	6	20	3	50
31	6	15	3	45
32	6	5	3	45
33	5	55	3	35

Table 19.8 Sedimentation times of soil particles settling through a depth of 5 cm (particle density 2.65 g/cc)

Temperature, °C	Particle diameter/settling time						
	2 μm		5 μm		20 μm		50 μm
	Time		Time		Time		Time
	h	min	min	s	min	s	s
20	3	50	37	30	2	20	22
25	3	30	33	20	2	5	20
30	3	3	29	10	1	50	17
35	2	28	25	50	1	40	15

the sample in a desiccator and weigh accurately. Calculate the dry mass of prepared soil (A). Also calculate the pretreatment loss, if required.

(A check should be made for the reaction with HCl by dropping a few drops of 1 N HCl on a small portion of the sample. If there is no effervescence, acid treatment is not required. A visible reaction indicates the presence of calcareous compounds, which could act as a cementing agent, preventing separation of individual grains.)

Stage IV: Dispersion of Soil

7. The hydrogen-saturated soil is transferred to a 2,000-mL glass bottle, and 25 mL of sodium hexametaphosphate is added and stirred for an about 2 h.
8. Next, the suspension is transferred to a 1,000-mL measuring cylinder and make up the volume up to the mark with distilled water or demineralised water. The measuring cylinder is then placed into a constant temperature water bath at 25°C, which may not be always possible. Better note the temperature of the suspension, and from the Table 19.6, note the time lapse required to collect the sample either for 10-cm depth or 5-cm depth.

Stage V: Separation of Silt + Clay

9. The special mechanical analysis pipette (10 mL) is used to withdraw 10-mL aliquot of the suspension from 10-cm depth after the desired lapse of time. *For example, at 25°C, 4 h 15 min is the elapsed time for silt and clay* (B) *and 7 h 5 min for clay only* (C). The pipette is lowered to 10-cm depth 5 s before the correct time (Tables 19.7 and 19.8)
10. The aliquot is taken in 100-mL beaker and evaporated to dryness at 105°C and reweighted.

11. Soil suspension is slowly decanted, and bottom sediment (*i.e. sand*) is collected, dried and reweighted (D).

For convenience of calculation of particle size distribution analysis, particle sizes of 50, 5 and 2 µm are taken for average sizes of sand, silt and clay particles, respectively.

19.3.5 Calculation

10-mL aliquot was in 1/100 of 1,000-mL total suspension:

(a) % Silit + Clay =
$$\dfrac{\text{Weight of } 20\mu m \text{ sample(B)} \times 100}{\text{Weight of H} - \text{saturated soil(A)}} \times 100$$

(b) % Clay =
$$\dfrac{\text{Weight of } 2\mu m \text{ sample(C)} \times 100}{\text{Weight of H} - \text{saturated soil(A)}} \times 100$$

(c) % of silt = [(a) − (b)]

(d) % of Sand =
$$\dfrac{\text{Weight of sample(D)}}{\text{Weight of H} - \text{saturated soil(A)}} \times 100$$

19.3.6 Comments

The application of Stokes' law requires that (1) all soil particles are completely dispersed in a uniform manner at the start of the sedimentation and (2) concentration of the suspension is sufficiently distributed so that the free fall of the same is not hindered by collision with other particles.

Soil texture determines water intake rate (absorption), water storage in the soil, the amount of aeration, which is vital for root growth, and influence the soil fertility.

Water infiltration is more rapid in coarse-textured soil as compared with fine-textured soil.

Example 19.1 Calculation of losses

Due to various losses in solution such as acidification, washing and sieving, the sum of sand, silt and clay fractions calculated will not be 100. Before reporting, convert the various fractions on the basis of 100 as follows. Suppose in a soil sample, the various fractions calculated as follows:

(continued)

Example 19.1 (continued)

Sand = 50%
Silt 30%
Clay = 15%
Total = 95%
Loss = 5%
Then in sand, loss is 5 × 50/100 = 2.5%
Silt = 5 × 30/100 = 1.5%
Clay = 5 × 15/100 = 0.75%
Total 4.75%

Differences (5.4.75) = 0.25%
This 0.25% will be added to the fraction, which is present in the greatest amount, that is, in sand in this case. Hence, the various fractions to be reported will be

$$\text{Sand} = 50 + 2.5 + 0.25 = 52.75\%$$
$$\text{Silt} = 30 + 1.5 \qquad = 31.50\%$$
$$\text{Clay} = 15 + 0.75 \qquad = 15.75\%$$

$$\text{Total} = \qquad\qquad = 100\%$$

Example 19.2 Determination of the texture class

Determine the textural name of the whole soil (coarse fragments + fine earth <2-mm diameter, with the following constituents):
Coarse fragments (2–50-mm diameter) = 500 g
Sand content (2–0.05-mm diameter) = 350 g
Silt content (0.05–0.02-mm diameter) = 100 g
Clay content (<0.002-mm diameter) = 50 g

Solution
Textural names consider only the <2-mm portion. The coarse fraction name is given if over 20% of the soil weight is coarse material.
Coarse fraction = (500 g/1,000 g) × 100 = 50%, and size is gravel.
Thus, term *gravel* is added to the textural class name as determined below.

1. Percentages of sand, silt and clay are based only on the <2-mm fraction.
   Portion of sand = (350/500) × 100 = 70%
   Portion of silt = (100/500) × 100 = 20%

(continued)

Example 19.2 (continued)

Portion of clay $= (50/500) \times 100 = 10\%$

2. From the textural diagram (Fig. 19.1) having 70% sand, 20% silt and 10% clay, the textural is *sandy loam.*
3. The correct textural name is *gravely sandy loam* (Ans).

19.4 Bulk Density and Pore Pace

19.4.1 Bulk Density

Soil under field conditions exists as a three-phase system viz. solid (soil particles), liquid (water) and gas (mostly air). The soil matter contained in a unit volume of the soil sample is called its bulk density. Bulk density of soil is quite variable. It depends on the texture, structures and organic matter status of the soils. High organic matter content lowers the bulk density, whereas compaction increases the bulk density. To determine the bulk density of the soil sample, the oven dry weight (heated at $105°C$ for 24 h) of a known volume of soil sample is determined, and the mass per unit volume is calculated.

19.4.1.1 Equipment
1. Oven
2. Soil sampling tube, etc.

19.4.1.2 Procedure (Core Method for Undisturbed Soil)
The bulk density of soil was measured by taking an undisturbed block of soil (soil core). The soil is dried at $105°C$ for 12 h and weights it. The exact volume of soil was determined by measuring the cylinder volume.

19.4.1.3 Calculation

$$\text{Bulk density} = \frac{\text{Weight of oven} - \text{dried soil(g)}}{\text{Volume of soil core(cm}^3)}$$

(Volume of soil core $= 3.14.r^2.h$; $r =$ inside radius of cylinder, cm)

Comments

For good plant growth, bulk density should be around 1.4 g/cm^3 for clay soils and 1.6 g/cm^3 for sandy soil. Bulk density greater than 1.8 g/cm^3 is considered as extremely bad for plant growth. In coal mine spoil dumps, bulk density is generally high in comparison to natural soil in the range of 1.7–2.1 (Maiti 1995). For example, in forest soil, bulk density is generally found 1.13–1.20 g/cm^3, grass land 1.20–1.28 g/cm^3 and cultivated land 1.35–1.48 g/cm^3 (Maiti 1995).

High bulk density favours the growth of weeds because most of weeds have superficial root system and restricted to 10-cm depth. Higher bulk density also reduces soil infiltration and enhances surface run-off thus increases soil erosion.

Example 19.3 Calculation of Bulk Density

A metal cylinder pushed into a loam soil is removed from the field, and the soil it contains is dried in an oven. The measured data are shown below:

Cylinder height $= 5.0$ cm

Cylinder inside diameter $= 4.4$ cm

Oven-dried soil weight $= 87.6$ g

Solution

The volume of the soil sample equals the volume of the cylinder. Volume of the cylinder $= 3.14 \, (4.4/2)^2 \times 5 = 76.0$ cm^3

Bulk density $=$ Soil mass/ soil volume
$= 87.6 \,\text{g}/76 \,\text{cm}^3 = 1.15 \,\text{g/cm}^3$

19.4.2 Pore Space

The pore space of a soil is that portion of the soil volume occupied by air and water.

% of pore space + % of solid space $= 100\%$

Hence, % of pore space (void space) $= 100\%$ - % of solid space

$$= 100\% - \frac{\text{Bulk density, g/cm}^3}{\text{Particle density, g/cm}^3} \times 100$$

This soil would be a clay loam or other clayey soil. A sandy soil would have a bulk density closer to 1.4–1.5 g/cm^3 and a pore space percentage close to 45–50%.

Example 19.4 Calculation of Pore Space Percentage

A soil was taken for determination of bulk density, and measurements were as follows:
Cylinder volume $= 73.6$ cm^3
Dry soil weight $= 87.8$ g
Standard particle density $= 2.65$ g/cm^3
Calculate the percentage pore space.

Solution

Bulk density $=$ Soil mass/soil volume
$\qquad = 87.8$ g/73.6 cm$^3 = 1.19$ g/cm^3
% pore space $= 100\% - [(1.19/2.65) \times 100]$
$\qquad = 100 - 44.9\% = 55.1\%$

19.5 Moisture Contents (*Gravimetric Method*)

When moist soil is heated at 105°C for about 24 h, only the water, which had been absorbed or held within the soil pores, evaporated. There is no loss of water of crystallisation, and there is not much oxidation of organic matter at this temperature.

19.5.1 Equipment

1. Drying oven
2. Dedicator container, containing anhydrous silica gel
3. Top-pan balance reading to 0.01 g
4. Container or moisture cans (numbered) (tin or aluminium moisture cans with lid)

19.5.2 Procedure

1. *Weighing container*: Weigh the empty moisture can nearest to 0.01 g.
2. *Selection of sample*: The test sample must be selected so that it is properly representative of the soil sample from which it is taken. For the measurement of natural moisture content, the approximate mass of specimen is about 100 g [for homogeneous clay and silts—30 g; medium-grained soils—300 g].
3. *Wet weighing*: Put soil sample immediately in the moisture can and close it to prevent loss of moisture by evaporation. Bring the can containing the moist soil to the laboratory and weigh immediately [if weighing is likely to be delayed, the lid must be fitted tightly and the containers left in a cool place].
4. *Oven drying*: Remove the lid from the container and place moisture can to an oven at 105–110°C. This takes approximately 24 h. Drying in the oven should be continued until the specimen has dried out to reach a constant mass.
5. *Cooling in desiccators*: Allow the sample to cool for some time in the oven. Then close the cans and put them in desiccators for further cooling.
6. *Drying weighing*: When cool, replace the lids on the container and weigh.

Observations
Weight of the empty moisture can $= $ m1
Weight of moisture can + moisture soil $=$ m2
Weight of moisture can + oven dry soil $=$ m3

19.5.3 Calculation

Moisture Content (%) $=$
$$\frac{\text{Loss of moisture } (m2 - m3), \text{g}}{\text{Weight of oven} - \text{dried sample } (m3 - m1), \text{g}} \times 100$$

Report the moisture in percentage, dry mass.

Example 19.5 Determination of moisture content

A laboratory test was conducted according to the procedure described previously. The following data were obtained:
Mass of container $= 60.02$ g
Mass of container and moisture soil $= 240.25$ g
Mass of container and oven-dried soil $= 216.25$ g

Solution

(continued)

Example 19.5 (continued)

Loss of moisture $= 240.25\,g - 216.25\,g = 24.00\,g$

Weight of oven dried sample $= 216.25\,g - 60.02\,g$
$$= 156.23\,g$$

$$\text{Moisture content} = \frac{24}{156.23} \times 100$$
$$= 15.362\% \,(\text{Ans.})$$

19.6 Measurement of Field capacity (*Field Method*)

19.6.1 Principle

The field capacity is defined as 'the amount of water held in the soil after the excess gravitational water has drained away and after the rate of downward movement of water has materially decreased'. Such a stage is generally reached 24–48 h after saturation.

Available water capacity or available water content (AWC) is the range of available water that can be stored in soil and available for growing plants. It is assumed that the water readily available to plants is the difference between water content at field capacity and permanent wilting point.

Plant available water = water at field capacity—water at permanent wilting point.

19.6.2 Equipments and Materials

1. Polythene sheet or straw mulch
2. Spade
3. Soil auger
4. Moisture cans
5. Balance
6. Drying oven
7. Dedicator
8. Water, etc.

19.6.3 Procedure

1. Select a uniform plot measuring 1 m × 1 m.
2. Remove weeds, pebbles, etc. if any.
3. Bind it from all side so that water can retain inside the plot.
4. Fill sufficient water in the plot to completely saturate the soil to the desired depth.
5. Cover the plot area with a thick straw mulch or polythene sheet to check evaporation.
6. Take soil sample from the centre of the plot from the desired layer starting after 24 h or saturation, and determine moisture content daily till the values of two successive days are nearly equal.
7. Plot the daily readings on graph paper. The lowest reading is taken as the value of field capacity of the soil.

19.6.4 Observation

Weight of the empty moisture can = m1.
Weight of moisture can + moisture soil = m2.
Weight of moisture can + oven dry soil = m3.
Repeat above on next day and so on until a constant value is reached.

19.6.5 Calculation

Percentage moisture in soil
$$= \frac{\text{Loss of moisture } (m2 - m3),\, g \times 100}{\text{Weight of oven} - \text{dried sample } (m3 - m1),\, g}$$
$$\times 100 \,(\text{e.g a1})$$

Percentage moisture in soil is calculated on succeeding days (a2) (a3)... and so on. The lowest reading is taken as the value of field capacity of the soil.

19.7 Water Holding Capacity (WHC)

Water holding capacity of soil usually refers to the amount of maximum water which can be held in the saturated soil. It is generally measured as the amount of water taken up by unit weight of dry soil when immersed in water under standardised conditions.

The field capacity of a soil is defined as the amount of water held in soil after the excess of gravitational water has drained away under free drainage and minimum evaporation.

19.7.1 Equipment

1. Oven
2. Perforated circular soil boxes 'keen-box' (5.6-cm diameter, 1.6-cm height, bottom perforated with holes of 0.75-mm diameter)
3. Filter paper (Whatman no. 42)
4. Petridish and balance

19.7.2 Procedure

1. Place a filter paper in the keen-box of an appropriate dimension so as to cover the whole perforated bottom of the box. Take the weight of the box plus filter paper (W1).
2. Now, transfer the crushed sample dried in an oven at 105°C. The dried sample is placed inside the perforated bottom of the circular soil box and weighted (W2).
3. The box was placed in a petridish of 10-cm diameter containing water and kept for overnight, so that water enters the box and saturated the soil. In the next day, the soil box is taken out from the water, whipped and recorded the weight (W3).

19.7.3 Calculation

Weight of box $+$ filter paper $=$ W$_1$

Weight of Box $+$ filter paper $+$ soil $=$ W$_2$

Weight of the soil $=$ (W$_2$ $-$ W$_1$)

Weight of the water absorbed $=$ (W$_3$)

$$\text{WHC}(\%) = \frac{W_3 - W_2}{W_2 - W_1} \times 100$$

Example 19.6 Calculation of Water holding capacity

Calculate the water holding capacity of a soil for the following data:
1. Weight of box + filter paper = 20 g
2. Weight of box + filter paper + soil = 30 g
3. Weight of the soil = (30–20) g = 10 g
4. Weight of keen-box after overnight water absorbed = 35 g

Solution

$$\textbf{WHC } (\%) = \frac{(35 - 30)g}{10g} = \frac{5g}{10g} x100 = 50 \,(\text{Ans.})$$

19.8   Wilting Point (*By Plant Method*)

It is used to determine the lower limit of available water. The turgidity of leaves depends upon the amount of moisture in them, which in turn depends upon the amount of moisture in the soil. As the amount of water decreases in the soil, the turgidity of plant leaves also decreases, and finally, they may start wilting. The amount of moisture in the soil, when the plants wilt (*plants do not recover even after addition of water*), is the permanent wilting point. The determination of wilting point water is based on the method suggested by Dakshinamurthi and Gupta (1968).

19.8.1 Materials and Equipments

1. *Metal containers*, 3 in. in diameter and 6 in. in height, having a tight-fitting lid with cotton-plugged hole in the centre
2. Dwarf sunflower seeds
3. Oven

19.8.2 Procedure

1. Take about 600 g of air-dried, <2-mm soil. Make a moisture content determination on a separate sample so that the oven-dried weight of soil is known.

2. Add nitrogen and other nutrients as needed to produce good plant growth.
3. Plant several seeds of draft sunflower (*Helianthus annuus*).
4. Moisten the soil to a water content suitable for plant growth and reweight the container,
5. After the seeds have germinated, cut off the weak plants and leave the few best-grown saplings (healthy sapling).
6. After the third pair of leaves has developed, irrigate the soil with enough water to bring the water content to the moist starting value.
7. When the lowest one or two pairs of leaves wilt, transfer the container and plant to a dark, humid chamber. If the leaves again show turgidity during an overnight period (14–19 h), return the experimental sample to the growth locations. *Repeat the process of wilting and humidification until the lowest pair of true leaves fails to recover in the humid chamber.*
8. When the stage of wilting has been reached, cut off the plant at soil level and discard the plant. Then determine the moisture content of the soil by oven-drying method (at 105°C for overnight).

19.8.3 Calculation

$$\text{Wilting point}(\%) = \frac{\text{Moisture at wilting point}}{\text{Initial moisture}} \times 100$$

19.9 Infiltration Rate (Double ring Infiltrometer Method)

Water infiltration is the process of water entry into soils though the surface. Infiltration or water intake rate of pounded water is measured in the field with the help of ring infiltrometer. Infiltration rate is the distance travelled by water through a soil column and is usually expressed as cm/h. It is initially high, but decreases with time and tends to approach a steady infiltration rate. Results of infiltration measurements can be presented by plotting infiltration rate as well as cumulative infiltration as a function of time.

The rate at which water can soak into the soil depends particularly on porosity. The other factors that control the rate of water movement into soil include (Donahue et al. 1990)

- *Percentage of sand, silt and clay*—Coarse sands permit rapid infiltration.
- *Soil structure.*
- *Organic matter*—The greater the organic matter, the higher the infiltration.
- The *depth of soil* hardpan, bedrock or other impervious layer.
- *Moisture*—amount of water in the soil.
- *Temperature*—warm soil takes more water.
- *Compaction*—which usually reduces pore space and slows infiltration.

Four infiltration rates have been classified by the National Co-operative Soil survey (Donahue et al. 1990) as shown in Table 19.9 and for mine soil as shown in Table 19.10.

19.9.1 Principle

Infiltration into soil is commonly measured by the cylinder (double-ring or single-ring) infiltrometer. The cylinder infiltrometer is the most common method used because it is relatively inexpensive and simple. It consists of two rings, one 50 cm *in diameter (outer ring)* and another 30 cm *in diameter (inner ring),* installed in the surface soil. Water flow assumes to be one-dimensional (only in the vertical direction), though flows in reality are three-dimensional. User often 'buffers' the effect of horizontal

Table 19.9 Classification of infiltration rate

Class	Rate of infiltration (cm/h)	Remarks
1. Very slow	Less than 0.25	Soil in this group are very high in percentage of clay
2. Low	0.25–1.25	Most of these soils are shallow, high in clay or low in organic matter
3. Medium	1.25–2.50	Soils in this group are loams and silts
4. High	Greater than 2.50	These are deep sands, deep and well-aggregated silt loams and some tropical soils with porosity

infiltration by using a double ring (Fig. 19.2). The larger ring is placed around the inner ring, and the same water level is maintained in the inside and outside rings.

19.9.2 Equipments

1. Two metal rings, 20 cm in length with one edge sharp, one 30 cm in diameter and another 50 cm in diameter
2. Stopwatch
3. Hammer (12 kg)

Table 19.10 Categorisation of infiltration rate for mine spoils (Hu et al. 1992)

Category	Rate of infiltration (cm/h)
1. Very slow	Less than 0.127
2. Slow	0.127–0.508
3. Moderate slow	0.508–2.000
4. Moderate	2.000–6.350
5. Moderately rapid	6.350–12.700
6. Rapid	12.700–25.400
7. Very rapid	Greater than 25.4

4. Driving plate; steel plate ½-in. thick, 60 cm in diameter with lugs to keep the plant on the rings

19.9.3 Procedure

1. The 30- and 50-cm rings are driven in the ground by hammering on the driving plate placed on the rings, until 10 cm of each ring goes into the soil.
2. The constant head water supply is maintained in the inner ring (head of 5 cm on the soil surface). The 5-cm water head is also maintained in the outer ring.
3. Then water is added to the inner ring to a depth of 5 cm and maintained a constant head of 5 cm from the soil surface.
4. The amount of water going into the soil from the inner ring at the end of 5, 10, 20, 30, 45, 60, 90, 120, 160 and 180 min is recorded. A typical infiltration test data sheet is presented in Table 19.11. The cumulative infiltration is also calculated as shown in Table 19.12.

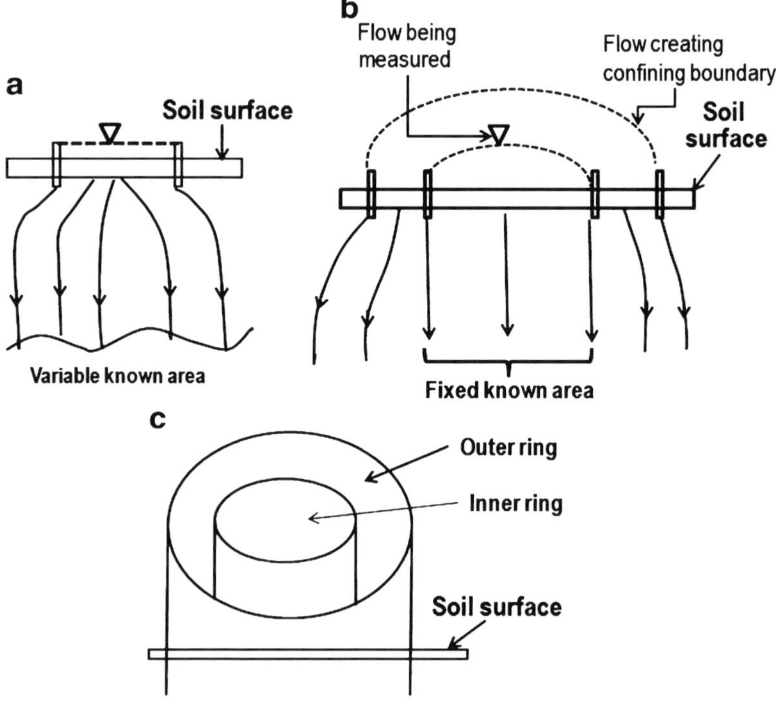

Fig. 19.2 (a) Divergence of stream lines during infiltration, (b) divergence of flow buffered by using a double-ring infiltrometer, (c) a typical double-ring infiltrometer

Table 19.11   Presentation of infiltration test data (Maiti 1995)

Elapsed time (min)	Volume of water intake (cc)	Infiltration amount (cm)	Cumulative infiltration amount (cm)	Infiltration rate (cm/h)	Cumulative infiltration rate (cm/h)
05	530	0.750	0.750	8.99	8.99
15	280	0.396	1.146	2.37	11.36
40	285	0.403	1.549	0.96	12.32
70	200	0.283	1.832	0.56	12.88
100	170	0.240	2.072	0.48	13.36
120	110	0.156	2.228	0.31	13.67
150	100	0.141	2.369	0.28	13.95
180	95	0.134	2.503	0.27	14.21

Moisture content = 3.7%

Table 19.12   Calculation of hourly infiltration rate and amount (Maiti 1995)

Time (h)	Infiltration rate (cm/h)	Infiltration amount (cm)
1	12.88	1.832
2	0.79	0.396
3	0.54	0.275

5. Once the amount of water going into the soil per unit time becomes constant, the recording of data is stopped.
6. The temperature of the water is also recorded. The initial soil moisture content is determined from a nearby area.

19.9.4 Observation

Observation sheet for collection of infiltration test data is given in Table 19.13.

19.9.5 Calculation

Infiltration rate

$$= \frac{\text{Vol.of water entering the inner ring(mL)}}{\text{Cross sectional area of the inner ring} \times \text{time}}$$

The cumulative water intake versus time is plotted and the slope of which will give rate of infiltration.

19.10 Test on Analysis of Soil Physical Parameters

1. List soil physical parameters that are analysed in the field and laboratory.
2. What size fraction of particle is considered as soil?
3. What is the size of gravel, cobbles and stones? How does it affect the soil quality?
4. Define soil texture. What is dispersing agents and why is it used in soil texture analysis?
5. What is H-saturated soil?
6. What is the principle of texture analysis?
7. List three importances of texture analysis data.
8. How is bulk density determined in the field? How is the bulk density corrected? How does it affect plant growth?
9. What is plant available water? How is it determined?
10. What is the importance of determining the wilting point? Which plat is used for wilting point determination?
11. List the important soil characteristics that influence the infiltration rate.
12. How is cumulative infiltration rate calculated?
13. 'Minesoil has either high infiltration rate or low infiltration rate'—why?

Table 19.13 Observation sheet for infiltration test data

Name of Project/ Location:				Date & Time:	
Elapsed time (min)	Volume of water intake (cc)	Infiltration amount (cm)	Cumulative infiltration amount (cm)	Infiltration rate (cm/h)	Cumulative infiltration rate (cm/h)
30 sec					
1 min					
2 min					
3 min					
5 min					
15 min					
30 min					
45 min					
60 min					
90 min					
120 min					
150 min					
180 min					

Moisture content = %; Temperature = °C

References

Dakshinamurthi C, Gupta RP (1968) Practicals in soil physics. IARI, New Delhi

Donahue RL, Miller RW, Shickluna JC (1990) Soils: an introduction to soils and plant growth, 5th edn. Prentice Hall of India Private Ltd., New Delhi

Hu Z, Candle RD, Chong SK (1992) Evaluation of firm land reclamation effectiveness based on reclaimed mine soil properties. Int J Sur Min Reclam 6:29–135

Jackson ML (1973) Soil chemical analysis. PHI Pvt Ltd New Delhi

Maiti SK (1995) Some experimental studies on ecological aspects of reclamation in Jharia coalfield. Ph.D. dissertation, Indian School of Mines, Dhanbad

Maiti SK, Karmakar NC, Sinha IN (2002) Studies into some physical parameters aiding biological reclamation of mine spoil dump – a case study from Jharia coalfield. IME J 41(6):20–23

Contents

S.K. Maiti, *Ecorestoration of the Coalmine Degraded Lands*,
DOI 10.1007/978-81-322-0851-8_20, © Springer India 2013

The following chemical parameters are analysed for soil/spoil in the laboratory:

1. pH
2. Lime requirement
3. Electrical conductivity
4. Organic carbon
5. Ignition loss
6. Total nitrogen
7. Available nitrogen
8. Available phosphorus
9. Total phosphorous
10. Available sulphur
11. Exchangeable potassium and sodium
12. Exchangeable calcium and magnesium
13. Cation exchange capacity (CEC)
14. Micronutrients and heavy metals (pollutants)

20.1 pH (*Pouvoir Hydrogene or Hydrogen Power*)

The pH of the soil is the measure of H^+ ion activity of the soil water system, indicating whether the soil is acidic, neutral or alkaline in reaction. The pH is defined as the negative logarithm to the base 10 of the hydrogen ion activity (moles/L) in soil solution. As the scale is logarithmic, so a change in pH of one unit represents a *tenfold* change in hydrogen ion concentration. Since the growth of plants suffers much both under very low (strongly acidic) as well as high (alkali) pH, correction of pH becomes necessary for commencement of bioreclamation work. Soil pH values below 7 are referred as 'acidic', pH value above 7 as 'alkaline' and at pH 7 as 'neutral'.

Acidity in soil: Below pH 4, the concentration of H^+ is directly harmful to the plants, while at pH less than 5, phosphorus becomes insoluble and not available to the plants. Soil acidity can be divided into three types: *active, reserve and potential*.

- Active acidity is simply the free H^+ concentration determined in the soil solution.
- Reserved acidity is related to the H^+ ions which are stored on the exchange complex of the soil.
- Potential acidity is the acid-generating power of soil; many mine spoils contain chemical elements such as sulphides (iron pyrite, FeS_2) that, on exposure to air, weather and produce acids.

Factors Affecting Soil pH Measurement

Soil samples directly in the field-moist condition may be considered the most valid in terms of the existing soil-biological environment. Measurement of air-dried soil samples is the most conveniently and generally used, and perhaps could be considered the standard procedure.

(*A*). *Equipment and Reagents*
- pH metre with electrode
- Electronic balance
- Mechanical shaker
- Beaker
- Distilled or deionised water
- Appropriate buffer solution (pH of 4, 7 and 9.2) for calibration of pH metre

(*B*). *Procedure*
1. *pH in Soil to Water Ratio of 1:2.5*
   Soil pH is normally measured in soil/water slurry in the ratio of 1:1, 1:2 or 1:2.5 (w/v). However, for reclamation point of view, soil/water ratio of 1:1 (pH paste) is recommended because it will provide close to field situation where plant roots are exposed. For pH measurement at 1:2.5 (w/v), take *20 g* of soil in a 100-mL beaker and *50 mL of distilled water*. The suspension is stirred at regular intervals for 30 min, and pH is recorded. The suspension must be stirred well just before electrode is immersed. The pH value obtained usually increases with increased volume of water used. The increase in pH value is caused by the dilution of the H^+ ion concentration in the solution.

2. *Measurement of Soil pH in a KCl Solution (Buffered pH) or pH$_{KCl}$*
   The soil is mixed with 1N KCl in 1:1 or 1:2 ratio. The pH is measured directly in the soil

suspension. This type of measurement will yield a more stable result than soil suspension method. It is sometimes referred as the *buffered pH*. The use of pH_{KCl} will provide better information concerning the chemical properties of the system.

Take 10 g soil in a 50-mL beaker and add 25-mL volume of 1-N KCl, and soil is stirred intermittently for 1 h. Measure the pH.

3. *Measurement of Soil pH in a 0.01-M CaCl₂ Solution (Soil–Water Ratio = 1:2)*

The presence of soluble salt in soil sample influences the pH measurement. To overcome the influence of soluble salts, some scientists prefer to measure soil pH in a mixture of soil and 0.01 M $CaCl_2$, because excess salt in the solution masks the effects of differential soluble salt concentration. Hence, to mask the variability of salt content of soils, to maintain the soil in a flocculated condition and to decrease the junction potential effect, the soil pH measurement is made in 0.01 M $CaCl_2$ solution. The advantages of use of 0.01M $CaCl_2$ for pH measurement are given in Box 20.1.

Box 20.1 Advantages of use of 0.01M $CaCl_2$ solution for pH measurement (Hendershot et al. 1993)

---

1. The pH is not affected within a range of the soil-to-solution rations used;
2. The pH is almost independent of the soluble salt concentration for nonsaline soils;
3. This method is a fairly good approximation of the field pH for agriculture soils;
4. Because the suspension remains flocculated, errors due to liquid junction potential are minimized;
5. No significant differences in soil pH determination are observed for moist or air-dried soil; and
6. One year of storage of air-dried soil does not affect the pH.

In this method, 25 g of soil is suspended in 50 mL of 0.01 M $CaCl_2$ solution (*completely dissolve 1.47 g CaCl₂.2H₂O in 1-L distilled water*), and suspension is stirred thoroughly for 15 min and stand for 30 min, stirring occasionally. The pH measurement is made in the usual way. The soil pH scale is shifted downwards in this solution. In acidic soil range from pH 4–5, pH 6 is the optimum pH of limed soils, and for calcareous soils, pH of 6.9–7.1 in the $CaCl_2$ solution.

Important Notes on pH Measurement

- The pH metre is switched on, and sufficient time (10–15 min) is given for warming up.
- pH metre should be calibrated at two-point scale before taking measurement.
- One point of calibration should be at 7.0, while another should be chosen based on the probable encountered pH (e.g. for acidic soil, pH 4, and alkaline soil, 9.2).
- An electrode should not be left to dry out for long periods, thus, it is always advisable to immerse it in distilled water or pH 7 buffer.

(C). Factors Affecting pH Measurement

The following factors may influence the measurement of pH:

- The nature and type or inorganic and organic constituents
- The soil/solution ratio
- The salt or electrolytic content
- The CO_2 content
- Errors associated with equipment standardisation

(D). Maintenance of Glass Electrode

A glass electrode frequently becomes sluggish in its operation as evidenced by slow changes or drift during the pH measurement. Such drift or slow response may be caused by a dried layer of clay or precipitated carbonate on the glass bulb that cannot be removed by ordinary washing with water. Or sluggish response may be caused by an ageing of the glass surface of the bulb. Rejuvenation of the glass surface by immersion in a dilute solution of HF for 10–15 s usually corrects the problems.

(E). Comments

pH determination of fresh sample is preferable. Soil pH is one of the most indicative measurements of chemical properties of a soil. According to Maiti (1995), three soil pH ranges are particularly informative for reclamationist:

- A pH <4 indicates the presence of free acid generally from oxidation of sulphides.
- A pH <5.5 suggests the likely occurrence of exchangeable aluminium.
- A pH from 7.8 to 8.2 indicates the presence of $CaCO_3$.
- Nearly all soils with pH values above 8 have a high percentage of Na^+ ions on their exchange sites.

(F). Exercise on pH Measurement

1. List three importance of soil pH measurement.
2. What are the important factors that influence the pH measurement?
3. What is buffered pH and how is it measured?
4. 'Sometimes soil pH is measured in $CaCl_2$ solution'—why?
5. How is pH measured in the laboratory?
6. The pH of the soil is <4.0, <5.5 and >8; what does it indicate?

20.2 Lime Requirement of Acidic Spoil/Soil

(A). Principles

The lime requirement (LR) of an acidic spoil is the amount of lime or other base required to neutralise the acidity (i.e. total soluble and exchangeable H^+ and Al^{3+} ions) from the initial acid condition to a selected less acid condition. Numbers of laboratory methods are available for this purpose. The procedure given by Shoemaker et al. (1962) is being widely used, known as SMP (Shoemaker, McLean and Pratt) single-buffer method.

For mine spoils, this method gives an estimate of the lime requirement to neutralise the actual and reserve acidity based on a change in pH of the standard buffer solution. Only mine spoils/soil having pH less than 6.0 needs to be tested.

(B). Instruments and Reagent

1. pH metre
2. Mechanical shaker
3. Electronic balance
4. Standard buffers, pH 7 and 4
5. Distilled water
6. SPM buffer solution

Preparation of Extracting SMP Buffer

Dissolve the following five chemicals (all chemically pure) in 1,000 mL of distilled water:

1. Nitrophenol 1.80 g
2. Triethanolamine 2.25 mL
3. Potassium chromate (K_2CrO_4) 3.0 g;
4. Calcium acetate 2.0 g
5. Calcium chloride dihydrate ($CaCl_2.2H_2O$) 53.1 g

First, add nitrophenol, potassium chromate and calcium chloride in 500–600 mL of distilled water and mix. Prepare calcium acetate solution separately, then mix. Shake periodically until the mixture is completely dissolved and dilute to 1,000 mL in a volumetric flask. Adjust to pH 7.5 with 15 % NaOH using the standardised pH metre. Filter through a fibreglass sheet or cotton mat. Since SPM buffer is adjusted to pH 7.5 and is protected from contamination, chances of pH change are very little from day to day.

(C). Procedure

1. Take 5 g of air-dried sieve soil (<2-mm size) in a dry 50-mL beaker.
2. Add 5 mL of distilled water and 10 mL of SMP extracting buffer and stir continuously for 10 min or intermittently for 20 min.
3. The pH of the suspension is determined on the basis of which requirement of lime is read from the Table 20.1. The values are given in *tonnes of pure calcium carbonate per acre* to bring the soil to the pH indicated and are to be converted to their equivalents of the form of agricultural lime to be used. For expressing *in metric unit, that is, tonnes/ha, the figures are to be multiplied by 2.43.*

(D). Comments

The SPM method is designed for soils that are particularly high in lime requirement (LR) and that have considerable exchangeable aluminium.

It is recognised that the SPM method is not very accurate when used on soils of very low LR. However, inclusion of the double-buffer features improves accuracy for low LR soils.

Calcite ($CaCO_3$) and dolomite ($CaCO_3.MgCO_3$) are two of the minerals whose impure limestone is commonly used as liming material.

(E). *Self Test on Lime Requirement*
1. Define lime requirement. Why is it measured, and below what pH is it measured?
2. What is SMP buffer?
3. List two liming materials used for correction of acidity of soil.

20.3 Soluble Salts (Electrical Conductivity)

Soluble salts for soil are technically defined as those dissolved inorganic solutes that are more soluble than gypsum ($CaSO_4.2H_2O$, solubility of 0.024 g/100 mL at 0°C). The most soluble salts in soils are Ca^{2+}, Mg^{2+}, Na^+, Cl^-, SO_4^{2-} and HCO_3^-. Smaller quantities of K^+, NH_4^+, NO_3^- and CO_3^{2-} are also found in most soil. Since ions are the carrier of electricity, the electrical conductivity (EC) of the soil water system rises according to the content of soluble salts. The measurement of EC can be directly related to soluble salts concentration of the soil at any particular temperature.

At normal concentration, soluble salts have little harmful effect on plant growth; however, if excessive salts exist, plant injury such as reduction of seed germination, leaf burning and death may occur (due to excessive salinity and/or exo-osmotic effect). In the laboratory, EC is measured by extracting the ions in a solution from a saturated soil sample (1:1; w/v; soil: water). This solution is called the saturation extract (ECe). The greater the concentration of ions or salts in the saturation extract, the higher is the EC.

In International System of units (SI), the reciprocal of 'Ohm' is the Siemens (S), and conductivity is reported as milliSiemens per metre (mS/m). The relationship between Siemens and mhos is given below:

Table 20.1 Lime requirement to bring the soil to the desire pH level according to pH value of the soil-buffer suspensions

pH of soil-buffer suspension	Lime required to bring the soil to indicated pH (in tonnes/acre of pure $CaCO_3$)		
	pH 6.0	pH 6.4	pH 6.8
6.7	1.0	1.2	1.4
6.6	1.4	1.7	1.9
6.5	1.8	2.2	2.5
6.4	2.3	2.7	3.1
6.3	2.7	3.2	3.7
6.2	3.1	3.7	4.2
6.1	3.5	4.2	4.8
6.0	3.9	4.7	5.4
5.9	4.4	5.2	6.0
5.8	4.8	5.7	6.5
5.7	5.2	6.2	7.1
5.6	5.6	6.7	7.7
5.5	6.0	7.2	8.3
5.4	6.5	7.7	8.9
5.3	6.9	8.2	9.4
5.2	7.4	8.6	10.0
5.1	7.8	9.1	10.6
5.0	8.2	9.6	11.2
4.9	8.6	10.1	11.8
4.8	9.1	10.6	12.4

1 dS/m (deciSiemens/metre)
$\quad$ = 1 mmhos/cm (millimhos/centimetre)
$\quad$ = 1,000 μmhos/cm.
1 mS/m = 10 μS/cm = 10 μmhos/cm.
1 μS/cm = 1 μmhos/cm.

(A). *Instrument*
Conductivity metre

(B). *Reagents*
Standard potassium chloride solution: 0.7156 g of dry reagent grade potassium chloride is dissolved in freshly prepared double distilled water and made to 1,000 mL. At 25°C, it gives an EC of 1,412 μmhos/cm (or 1.412 mmhos/cm = 1.412 dS/m). The conductivity bridge is calibrated, and the cell constant is determined with the help of this solution.

(C). Procedure

In the laboratory, EC is measured in a soil–water extract based on a fixed soil/solution ratio (e.g. 1:2; 1:2.5 or 1:5). Measurement of EC at 1:1 ratio (ECe) is most desirable for reclamation work because it gives very close to field moisture conditions. Two methods are discussed below:

EC Measurement in Soil to Water Ratio of 1:2 (w/v)

- Take 20 g of soil in 150 mL of conical flask and add 40 mL of distilled water, shake intermittently for 1 h and allow to stand. Alternatively, the clear extract after pH determination can be used for electrical conductivity measurement. The conductivity of the supernatant liquid is measured with the help of a conductivity metre.

EC Measurement in Soil to Water Ratio of 1:1 (w/v)

Preparation of Saturation Extract (ECe)

1. Weigh 200–400 g of air-dried soil of known water content in a plastic container having the lid.
2. Weigh the container plus soil.
3. Add distilled water to the soil with stirring until it is heavily saturated.
4. Allow the mixture to stand covered for several hours (4 h to overnight) to permit the soil to imbibe the water. Then add more water to achieve a uniformly saturated soilwater pest.
5. After mixing, allow the sample to stand preferably overnight (at least 4 h).
6. If pest is too wet, add additional dry soil to the pest mixture.
7. Upon attainment of saturation, reweigh the container plus content.
8. *Calculate the saturation water percentage from the weight of oven-dried soil and sum of the water added and that initially present in the air-dried sample.*
9. After allowing the saturated soil paste to stand 4 h or more, transfer it to the Buchner funnel fitted with filter paper.
10. Apply vacuum, and collect the filtrate in a test tube of bottle.

Table 20.2 Salinity hazards viz. ECe (mmhos/cm)

Salinity hazards	ECe (mhos/cm)
1. Low	<0.75
2. Medium	0.75–1.5
3. High	1.5–3.0
4. Very high	>3.0

11. If the initial filtrate is turbid, refilter it.
12. Terminate the filtration, when air begins to pass through the filter.

The electrical conductivity of saturation extract (ECe) is frequently used as an index of salinisation hazards. The relationship between salinity hazards corresponding ECe is given in Table 20.2.

SAR (sodium absorption ratio): The SAR is used to estimate what is the exchangeable Na percentage of a soil. The SAR is defined as

$$SAR = \frac{[Na^+]}{\sqrt{\dfrac{Ca^{+2} + Mg^{+2}}{2}}}$$

The concentration of Na^+, Ca^{2+} and Mg^{2+} is expressed as millimole per litre or in milliequivalent per litre.

(E). Exercise   on   Electrical   Conductivity Measurement

1. List the importance of conductivity measurement.
2. Define EC.
3. If conductivity is high, what will be the impacts?
4. Are there any differences between EC and ECe? Which value will be higher and why?
5. How is saturation extract for ECe measurement prepared?

20.4 Organic Carbon (OC)

(Walkley and Black 1934 by rapid dichromate oxidation techniques)

Organic matter is critical for soil health and for soil productivity. It provides energy for soil microbes, builds soil microbial biodiversity, supports and stabilises soil structure, reduces bulk density, increases water storage, enhances water infiltration, stores and supplies nutrients and enhances cation exchange capacity. Soil organic matter (SOM) is made up of living plants and animals (roots, fungi, bacteria, macrofauna and microfauna), plant litter and all the degraded material from decomposing plant and animal material. Organic carbon is contained in the soil organic fraction, which consists of cells of microorganisms, plant and animal residues at various stages of decompositions.

(A). Principles

The organic matter (humus) in the soil gets oxidised by chromic acid (potassium dichromate plus concentrated H_2SO_4) utilising the heat of dilution of H_2SO_4. The unreacted dichromate is determined by back titration with ferrous (ammonium) sulphate (redox titration).

$$2Cr_2O_7{}^{2-} + 3\ C^0 + 16H^+$$
$$= 4\ Cr^{3+} + 3CO_2 + 8\ H_2O$$

The amount of $Cr_2O_7{}^{2-}$ remaining after reaction with soil organic matter, can be estimated by colorimetrically (intensity of green colour) after the removal of soil by filtration or centrifugation.

Dichromate oxidation method uses heat of dilution, which do not give complete oxidation of organic compound in soil, although the most active forms of organic carbon are converted to CO_2. A temperature of approximately $120\,^{\circ}C$ is obtained, which is sufficient to oxidise the active form of organic carbon.

Main Interference

There are three major sources of errors with rapid dichromate oxidation technique:

1. Presence of chloride: If chloride is present, it reduces $Cr_2O_7{}^{2-}$, which leads to higher result of organic carbon. The chloride interference can be removed by adding Ag_2SO_4 to digestion acid (concentrated H_2SO_4).

2. Presence of Fe^{2+}: When Fe^{2+} presents higher amount in soil (in case of OB dumps), Fe^{2+} will be oxidised to Fe^{3+} by $Cr_2O_7{}^{2-}$, resulting

a positive error in the analysis, that is, giving higher values for organic carbon content:

$$Cr_2O_7{}^{2-} + 6\ Fe^{2+} + 14H^+$$
$$= 2Cr^{3+} + 6Fe^{3+} + 7H_2O.$$

So, by sufficient air-drying of soil (i.e. about 1–2 weeks), Fe^{2+} will oxidised to Fe^{3+} and OC can be estimated accurately.

Higher oxides of Mn compete with $Cr_2O_7{}^{2-}$ for oxidation of organic matter, resulting to lower value of OC. Usually this is not a serious error.

3. Differences in digestion conditions and reagent composition and from the variable composition of organic matter itself.

(B). Apparatus

1. Balance accurate to 0.001 g
2. Volumetric flasks—500 mL
3. Burette—50 mL
4. Pipette—10 mL, 1 mL with rubber teat
5. Graduated measuring cylinder—200 mL and 20 mL
6. Sieves—500 μm
7. Pestle and mortar
8. Wash-bottle with distilled water
9. Drying oven (105–110 °C) and desiccator.

(C). Chemicals and Reagents

Chemicals

1. Potassium dichromate, $K_2Cr_2O_7$ (1N)—49.04 g/L
2. Diphenylamine indicator, $(C_6H_5)_2NH$—0.5 g
3. Orthophosphoric acid (H_3PO_4, 85 %)
4. Ferrous ammonium sulphate, $Fe(NH_3)_4SO_4$—(0.5N approx.)—196 g/L
5. Sulphuric acid (H_2SO_4)-concentrated 96 %
6. Sodium fluoride (NaF)—solid

Preparation of Reagents

1. *Potassium dichromate, 1N*: Dissolve 49.04 g of AR (analytical grade) $K_2Cr_2O_7$ (dried at 105 °C) in 1 L of distilled water.
2. *Ferrous ammonium sulphate, 0.5N (approx.)*: Dissolve 196 g of the hydrated crystalline salt of Ferrous ammonium sulphate per L

containing 20 mL of concentrated H_2SO_4. This solution is relatively stable and convenient to work than that of ferrous sulphate.

3. *Diphenylamine indicator*: 0.5 g diphenylamine dissolve in a mixture of 20 mL of water and 100 mL of concentrated H_2SO_4.
4. *Concentrated sulphuric acid* (sp. gravity 1.84, not less than 96 %), containing 1.25% silver sulphate (Ag_2SO_4): Add 5 g of silver sulphate/L of H_2SO_4. (In case of soils free from chlorides, use of Ag_2SO_4 can be avoided.)
5. *Orthophosphoric acid* (85%) and/or *sodium fluoride* (chemically pure).

(D). Procedure

The procedure described below is divided into five parts:

- Standardise ferrous ammonium sulphate (FAS) solution
- Sample preparation
- Test for organic carbon
- Calculation
- Reporting of results

(i) *Standardisation of Ferrous Ammonium Sulphate Solution*

Take 10 mL of 1-N potassium dichromate solution in a 250-mL conical flask. Then very carefully add 20-mL concentrated H_2SO_4. This will generate heat. Swirl the mixture and allow to cool. Add 200 mL of distilled water. Then add 10 mL of orthophosphric acid and 1-mL indicator solution and mix thoroughly.

Add ammonium ferrous sulphate from the burette, swirl the flask until the colour changes from blue to green.

This data is also used as blank reading.

(ii) *Test for Organic Carbon*

1. The oven-dried soil is ground and completely passed through 0.2-mm sieve, and 0.50-g sample is placed at the bottom of dry 500-mL conical flask (corning/Pyrex).
2. Add 10 mL of 1N potassium dichromate in the conical flask and swirl the flask gently to disperse the soil in the dichromate solu-

tion. The flask should be kept on asbestos sheet.

3. Add 20 mL of concentrated H_2SO_4 (containing 1.25% Ag_2SO_4) very carefully from a measuring cylinder. Swirl two to three times. Avoid excessive swirling that would result in organic particles adhering to the sides of the flask out of the solution. The flask is allowed to stand for 30 min. Protect the flask from draughts.
4. Add 200 mL of distilled water and 10 mL of orthophosphoric acid to get a sharper end point of the titration.
5. Add 1 mL of diphenylamine indicator (approx. 10 drops). The indicator should be added just prior to titration to avoid deactivation by adsorption onto clay surfaces.
6. The contents are titrated with FAS solution until the colour flashes from blue-violet to green.
7. Simultaneously, a blank is run without soil. The blank is used to standardise the FAS solution daily.
8. If more than 7 mL of dichromate solution is consumed, the determination is repeated with a smaller quantity (0.25 to <0.5 g) of soil.

(D). Calculation

(a) Oxidisable organic carbon $(\%) = \frac{10 \ (B-T)}{B} \times 0.003 \times \frac{100}{\text{wt. of}}$ soil

where,

B = volume (mL) of ferrous ammonium sulphate required for blank titration

T = volume of ferrous ammonium sulphate needed for soil sample

Wt. of soil in g

Note: 1 L of 1N $K_2Cr_2O_7$ will be equal to 12/4 g carbon = 3 g carbon

or 1 mL of 1N $K_2Cr_2O_7$ will be equal to 3 g C $\times 10^{-3} = 0.003$ g C

To convert easily oxidisable organic carbon to total carbon, divide by 0.77 or multiply by 1.334 factor.

(b) Total organic carbon $(\%) =$ oxidisable organic carbon $(\%) \times 1.334$

(c) Organic matter (%) = total organic carbon (%) × 1.724

Note:

1. For soil high in organic matter (1% oxidisable organic carbon or more), more than 10-mL potassium dichromate is needed.
2. The factor 1.334 and 1.724 used to calculate TOC (total organic carbon) and OM (organic matter) are approximate; they may vary with soil depth and between soil.
3. Soil containing large amount of chloride (Cl^-), manganese (Mn^{2+}) and ferrous iron (Fe^{2+}) will give higher results. The chloride interference is removed by adding silver sulphate in the oxidising agent. The inference of Fe^{2+} is reduced by air-drying of soil samples. The presence of $CaCO_3$ up to 50% causes no interference.

(E). Exercise on Organic Carbon

1. What are the sources of organic carbon in soil?
2. Write the principle of organic carbon estimation?
3. *'Fe^{2+} is one of the most positive interfering elements in OC determination.'*—how to overcome the interference of Fe^{2+}.
4. What is the function of potassium dichromate in OC estimation?
5. In concentrated H_2SO_4, why is 1.25% Ag_2SO_4 added?

20.5 Organic Matter (OM) by Loss of Ignition (LOI)

(A). Introduction

The organic matter (OM) content influences many soil properties like capacity of soil to supply N,P,S and trace elements; infiltration and retention of water; degree of aggregation and overall soil structure; cation exchange capacity and soil colour.

The OM content of soil may be indirectly estimated through multiplication of organic carbon (OC) concentration by the ratio of OM to OC commonly found in the soil. *The factor is 1.724, that is, if, OC is 1%, then OM = 1.724%.* However, a number of studies suggested that this factor is too low for many soils, and consequently the organic matter content is underestimated. It has been concluded that *conversion factors of 1.9 and 2.5* would be appropriate for surface soil and subsoils, respectively.

It is evident that estimation of OM from OC is not highly accurate, because the ratio of OM to OC varied from soil to soil and with profile depth. As per Nelson and Sommers (1996), a factor of 2 appears to be most universally acceptable.

(B). Principles

To achieve a direct determination of OM, one must separate it from inorganic material, which in most soils makes up 90% or more of the weight of the soil. The OM is destroyed after the loss in weight of the soil is taken as a measure of the OM content.

The ignition of soil at high temperature gives quantitative value of OM, but inorganic constituents of the soil, chiefly the hydrated aluminosilicates, lose structural water, and carbonate minerals are decomposed, upon heating, thus resulting in weight losses considerably in excess of the actual organic matter content. However, if the soil is pretreated with a mixture of HCl and HF to remove the hydrate mineral water, loss of ignition gives a valid estimate of organic content

Some workers have attempted to avoid the cumbersome pretreatment of soil with HCl–HF mixtures before carrying out the loss on ignition procedure. Combustion of samples at 350–440°C has been found to destroy OM without removing structural water from inorganic soil components, thereby eliminating need for pretreatment.

It has been shown that a 24-h heating at 430°C did not destroy $CaCO_3$ in soil and led him to propose that OM content may be estimated from loss on ignition at 430°C.

(C). Apparatus

1. Porcelain crucible—20 mL.
2. Analytical balance sensitive to ± 1 mg in draught-free environment.

3. Muffle furnace—heating up to 650°C
4. Desiccator
5. Hot air oven
6. Gloves, forceps, etc.

(D). Procedure
1. Take accurately 10 g of sample (<2 mm size) into tarred crucibles.
2. Dry at 105°C for approximately 18 h and record the weight nearest to $\pm$ 1 mg. (Usually 2 h drying at 105°C could be sufficient.)
3. The crucible containing the sample is then placed in a muffle furnace, and heat was increased slowly (2°C/min) to 430°C and kept overnight. (Usually, 2-h heating after reaching temperature 430°C could be sufficient.)
4. Then the crucible is removed and cooled in a desiccator and reweight in a draught-free environment nearest to ±1 mg.

(E). Calculation

$$\% \text{ Loss of weight n ignition } (430°C)$$
$$= \frac{(\text{wt at } 105°C) - (\text{wt at } 430°C)}{\text{wt at } 105°C} \times 100$$

Estimation of Organic Matter (OM)
Estimation of OM from LOI is done by regression analysis method. Select mine spoils covering the various ranges in OM expected by considering different age of plantation, age of spoil, grass or herbaceous vegetation cover and litter accumulation. Determine the OM (%) by Walkley–Black method as described. Then carry out the regression analysis by plotting OC viz. LOI. Use the resulting equation to convert LOI to OM (%).

(F). Exercise on Organic Matter (OM) Estimation
1. Discuss the role of OM in soil.
2. How is OM in soil estimated by using OC data?
3. List the difficulties faced in the laboratory during OM estimation.
4. 'Temperature in the furnace is very crucial for OM estimation'—why?

20.6 Total Nitrogen

The total N content of soils ranges from <0.02% in subsoils to > 2.5% in peats. The surface layer of most cultivated soils contains between 0.06 and 0.5% nitrogen. Two methods have gained acceptance for determination of total N: (a) *the Kjeldahl method* which is essentially a *wet oxidation* procedure, and (b) the *Dumas method* which is a fundamentally a *dry oxidation* (i.e. combustion) technique.

(A). Principles of Kjeldahl Method
The dried and homogenised soil or mine spoils sample is digested in a suitable Kjeldahl apparatus with concentrated H_2SO_4 which decomposes organic substances by oxidation, and liberated nitrogen as ammonium sulphate. In this step, to rise the temperature, digestion catalyst mixture (Potassium sulphate, copper sulphate and Se powder) is added. Copper sulphate acts as a catalyst to accelerate the decomposition, while potassium sulphate enhances the temperature of digestion. Chemical decomposition of the sample is complete when the initially very dark-coloured medium has become clear and colourless. After adding sodium hydroxide to the digestion solution, the produced ammonium from all nitrogen species is evaporated by distillation as ammonia gas. This is condensed in a conical flask with boric acid solution, and the amount is titrated against the indicator with a standard acid (H_2SO_4 of HCl).

In a nutshell, it measures only organic and ammonium forms of nitrogen excluding nitrate nitrogen. Organic and ammoniacal nitrogen is converted to ammonium sulphate, and *ammonia gas* is distilled into boric acid and titrated with dilute acid.

Basically, the Kjeldahl procedure employed for determination of total N involved two steps:
- Digestion of sample to convert organic N to a NH_4^+–N
- Determination of ammonium N in the digest

The chemical reactions involved in TKN estimation is given below:

Degradation: Sample $+ H_2SO_4 \rightarrow$

  $(\mathbf{NH_4})_2\mathbf{SO_4(aq)} + CO_2(g) + SO_2(g) + H_2O(g)$

Liberation of NH_3: $(NH_4)_2SO_4(aq) +$

  $2NaOH \rightarrow Na_2SO_4(aq) + 2H_2O(l) + \mathbf{2NH_3(g)}$

Capture of NH_3 : $B(OH)_3 + H_2O + NH_3(g) \rightarrow$

  $\mathbf{NH_4^+} + B(OH)_4^-$

Back titrationwith H_2SO_4 (or HCl)

Interferences and Sources of Errors

The Kjeldahl method in principle does not capture all nitrogen compounds. The nitrogen that occurs in N–N and N–O linkages is not completely recorded. Furthermore, the inorganic fraction, nitrate and nitrite, is not determined. Other sources of error include fluctuations in residual water content of samples, impurities in the apparatus and fluctuations during titration. Therefore, the apparatus has to be rinsed after each analytical series, and blank determinations have to be carried out. The amount of sulphuric acid has to be investigated with different materials to yield the highest digestion efficiency.

(B). Apparatus: Kjeldahl Distillation Assembly
(C). Chemicals and Reagent
Chemicals

1. Hydrochloric acid, HCl	7. Sulphuric acid, H_2SO_4 , concentrated
2. Cupric sulphate pentahydrate, ($CuSO_4$. $5H_2O$)	8. Sodium carbonate, Na_2CO_3
3. Selenium powder	9. Potassium sulphate, K_2SO_4
4. Sodium hydroxide, NaOH	10. Boric acid
5. Methyl red	11. Bromocresol green
6. Ethyl alcohol	

Reagents

1. *Concentrated sulphuric acid* (sp. gravity 1.84)
2. *Digestion catalyst*: Grind together 20-g cupric sulphate ($CuSO_4$. $5H_2O$), 2-g selenium powder and 200-g potassium sulphate (K_2SO_4). Powder the reagents separately before mixing.
3. *Sodium hydroxide 40%*: Dissolve 40 g of NaOH in 100 mL of distilled water.
4. *Boric acid solution 4%*: Dissolve 4-g boric acid in 100 mL of distilled water.
5. *Boric acid + mixed indicator solution*: Dissolve 0.066-g methyl red plus 0.099-g bromocresol green in 100 mL of 95% ethyl alcohol. Add 5 mL of this solution in 100 mL of boric acid solution. (In case the solution turns bluish, add drop wise 0.01N HCl until the colour just turns pink to brown.)
6. *Sulphuric acid* (H_2SO_4) or *hydrochloric acid* (HCl); 0.01N.
7. *Hydrochloric acid (0.01N)*: Take 8.34 mL of concentrated HCl (12N), and dilute it with distilled water to prepare 100 mL of 1-N HCl. Take 1 mL of 1-N HCl, and dilute it to 100 mL with distilled water, and get 0.01-N HCl.

(D). Procedure
I. *Pretreatment of Sample*

Soil samples to be analysed for total N are usually dried, grounded and sieved before analysis. They are often stored for analysis in paper bags or other containers that are not airtight. Jackson (1973) recommended that soil samples for Kjeldahl analysis be grounded to pass through a *0.15-mm (100-mesh)* screen to ensure complete oxidation of organic matter within small aggregates.

II. *Digestion stage*

Step-1 Take 5 g of sample in a 500 mL of Kjeldahl flask, and add 25 mL of distilled water.

Step-2 Then *20 g of digestion catalyst mixture, 35 mL of concentrated sulphuric acid* and liquid paraffin are added and mixed thoroughly.

Step-3 The contents are heated on low heat for the first 10–30 min until the frothing stopped and raised the heat thereafter about an hour to release all the residual nitrogen.

Step-4 The digest is cooled, and 100 mL of distilled water (very slowly) is added and mixed thoroughly.

Step-5 The supernatant is transferred to a distillation flask, leaving as much soil as possible.

Step-6 The supernatant is put into a Kjeldahl assembly.

Step-7 Wash the residue further with a little distilled water several times (3–4 times) and transfer the supernatant each time to the distillation flask.

III. *Distillation in Kjeldahl Assembly.*

1. Add 100 mL of NaOH (40%) and few granules of Zn in the Kjeldahl flask. Take 25 mL of boric acid cum indicator solution (Pink colour solution) in 250 mL of flask, and place it below the condenser of the distillation assembly so that the lower open end of the condenser is dipped in the solution.

2. Heat the distillation flask on hot plate or heating mantle and collect about 150 mL of distillate in flask (pink solution turns to green due to absorption of ammonia). (It is good to put a mark at the level of 170 mL in the flask to know easily when 150-mL distillate is collected).

3. Remove the flask with distillate.

4. Titrate the distillate with 0.01N HCl to its original colour (i.e. green turns to pink colour indicates the end point of titration).

5. Run a distilled *water blank* in the same manner.

(*E*). *Calculation*

$$\text{Kjeldahl Nitrogen (mg/g)} = \frac{(T_1 - T_2) \times N \times 14}{W}$$

$$\text{Kjeldahl Nitrogen (\%)} = \frac{(T_1 - T_2) \times N \times 1.4}{W}$$

Where,

$T_1 =$ Volume of titrant used against sample (mL);

$T_2 =$ Vol. of titrant used against distilled water (Blank) (mL);

$N =$ Normality of titrant (0.01N HCl);

$W =$ Weight of soil/spoils (g);

If 0.01M H_2SO_4 is used as a titrant, 1 mL of 0.01 M H_2SO_4 is equivalent to 0.28 mg of N

$$\text{Kjeldahl Nitrogen (\%)} = \frac{(T_1 - T_2) \times M \times 2.8}{W}$$

where $M =$ molarity of titrant (0.01 M H_2SO_4).

(*F*). *Comments*

- *Hg as catalyst*: There is considerable evidence that Hg is the most effective single catalyst. However, use of Hg has been discouraged because Hg^+ reacts with alkali; some of the NH_4^+ in the digest react with the HgO and precipitated by the alkali to form a Hg-NH_4^+ complex, and the NH_4^+ in this complex is not readily liberated by distillation of alkali.

- Temperature is the most important factor in Kjeldahl digestion of soil. K_2SO_4 increases the temperature. Loss of nitrogen occurs when digestion temperature exceeds 400°C. Se acts as a catalyst.

- During the process of digestion, the Kjeldahl flask should be swirled at an interval to dislodge any material adhering to the walls and bring it into the contact with the acid.

- Bumping is commonly encountered during Kjeldahl *digestion* of soil (particularly sandy soil), but this problem can usually be eliminated by adding glass bids.

- The sharpness of an end point depends upon the quality of boric acid, concentration of boric acid and indicator. The better the quality of boric acid, and the more dilute the boric acid solution consistent with complete retention of NH_3, the sharper is the end point.

- The 5 mL of 2% boric acid solution used in the method described should effectively absorb about 5 mg of NH_3-N.

- Control (blank) should be performed in each estimation.

Experiences in laboratory have shown that a 60-mesh sieve sample should be use. More finely sieved materials have the higher accuracy of the results. To ensure precise results, *it is recommended that soil containing <1% N be ground to pass through a 100-mesh (0.14-mm) screen, that soils containing >1% N be ground to pass through a 150-mesh (0.105-mm) screen* and that the sieved material be thoroughly mixed before analysis.

(G). Exercise on Total Nitrogen Estimation
1. What is TKN?
2. What is the range of total N found in the soil?
3. What is the principle of Kjeldahl method of N estimation?
4. Why is pretreatment of the soil sample in necessary for total N estimation?
5. What is digestion catalyst? Why is it used during total N estimation?
6. Why has use of Hg as a catalyst been abandoned?

20.7 Available Nitrogen

(Easily Mineralisable Nitrogen-by Alkaline Permanganate Method, Subbiah and Asija 1956)

(A). Introduction
In soil/spoil, nitrogen presents as organic nitrogen, ammoniacal nitrogen, nitrate and nitrite nitrogen. Major portion (90%) of soil nitrogen exists in combination with the organic matter. Only a negligible fraction of soil-N, which is present as inorganic form, available to the plant. As, organic N mineralised to inorganic form then available to plant, therefore, *estimation of organic carbon is usually used as a measure of available N in soil.*

(B). Principles
Nitrogen mineralisation test gives a measure of the amount of nitrogen that may become available through microbial decomposition of the total organic nitrogen present.

$$\text{Organic matter} \xrightarrow[\text{Mineralisation process}]{} NH_4^+$$

The procedure involves distilling the soil with alkaline potassium permanganate solution and determination of the *ammonia* liberated which serves as an index of the available/mineralisable nitrogen status.

(C). Apparatus: Kjeldahl Distillation Assembly
(D). Chemicals and Reagents

Chemicals
1. Potassium permanganate, $(KMnO_4)$—3.2 g/L
2. Sulphuric acid, H_2SO_4—0.02N
3. Sodium hydroxide, NaOH—25 g/L
4. Boric acid, 2%—20 g/L
5. Bromocresol green
6. Sodium carbonate, Na_2CO_3
7. Methyl red
8. Ethyl alcohol

Preparation of Reagents
1. *Potassium permanganate $(KMnO_4)$ solution; 0.32% (3.2 g/L)*: Dissolve 0.32 g in 100 mL of distilled water.
2. *Sodium hydroxide, 2.5% (25 g/L):* Dissolve 2.5-g NaOH in 100 mL of distilled water.
3. *Sulphuric acid (0.02N):* Prepared a stock solution of approximately 0.1N H_2SO_4 by diluting 3 mL of concentrated H_2SO_4 (sp.gr. 1.84) to 1,000 mL, standardised against 0.1-N Na_2CO_3 solution. (Dissolve 1.060 g of anhydrous Na_2CO_3 oven-dried at 140°C and diluting to 1,000 mL with distilled water.) Next, dilute 200 mL of 0.1N H_2SO_4 to 1,000 mL with distilled water will give 0.02-N H_2SO_4.
4. *Boric acid (H_3BO_3) solution, 2% (20 g/L):* Dissolve 2 g boric acid in 100 mL of distilled water and add containing 20 mL of mixed indicator per L.
5. *Mixed indicator:* 0.066 g methyl red plus 0.099 g bromocresol green dissolved in 100 mL of 95% ethyl alcohol.

(E). Procedure
1. Take 20 g of soil in an 800-mL Kjeldahl flask and add 20 mL of water and 100 mL 0.32% $KMnO_4$ and 100 mL 2.5% NaOH solutions. The frothing during boiling is prevented by adding a few glass beads.
2. The content is distilled in a Kjeldahl assembly at a steady rate, and the liberated ammonia gas is collected in a 250-mL conical flask

containing 20 mL of boric acid cum mixed indicator solution. The lower open end of the condenser must be dipped into the solution.

3. With the absorption of ammonia gas, the pinkish colour boric acid solution turns to green.
4. Nearly 100 mL of distillate is collected in about 30-min time and titrated with 0.02N H_2SO_4.
5. The blank correction (without soil) is made for the final calculation.

(*F*). *Calculation*

1. Available N (ppm) $= \frac{(A-B)\times N \times 14 \times 10^3}{W}$
   where,
   A = volume of H_2SO_4 consumed for blank, mL
   B = volume of H_2SO_4 consumed for sample, mL
   N = normality of H_2SO_4 acid
   W = wt. of the soil sample
2. *in % = nitrogen in ppm/10,000*
3. *in kg/ha = % of nitrogen × 22,500* (bulk density = 1.5 g/cm^3.)

(*G*). *Comments*

Alkaline permanganate oxidisable nitrogen is a good measure of organic N or total available N.

A soil having 2% of organic carbon considered to have 0.2% (i.e.1/10 of organic carbon) organic nitrogen, which is equal to 4.4 million g-N/ha (i.e. 4.4 × 10^6 = 4,400 kg-N/ha), constitutes the total N in the soil that will be available to plants growing in it. In fact, not more than 1–2% of this amount (i.e. 4,400 kg-N/ha), which is about 44–88 kg-N/ha is present in an instantly available form. This availability is the result of nearly 2.5% mineralisation of organic N that is taking place per year. The relationship between total nitrogen and available nitrogen as reported in some of the Indian soils are given below:

Total nitrogen (g/kg)	Available N (g/kg)	Ratio (TN: Av-N)
2.20 (or, 4950 kg/ha)	0.16 (or 360 kg/ha)	13.75

(*H*). *Exercise on Easily Mineralisable Nitrogen*
1. What is easily mineralisable nitrogen?
2. What is relationship between easily mineralisable N and total N?
3. What is alkaline permanganate method?
4. How to convert N concentration between ppm, % and Kg/ha.

20.8 Available Phosphorous

Phosphorus is the second key important nutrient. The amount of plant-available P is generally not exceeded 0.01% of the total P. The total P in soil found between 0.02 and 0.10% by weight. Two forms of P are found in soil: *Organic P* (20–28%), and the rest is *inorganic P*. There has been no report of plant absorbing organic P, either from solid or solution phase of the soil. Inorganic P occurs as orthophosphate ($H_2PO_4^{-}$ and HPO_4^{-2}) in several forms and combinations, and only a small fraction of the total amount present may be available to plants which are of direct relevance in assessing the phosphorus fertility level.

In soil testing work, the available phosphorus content is readily determined by the following:

- Extraction with suitable reagents according to specified soil to solution ratio and time of shaking.
- In the filtered extract, phosphorus is estimated calorimetrically by adding ammonium molybdate and thereafter reducing the molybdenum phosphate complex in acidic medium.
- The Mo blue method is the most sensitive and, as a result, is widely used for soil extract containing small amount of P.

This method is based on principles that, in an acidic molybdate solution containing orthophosphate ions, a phosphomolybdate complex form that can be reduced by $SnCl_2$ and other reducing agent to a Mo blue colour. The intensity of blue colour on reduction provides a measure for the concentration of P in the test solution.

Bray and Kurtz (1945) suggested a combination of HCl and NH_4F to remove easily acid-soluble P forms, largely Al and Fe phosphates popularly known as Bray's Phosphorus. In 1953, Mehlich introduced a combination of

HCl and H_2SO_4 acids (Mehlich 1) as extractants to extract P from soils. This acid solution can dissolve Al and Fe phosphates in addition to P adsorbed on colloidal surfaces in soils. Later on in 1984, Mehlich changes the composition of extractant, which is a combination of acids (acetic acid and nitric), salts (ammonium fluoride and ammonium nitrate [NH_4NO_3]), and the chelating agent ethylenediaminetetraacetic acid (EDTA) known as Mehlich III. Olsen and Sommers (1982) used 0.5-M sodium bicarbonate ($NaHCO_3$) solution at a pH of 8.5 to extract P from calcareous, alkaline and neutral soils. This extractant decreases calcium in solution (through precipitation of calcium carbonate), and this decrease enhances the dissolution of Ca phosphates. Moreover, this extracting solution removes dissolved and adsorbed P on calcium carbonate and Fe-oxide surfaces.

The two common procedures applied are:

1. Bray's No.1 method (pH around 5.5 or less)—for acidic soil (Bray and Kurtz 1945)
2. Olsen's method (slightly acidic, neutral and calcareous soil) (Olsen and Sommers 1982)

20.8.1 Available P by Bray's Method

(A). Principles
This method has been widely used *as an index of available P in soil*. The combination of HCl and NH_4F is designed to remove easily acid-soluble P forms, largely calcium phosphate, and a portion of the aluminium and iron phosphates. The NH_4F dissolves aluminium and iron phosphate by its complex ion formation with these metal ions in acidic solution.

(B). Equipment: Spectrophotometer
(C). Chemicals and Reagents

Chemicals

1. Ammonium fluoride, NH_4F—1.11 g/L	4. Concentrated hydrochloric acid, HCl
2. Ammonium molybdate (AR grade)	5. Stannous chloride, $SnCl_2$
3. HCl—0.025, 10N	6. KH_2PO_4

Reagents

1. *Bray's No.1 reagents (for extraction)—2.5 L: Consists of 0.03-N NH_4F in 0.025N HCl solution. Dissolve 2.775 g of NH_4F to 2.5 L of* 0.025-N HCl (check against standard NaOH solution).
   - *Preparation of 500-mL extracting reagent*: Dissolve *0.555 g of NH_4F in 500 mL of 0.025N HCl. (HCl solution—0.025N*: Add 41.5 mL of concentrated HCl in 500 mL of distilled water to get 1-N HCl. Next, add 12.5 mL of 1N HCl in 500 mL of water to get 0.025N HCl). It will keep in glass bottle more than 1 year.
2. *Dickman's and Bray's reagent (for Bray's No.1 method)—1 L*: It is used for development of colour. Dissolve *15 g* of ammonium paramolybdate [$(NH_4)_6MO_7O_{24}. 4H_2O$] (AR) in 300 mL of distilled water, warm to about 60°C and filter, if necessary, after cooling. To this, add *350-mL 10N HCl* and make up to 1,000 mL. The normality of the HCl should be adjusted correctly by titration. *Store the solution in a black glass-stoppered bottle. Prepare a fresh solution every 2 months.*
   - *For 100 mL of reagent: Dissolve 1.5 g of* ammonium molybdate (AR) in 30 mL of distilled water; warm to about 60°C and filter, if necessary, after cooling. To this, add *35-mL 10N HCl.*
3. *Preparation of 10N HCl:* Concentrated HCl is available with sp. gr. 1.174–1.189, active ingredients 36–37% and normality in between 11N and 12N. Add 909 mL of 11N HCl in 1,000 mL that will give 10N, or 833.34 mL of 12N HCl in 1,000 mL will give 10N.
4. *Stannous chloride solution ($SnCl_2$, $2H_2O$)*: *10 g* of crystalline stannous chloride (LR) is dissolved in *25 mL* of concentrated HCl by warming and stored in an amber-coloured bottle, carefully avoiding all contact with air. *This is 40% $SnCl_2$ stock solution.* Keep the solution in a black glass-stoppered bottle. Prepare a fresh solution every 6 weeks.
5. *Stannous chloride ($SnCl_2$), dilution solution*: Just before use, 0.5 mL is diluted to 66 mL with distilled water. Make a fresh solution

Table 20.3 Standard curve of phosphorus (Phosphorous standard solution 1 mL = 2 µg p)

Std. solution. mL	Bray's extracting reagents, mL	Dickman–Bray's reagents, mL	SnCl$_2$ solution, mL	Distilled water, mL	Final volume, mL	Con. µg/mL or in PPM
1	5	5	1	13	25	0.08
2	5	5	1	12	25	0.16
3	5	5	1	11	25	0.24
4	5	5	1	10	25	0.32
5	5	5	1	9	25	0.40
10	5	5	1	4	25	0.80

The colorimeter reading was taken against 660 nm filter just after 10 min

every 2 h as needed. A piece of tin metal (AR) is added to keep the stock solution for longer use.

6. *Standard phosphorous solution (100 ppm or 1 mL = 100 µg)*: Analytical grade dihydrogen orthophosphate (KH$_2$PO$_4$) is dried in air oven at 60°C for 1 h, and after cooling (desiccator), exactly *0.439 g* is dissolved in about 500 mL of distilled water. Then, 25 mL of 7N H$_2$SO$_4$ (approx.) is added and made up to 1,000 mL with distilled water. (7N H$_2$SO$_4$: Add 19.45 mL of concentrated H$_2$SO$_4$ (sp. gr. 1.84) in 100 mL of distilled water). *This gives a 100-ppm stock solution of phosphorus (100 µg P/mL).*

7. *Standard working phosphorous solution (2 ppm or 1 mL = 2 µg p)*: Dilute 20 mL of standard phosphorous solution to 1,000 mL with distilled water. This solution contains *2 µg p/mL.*

(D). Procedure (for Determination of Bray's Phosphorous)

I. *Extraction*

   Take 5 g of soil and 50 mL of the Bray's reagent in a 100-mL conical flask, and shake for 5 min. Filter the mixture by using Whatman no. 42 paper. If the filtrate is not clear, filter it again.

II. *Colour Development and Estimation*

   1. Take 5 mL of the soil extract (filtrate) into a 25-mL volumetric flask to which add 5 mL of Dickman's and Bray's reagent.
   2. The neck of the flask is washed down, and the contents are diluted to about 22 mL.
   3. Then 1 mL of the dilute stannous chloride solution is added, and volume is made up to the mark.

4. The intensity of the blue colour is measured (using 660 nm filter) just after 10 min, and the concentration of phosphorus is determined from the standard curve.

5. This is very important as the colour starts fading after some time.

6. With each set of samples, a blank is (without soil) also run

III. *Development of Standard Curve for Phosphorus*

   1. For preparation of the standard curve different concentration of phosphorus (1,2,3,4,5 and 10 mL of 2-ppm phosphorus solution) are taken in 25-mL volumetric flasks. Prepare the standard concentration of phosphorous in the range of 0.08–0.80 µg/L as follows (Table 20.3):

   The curve was plotted taking the colorimeter reading on the vertical axis and the amount of phosphorus (in µg P/mL) in the horizontal axis.

(E). Calculation

The available P in kg/ha can be calculated using the formula:

$$Bray's\ P\ (kg/ha) = R \times \frac{50}{5} \times \frac{1}{5} \times 2.24$$
$$= µg\ of\ P \times 4.48$$

where R = µg of phosphorus in aliquot (obtained from standard curve).

(F). Comments

Boric acid eliminates possible interference of fluorides in the colour development, but its necessity for most soils has not been established. Some acid sandy soils may show interference from fluorides. The soil test values (soil/solution, 1:7) are interpreted in general as follows: <3 ppm, very low; 3–7 ppm, medium; and >20 ppm, high (Olsen and Sommers 1982).

20.8.2 Available P by Olsen's Method

(*Phosphorous Soluble in Sodium Bicarbonate*)

This procedure is recommended for calcareous soil, particularly those containing >2% caCO$_3$. The extraction solution is a solution of weak sodium bicarbonate. The correlation between the amounts of P extracted from the soil with the P taken up by the plants ranges from 0.63 to 0.73 for the alkaline soil conditions.

(A). Principles

Phosphorous is extracted from soil with 0.5 M NaHCO$_3$, at a nearly constant pH of 8.5. In calcareous, alkaline or neutral soils, containing Ca phosphate, this extractant decreases the concentration of Ca in solution by causing precipitation of Ca as CaCO$_3$; as a result, the concentration of P in solution increases. The amount of P extraction varies with temperature and shaking time.

(B). Chemicals and Reagents
Chemicals
1. Sodium bi carbonate (NaHCO$_3$)—42 g/L
2. Sodium hydroxide (NaOH)—10% (10 g in 100 mL)
3. H$_2$SO$_4$ (concentrated)—141 mL.
4. Ammonium molybdate [(NH$_4$)$_6$MO$_7$O$_{24}$·4H$_2$O]—15 g
5. Hydrochloric acid (HCl, 10N)—400 mL
6. Potassium di hydrogen phosphate (KH$_2$PO$_4$)—0.4393 g for 1-L stock solution

Preparation of Reagents
1. *Sodium bicarbonate (NaHCO$_3$) solution, 0.05 M*: Dissolve *42.0 g* of NaHCO$_3$ (laboratory reagent) in 1,000 mL of distilled water. The pH is adjusted to 8.5 by adding small quantities of 10% NaOH solution. Add mineral oil to avoid exposure of the solution to the air. Prepare a fresh solution before use, if it has been standing over 1 month in a glass container. *Store the solution in a polythene container for periods more than 1 month, but check the pH of the solution each month.*

2. *Sulphuric acid (H$_2$SO$_4$), 5N:* Add 141 mL of H$_2$SO$_4$ to 800 mL of distilled water. Cool the solution and dilute to 1,000 mL with distilled water.

3. *Dickman's and Bray's reagent having excess of acid for Olsen's method*: 15 g of ammonium molybdate (AR) is dissolved in 300 mL of warm water (about 60°C), cooled and filtered if necessary. To this, 400 mL of 10N HCl is added and made up to 1 L.

4. *Standard phosphate solution*: Dissolve *0.4393 g* of potassium dihydrogen phosphate (KH$_2$PO$_4$) into a 1-L volumetric flask. Add 500 mL of distilled water, and shake the content until the salt dissolves. Dilute the solution to 1 L with distilled water. Add five drops of toluene to diminish microbial activity. This solution contains 0.1 mg of P/mL.

5. *Dilute phosphate solution*: Dilute 20 mL of standard phosphate solution to 1 L with distilled water. The solution contains 2 mg P/mL.

(C). Procedure
Preparation of Extract
1. Add *2.5 g* of soil and 50 mL of extracting solution (NaHCO$_3$) in a 250-mL conical flask.
2. Shake the flask for 30 min with a suitable shaker.
3. Filter the suspension through the Whatman no. 40 paper.
4. Add activated carbon (free of phosphorous), if necessary, to obtain a clear filtrate.
5. Shake the flask immediately before pouring the suspension into the funnel.

Colour Development
1. Take 5 mL of extract into a 25-mL volumetric flask, to which 5 mL of Dickman and Bray's is added[*] drop by drop with constant shaking till the effervescence due to CO$_2$ evolution ceases.

Table 20.4 Calibration of Bray-I, Melich III and Olsen phosphorous (in ppm)

P sufficiency level	Bray-I	Melich III	Olsen	P fertiliser recommended, Kg/ha
Very low	< 5	<7	<3	25
Low	6–12	8–14	4–7	15
Medium	13–25	15–28	8–11	8
High	>25	>28	>12	0

Source: Soil fertily and fertiliser by Tisdale et al. (1985), pp. 205

2. The neck of the flask is washed down and the contents are diluted to about 22 mL. (*Acidification to be checked, pH 5.0; if less, then acidify with 5N H_2SO_4 to pH 5.*)
3. Then 1 mL of diluted $SnCl_2$ is added and volume is make up to mark. The colour is stable for 24 h, and maximum intensity is obtained in 10 min.

Preparation of the Standard Curve (Like Bray's Method).

(*D*). *Calculation*: Olson's P (kg/ha)

$$\text{Olson's P (Kg/ha)} = R \times \frac{V}{v} \times \frac{1}{W} \times \frac{2.24 \times 10^6}{10^6}$$

$$= R \times \frac{50}{5} \times \frac{1}{2.5} \times 2.24$$

$$= \mu g \, P \times 8.96$$

where
V = total volume of extractant (50 mL)
v = volume of aliquot taken for analysis (5 mL)
W = weight of soil (2.5 g)
R = microgram P in the aliquot (from standard curve)

(*E*). *Comments*
- Possible sources of variation in the analytical results in the $NaHCO_3$, extraction method, are associated with the temperature of the extracting solution and the shaking period. The extractable P increase approximately 0.43, expressed as P in ppm soil, for each degree rise in temperature 20 and 30°C, forsoils testing between 5 and 40 ppm of P.

- Plastic containers are preferable to glass for storing the extracting solution. If glass is used, a fresh solution should be prepared every month since pH tends to increase with time. This changes cause higher values for extractable P.
- The relation between the P soluble in the extract and the expected yield response to applied fertiliser P is as follows: <5 ppm, a response; between 5 and 10 ppm, a probable response; and >10 ppm, a response unlikely. The relationships between Bray-I, Melich and Olsen P are given in Table 20.4.

(*F*). *Phosphorus Fertiliser*
The P content of fertiliser is expressed as the oxide P_2O_5 instead of elemental P. The conversion of % P to % P_2O_5 and vice versa is simple and given by

$$\% \, P = \% \, P_2O_5 \times 0.49$$
$$\% \, P_2O_5 = \% \, P \times 2.29$$

The conversion factors are derived from the ratio of molecular weights of P and P_2O_5

$$= \frac{2 \times \text{molecular wt of P}}{\text{Molecular wt of } P_2O_5} = \frac{2 \times 31}{142} = 0.43$$

(*G*). *Exercise of Phosphorus Estimation*
1. What is the original source of soil P?
2. What form of P is available to plant?
3. For acidic soil which method is used?
4. What is Bray's reagent?
5. A soil having pH >8, which method will be used for P estimation?

6. A fertiliser contain 46% P_2O_5. To what percent of P does this correspond?

20.9 Total Phosphorous (*Digestion Method*)

(*A*). *Principle*

Total P analysis of soils requires the conversion of insoluble materials to soluble forms suitable for colorimetric procedures. The two most widely used methods for extracting the total P from soils are:

- Digestion with $HClO_4$
- Fusion with Na_2CO_3

(*B*). *Reagents*

1. *Perchloric Acid ($HClO_4$), 60%*
2. *Ammonium paramolybdate-vandate*: Dissolve 25 g of ammonium paramolybdate $[(NH_4)_6Mo_7O_{24}.4H_2O]$ in 400-mL distilled water. Dissolve 1.25 g of ammonium metavanadate (NH_4VO_3) in 300 mL of boiling distilled water. Cool the solution, and add 250 mL of concentrated HNO_3. Cool the solution to room temperature. Pour the ammonium paramolybdate solution into the NH_4VO_3–HNO_3 solution, and dilute to 1 L with distilled water.
3. *Standard phosphate solution*: Dissolve *0.4393 g* of potassium dihydrogen phosphate (KH_2PO_4), and dilute the solution to 1 L with distilled water. Add five drops of toluene to diminish microbial activity. This solution contains 0.1 mg of P/mL.
4. *Dilute phosphate solution*: Dilute 20 mL of standard phosphate solution to 1 L with distilled water. *The solution contains 2 μg P/mL.*

(*C*). *Procedure*

1. Mix 2 g of finely ground soil (0.5 mm) with 30 mL of 60% $HClO_4$ in 250-mL volumetric or Erlenmeyer flask.
2. Digest the mixture at a temperature a few degrees below the boiling point on a hot plate in a fume hood until the dark colour due to organic matter disappears.
3. Then continue heating at the boiling temperature 20 min longer. At this stage, heavy fumes of $HClO_4$ appear, and the insoluble material becomes like white sand. If necessary, add 1 or 2 mL of $HClO_4$ to move down any black particles that stick to the sides of the flask.
4. If sample is high in organic matter, add 20 mL of HNO_3 and heat to oxidise the sample. Then add the $HClO_4$. The total digestion with $HClO_4$ usually requires about 40 min. Cool the mixtures. Add distilled water to obtain the volume of 250 mL and mix the content.

(*D*). *Colour Development*

1. Pipette an aliquot that contains 0.05–1.0 mg of P of the solution into a 50-mL volumetric flask.
2. Add 10 mL of the vanamolybdate reagent and dilute the solution to 50 mL with distilled water.
3. Measure the optical density after 10 min at wavelengths from 400 to 490 nm. The sensitivity varies tenfold between these wavelengths with the higher sensitivity at the lower wavelength.
4. Prepare a reagent blank and subtract its optical density from the optical density of the sample.

20.10 Phosphorous Fixing Characteristics of Soil

Soils vary appreciably in their P-fixing capacity. The P-fixing capacity can be determined as per procedure given by Ghosh et al. (1983).

(*A*). *Reagents*

Standards phosphorus solution: A stock solution of 2,500 ppm of P is prepared by dissolving 9.440 g of pure anhydrous monocalcium phosphate, Ca $(H_2PO_4)_2$, in distilled water and made to 1 L (old stock should not be used). It is desirable to check the concentration of P by Dickman and Bray's method after appropriate dilution. Out of this, a solution of different concentration of P is prepared such that by adding 1 mL each to 2.0-g soil, the resultant concentration of added P would be 0, 25, 50, 75,100, 125, 250, 375 and 500 ppm.

(B). Procedure
Two gram of soil is weighed out (in duplicate) in separate 100- mL conical flask. One milliliter of solution of different concentration of P is added to each so that they cover the range from 0 to 500 ppm as stated above. The flasks are plugged with cotton wool and allowed to incubate for 96 h at room temperature, and thereafter, the P is extracted by Olsen's or Bray's method and estimated colorimetrically.

The results of extracted P are plotted (*Y*-axis) on the graph sheet against added P (*X*-axis). The curve is examined carefully, and the point where there is an abrupt increase in extractable P is located. Form this inflexion point (on the curve), the corresponding value of *X* is found out, that is, approximate amount of P required to be added to more than overcome the effect of fixation. If the fixation curve is linear, its shape may be considered to work out the amount of P to be applied which would take into account the fixing capacity.

(C). Calculation (for Percentage P Fixation)

Amount of available P in control
  (without P added) $= a$ ppm
Amount of available P in test sample $= b$ ppm
Amount of available P extracted $(b - a) = \times$ppm
Amount of P added $= c$ ppm
Amount of P fixed $= (c - x)$ ppm
Percentage of P fixation (% of added P)
  $= [(c - x)/c] \times 100$

20.11 Exchangeable Potassium (K)

The elements Li, *Na and K* belong to group 1A of the periodic table and are part of the group known as the *alkali metals*. They are characterised by a single electron in their outermost cell. This electron is easily lost and thereby readily form stable monovalent ions.

Exchangeable K is the major source of K to plants. It is present in relatively large quantities in most soils, averaging about 1.9%. The various forms of K in soils can be classified on the basis of availability in three general groups:
- Unavailable—90 to 98% of the total soil K
- Slowly available—1 to 10%
- Readily available—0.1 to 2%

A relatively small portion of total K in soils is exchangeable (approximately 1%). Exchangeable K ranges from <100 to 2,000 ppm or more compared with total K values in the order of 1–2%.

Potassium fertility rating (Ghosh et al. IARI manual, 1983)
- High fertility $\geq$ 280 kg/ha (>= 125 ppm)
- Medium fertility = 120–280 kg/ha (54–125 ppm)
- Low fertility $\leq$ 120 kg/ha (<= 54 ppm)

(A). Principles
The term available potassium incorporates both exchangeable and water-soluble forms of nutrients present in the soil. The readily exchangeable plus water-soluble potassium is determined in the neutral normal ammonium acetate (1N NH_4OAc) extract of soil. But complications arise from the fact that plants can also utilise some of the non-exchangeable potassium. A recent study indicates that potassium concentration in soil solution (saturation extract) is a direct measure of potassium availability (Biswas and Mukherjee 1994).

(B). Instrument
1. Flame photometer with accessories
2. Whatman filter paper
3. Volumetric flask
4. Mechanical shaker (reciprocating)
5. Centrifuge

(C). Reagents
1. *Ammonium acetate solution*, 1(N): Dissolve 70 mL of reagent grade concentrated NH_4OH to 57 mL of glacial acetic acid (99.5%) and

Table 20.5 Preparation of different K standards (stock solution 1 mL = 1 mg K)

K- Stock solution, mL	Add NH$_4$OAc to make 100 mL final volume	Concentration of Na in ppm (mg/L)	Flame photometer reading
1	99	10 ppm	
2	98	20 ppm	
3	97	30 ppm	
4	96	40 ppm	
6	94	60 ppm	
Blank	100	0 ppm	

dilute to 1 L. Adjust the pH 7.0 with either addition of NH$_4$OH or glacial acetic acid.

2. *Standard potassium solution* (1,000 μg K/mL): Dissolved *1.90676 g of dried potassium chloride (KCl)* in distilled water, dilute the solution to *1000 mL* and mix it. The solution contains 1,000 ppm *of K (1 mL = 1 mg)*.

   (Molecular wt K = 39.0983; Cl = 34.453; Mol wt of KCl = 74.5513; thus 1 g K = 74.5513/59.0983 = 1.90676 g of KCl)

(D). Procedure

Since extraction with neutral 1N NH$_4$OAc is the procedure of preference by definition, two extraction ration of soil-ammonium acetate is discussed:

Extraction by Centrifugation and Decantation Procedure (1:10)

1. Place 10 g of soil (or use 5 g if the soil contains >500 ppm of K) in a 50-mL centrifuge tube.
2. Add 25 mL of NH$_4$OAc, and shake the tube for 10 min.
3. Centrifuge the tube until supernatant is clear.
4. Decant the supernatant liquid into a 100-mL volumetric flask.
5. Make three additional extractions in the same manner.
6. Dilute the combined extract to 100 mL with NH$_4$OAc.
7. Mix the solution and determine emission readings on a flame photometer at 767-nm wavelength.

Extraction (As suggested by Ghosh et al. 1983) (Soil: extractant = 1:5)

1. Weigh 5 g air-dried soil (<2-mm) that is shaken with 25 mL of neutral normal ammonium acetate solution for *5 min and filtered immediately* through a dry filter paper (Whatman no.1).
2. First few mL of the filtrate is rejected.
3. Potassium concentration in the extract is determined by flame photometer by using K filter.

Preparation of Standard Curve for K

From the stock solution, *10–60 ppm K* solution is prepared by adding ammonium acetate solution. After attaching the appropriate filter, gas and air pressure are adjusted in the flame photometer, and reading is adjusted to zero for the blank (ammonium acetate). Flame photometer reading is noted at different concentration for K solution. (10–60 ppm). The curve is obtained by plotting the flame photometer readings against the different concentration of K (Table 20.5).

Flame Photometer Calibration

Set the readout to zero using distilled water as a blank. Set the peak reading according to the instrument instructions using the most concentrated sodium solution (100 mg/mL). Measure the emission intensity of each of the remaining sodium standard solutions and of the sodium unknown solution. Check for accuracy and repeatability by measuring the standards several times. Be sure to aspirate deionised distilled water between measurements.

(E). Calculation

1. Available Potassium (in ppm) $= R \times \dfrac{V}{W}$

$$= R \times \frac{100 \text{ mL of extract}}{10 \text{ g of soil}} = R \times 10$$

2. Available potassium (mg of K/g of soil)

$$= R \times \frac{V}{W} \times \frac{1}{100} = R \times \frac{100}{10} \times \frac{1}{100} = \frac{R}{10}$$

Where,

R = K content of soil extract from standard curve, mg/l;

V = Volume of the soil extract (mL)- in the present case 100 mL was taken as final volume of extract for calculation;

W = Weight of air-dried samples taken for extraction in g; in the present case 10 g was taken as final volume of extract for calculation;

K meq/L = K mg/L × 0.02558

K mg/L = K meq/L × 39.10

For example, 2 meq K/100 g soil = 780 ppm of K or 1 meq K/100 g soil = 390 ppm K

(E). Determination of Non-exchangeable K

1. Take 5 g of sieved soil (2 mm) in a dry 100-mL conical flask.
2. Add 50-mL 1-N HNO_3 and heat content on a hot plate (low heat) under reflux (fitted with cork and long glass tube) for 10 min after boiling starts.
3. On cooling, the suspension is filtered through Whatman no. 1 filter paper.
4. Determine the K in flame photometer after appropriate dilution.

   Non-exchangeable K = HNO_3 soluble K— ammonium acetate K.

(F). Comments

Probably the most universally employed index to K availability is the sum of the exchangeable and water soluble, that is, the K extracted by neutral 1-N NH_4OAc. As a soil test, an equilibrium extraction at a 1:5–1:10 soil/solution ratio is commonly used rather than a repeated or continuous leaching extraction. However, a wide range of soil extractant has been used and is still being used: those are $NaNO_3$, double acid (HCl-H_2SO_4) and Morgan extractant ($NaOAc$-$HOAc$).

K Fertiliser

More than 90% of potassium fertilisers used are in the form of KCl, the result being sulphate and a small amount of mixed salt. In India, all potassium fertilisers are imported. The average content of K_2O in commercial fertiliser is the following:

- Murate of potash (KCl)—58%
- Sulphate of potash (K_2SO_4. $2MgSO_4$)—48%
- Scoenite (K_2SO_4, $MgSO_4$. $6H_2O$)—23%

   K contents of fertilisers are guaranteed in terms of its oxide (K_2O) equivalent. Conversion of % K and % K_2O and vice versa can be accomplished by the following relationships:

$$\%K = \% \ K_2O \ / 1.2$$
$$\% \ K_2O = \%K \times 1.2$$

(G). Exercise on K Estimation

1. What are the sources of K in soil?
2. In what form do plants absorb K?
3. Name different forms of K present in soil.
4. How is plant-available K from the soil extracted?
5. How is non-exchangeable K in soil?
6. Give K rating of Indian soil.
7. Name few common K fertiliser.
8. A fertiliser is guaranteed to contain 30% K_2O. To what percentage of K does this correspond?

20.12 Exchangeable Sodium (Na)

Most soil analysts interested in Na are concerned with the diagnosis of sodic and sodic-saline problems. There are widespread problems in the arid areas of the world and especially in arid and semi-arid regions in world where irrigation is practised.

(A). Apparatus: *Same as Potassium*

(B). Procedure

The estimation of Na in a soil sample consists of following steps:

1. Extraction of Na from soil sample.
2. Preparation of standard curve of Na.
3. Estimation of Na from the standard curve
4. Calculation of results.

(C). Extraction of Na from Soil Sample
The extraction solution of K can be used for analysis of Na.

(D). Preparation of Standard Curve of Na
Standard Stock Na Solution: Weigh *2.542 g* of sodium chloride (NaCl) that has been dried at 110°C. Dissolve the NaCl in distilled water and dilute the solution to a volume of *1,000 mL*. This solution contains 1,000 ppm (1,000 mg/L) of Na (1 mL = 1 mg Na = 1,000 μg Na). Prepare different concentration of Na, preferably in the range of 10, 20, 30, 40 and 60 ppm, as shown in Table 20.6. Estimate Na directly from the standard curve.

(E). Calculation
1. Exchangeable sodium (in ppm) $= R \times \dfrac{V}{W} =$

$$R \times \frac{100 \text{ mL of extract}}{10 \text{ g of soil}} = R \times 10$$

2. Exchangeable sodium (mg of Na/g of soil)

$$= R \times \frac{V}{W} \times \frac{1}{100} = R \times \frac{100}{10} \times \frac{1}{100} = \frac{R}{10}$$

Where,

R = Na content of soil extract from standard curve, mg/l;

V = Volume of the soil extract (mL)- in the present case 100 mL was taken as final volume of extract for calculation;

W = Weight of air-dried samples taken for extraction in g; in the present case 10 g was taken as final volume of extract for calculation.

$$Na^+ meq/L = mg/L \times 0.04350$$
$$Na^+ mg/L = Na^+ meq/L \times 22.99$$

Note: 1 meq/100 g soil = 1 Cmol (p+)/ kg of soil

(F). Comments
Since Na is not an essential element for most plants, there has been little concern about assessing its plant availability by soil test. The best index of availability of Na in nonsaline soil is the amount in the NH_4OAc extract. The effect of exchangeable Na on the physical properties of soil is more related to the percent saturation than to the total quantity present.

20.13 Exchangeable Calcium and Magnesium

Ca and Mg are essential nutrients to plants and are widely distributed and generally abundant elements in soil. The Ca and Mg share common chemical properties, like their natural occurrence as carbonate, phosphate, sulphate or silicates, and they precipitate in similar fashion.

Calcium is absorbed by plants as Ca^{2+} from soil solution. Ca deficiency is uncommon but can occur in highly leached and unlimed acid soil. Ca in soil solution ranges from 30 to 300 ppm. In soils of higher-rainfall, soil solution Ca^{2+} concentration will vary from 8 to 45 ppm (average 33 ppm). A level of 15 ppm Ca^{2+} in soil solution adequate. Even though, Ca^{2+} concentration of the soil solution is about 10 times greater than K^+, its uptake is usually lower than that of K^+. Ca is supplied by both dolomite and calcite limestone as well as gypsum.

Table 20.6 Preparation of different Na standards (stock Na solution 1 mL = 1 mg Na)

Na-stock solution, mL	Add NH_4OAc to make 100 mL final volume	Concentration of Na in ppm (mg/L)	Flame photometer reading
1	99	10 ppm	
2	98	20 ppm	
3	97	30 ppm	
4	96	40 ppm	
6	94	60 ppm	
Blank	100	0 ppm	

(A). Exchangeable Fractions (Plant-Available Fraction)

The exchangeable fraction is most commonly measured and reported as alkaline earth metals. The extractant, 1-N ammonium acetate (NH_4OAc) with pH 7.0 is a most widely accepted reagent for extraction of these elements from soil and widely regarded as *available to plants.* Another fraction such as that extracted by 0.025N $CaCl_2$ may be considered as *readily available.* When extracted with 1-N HCl, it may include *potentially available.*

(B). Acid-Soluble or Non-Exchangeable Fraction (Potentially Available)

The acid soluble fraction is not susceptible to exchangeable reactions, and it does not require complete dissolution of the sample.

Reagents

1. Hydrochloric acid (HCl), 1N
2. Hydrochloric acid (HCl), 0.1N

Take 2.5 g of <2-mm soil in a 250-mL Erlenmeyer flask. Add 100 mL of 1N HCl, and heat until boiling starts. Boil the suspension for 45 min. Remove the flask from the heat, and transfer the content to a funnel lined with Whatman no.1 filter paper. Received the filtrate in a 250-mL volumetric flask, and wash the soil with four 15-mL portions of 0.1N HCl. After the filtrate has cooled, dilute it to volume and mix thoroughly. Analyse the cation by a convenient method.

(C). Total Analysis

Both sodium carbonate fusion and HF–HCl dissolution have been used. Total analysis is not useful for determining their availability to plants.

(D). Analysis of Ca and Mg.

Once a fraction of Ca and Mg has been obtained from soil, these can be determined by

- Atomic absorption spectrometry (AAS)
- Ion selective electrodes
- EDTA-titrimety methods

Although AAS is most widely used, but EDTA-titrimety methods for Ca and Mg may be considered when an instrument is not available.

(E). Estimation of Ca and Mg by EDTA Titrimetry Method

a. Principles

Exchangeable Ca and Mg can be determined in NH_4OAc extracts of soils by titration with EDTA. Both Ca and Mg may be titrated at pH 10 using Eriochrome black T (EBT) as an indicator. Magnesium forms insoluble $Mg(OH)_2$ at pH 12 or higher if NH_4^+ salts are present, thereby allowing Ca to be titrated using *Murexide or Calgon* as an indicator.

Reagents

1. *Ammonium chloride–ammonium hydroxide buffer solution*: Dissolve *67.5-g* NH_4Cl in 570 mL of concentrated NH_4OH and make to 1 L. Store in a plastic or borosilicate glass container for no longer than 1 month. Stopper tightly to prevent loss of ammonia (NH_3) or pick up of carbon dioxide (CO_2) from air.
2. *Eriochrome black T indicator*: Mix 0.5 g dye with and 4.5 g of hydroxylamine hydrochloride in 100 mL of 95% ethanol *or* dissolve 0.5 g of indicator in 100 mL of ethyl alcohol.
3. *Murexide (ammonium purpurate) indicator*: *The indicator changes from pink to purple at the end point.* Mix 0.2 g of ammonium purpurate and 100 g of sodium chloride, and grind thoroughly to fine powder (i.e. one that passes through a 40-mesh sieve).
4. *Sodium hydroxide solution (4N)*: Dissolve 160 g of NaOH in 1 L of distilled water.
5. *Standard calcium solution (0.01N)*: Weigh accurately *0.5 g* AR grade $CaCO_3$ (dried at 180°C for 1 h before weighing) and dissolve in a minimum of 0.2N HCl. This solution is boiled to expel CO2 and is then diluted to 1 L. The solution is 0.01N with respect to Ca.
6. *Standard EDTA solution (0.01N)*: Dissolve 3.723-g EDTA-sodium salt and dilute to 1 L. Standardised against standard Ca solution; *1 mL = 1 mg CaCO₃. The reagent is stable for several weeks.*

b. Procedure

Standardisation of EDTA solution

Table 20.7 Format of table for recording the observations for estimation of Ca and Mg in soil and spoils samples (Maiti 2003)

| No of observation | Volume of a liquot taken, mL | Volume of EDTA titrant, mL | | Mean reading for Ca (B) | Mean reading for Mg (A–B) |
		A for (Ca + Mg)	B for (Ca only)		
1	10	10.2	4.9	4.9	5.35
2	10	10.3	4.9		

Estimation of total Ca^{2+} and Mg^{2+}

1. 10 mL of aliquot is taken in a 100-mL conical flask.
2. Next add 5 mL of buffer solution (which will raise the pH to 10 and prevent the precipitation of Mg).
3. Add 2–3 drops of *Eriochrome black T indicator*.
4. The content is titrated with standard EDTA solution to a pure blue end point.

Estimation of Ca^{2+}

1. Another 10 mL of aliquot is taken in a 100 mL of conical flask, and 5 mL of 4N NaOH is added to precipitated Magnesium as Mg $(OH)_2$.
2. Next, add a pinch of *Murexide indicator*.
3. The content is titrated against 0.01N EDTA to an orange red to purple or violet colour.

Observation
The observations are recorded as per Table 20.7

c. Calculation

1 mL of 0.01M EDTA = 0.4008 g Ca $^{2+}$

1 mL of 0.01M EDTA = 0.2432 g of Mg^{2+}

Hence, 4.9 mL of EDTA = 0.4008 × 4.9 mg of Ca^{2+} = 1.96392 mg of Ca^{2+}

In 10 mL of soil extract, Ca^{2+} present = 1.96392 g

In 1,000 mL of soil, Ca^{2+} present = 196.392 g/mL or 196.392 mg/L or ppm (Ans)

Again 1-mL 0.01-M EDTA = 0.2432 g of Mg^{2+}

5.35-mL of 0.01-M EDTA = 0.2432 × 5.35 = 1.30112 g of Mg^{2+}

In 10 mL of soil extract, Mg^{2+} present = 1.30112 g of Mg^{2+}

In 1,000 mL of soil, Mg^{2+} present = 1.30112 × 1,000/10 = 130.112 g/mL or 130.112 mg/L or ppm (Ans)

Results: Ca = 196.392 ppm and Mg = 130.112 ppm

Example 20.1 Conservation of meq to ppm

Convert 10 meq Ca/100 g soil = ppm? (Ans, *2,000 ppm*.)

How,

Remember that 1 ppm = 1 mg/kg; mg per 100 g soil × 10 = mg per kg

The meq wt of Ca^{2+} = 40/2 = 20 mg/meq

10 meq × 20 mg/meq = 200 mg/100 g soil × 10 = *2,000 ppm*

(*or* 1 meq Ca = 200 ppm of Ca)

(F). Exercise Ca and Mg Estimation

1. What forms of Ca and Mg are absorbed by plants?
2. Are deficiencies of Ca and Mg common?
3. How are plat available fraction and potentially available fraction of Ca and Mg in soil determined?
4. Why is total analysis of Ca and Mg not useful?
5. How are Ca and Mg analysed after the extraction into solution?

20.14 Cation Exchange Capacity (CEC)

(A). Introduction
Cation exchange capacity is the maximum quantity of total cations that a soil can hold at a given pH value, available for exchange with the soil solution. CEC is used as a measure of fertility, nutrient retention capacity and the capacity to

Table 20.8 Generalised relationship between soil texture and CEC

Soil texture	CEC (Cmol (P^+)/Kg of soil)
Sands	1–5
Fine sandy loams	5–10
Loam and silty loams	5–15
Clay loams	15–30
Clays	>30

Table 20.9 CEC of some Indian soil

Soil type	CEC (Cmol (P^+)/Kg of soil)
Alluvial soil	5.6–30.4
Black soil	44.2–61.3
Red soils	4.9–9.2
Lateritic soil	3.2–6.9
Forest soil	31.1–29.7
Sodic soil	8.5–11.6

protect ground water from cation contamination. Cations are positively charged ions consisting of Calcium (Ca^{2+}), Magnesium (Mg^{2+}), Potassium (K^+) and sodium (Na^+), and the acidic cations include hydrogen (H^+) and aluminium (Al^{3+}). These positive ions are absorbed on the negatively charged site of the soil colloids. The larger the CEC value, the more cations the soil can hold, and soil has good fertility. For example, a clay soil will have a larger CEC than a sandy soil. Increasing the organic matter content of any soil will help to increase the CEC value.

CEC is usually expressed in milliequivalents of hydrogen per 100 g of dry soil (meq+/100 g) or the SI unit is centi-mol per kg (cmol+/kg).

$$1 \text{ meq/100 g of soil} = 1 \text{ mmol/100 g of soil}$$
$$= 1 \text{ cmol } (P^+)/\text{kg of soil.}$$

For example, 200 meq/100 g of soil is same as 20 cmol (P^+)/kg of soil.

A soil having a CEC of 1 meq (1 meq/100 g soil) is capable of absorbing 1 mg of H or its equivalent per 100 g of soil. This is the total capacity of the soil to absorbed cations on to electrically charged sites, from where they are available to plants, but resistant to leaching. The CEC is dependent on soil texture (Table 20.8). Table 20.9 shows the CEC values of some of the Indian soil types.

Symbolically, the relationship is as follows:

$$CEC = Ca^{2+} + Mg^{2+} + K^+ + Na^+ + H^+$$
$$\text{or } T = S + H$$

where

T = total cations
S = sum of basic ions
H = hydrogen

$$\text{Base saturation } (\%) = \frac{S}{T} \times 100, \text{ and}$$

$$\text{Base saturation } (\%) = \frac{H}{T} \times 100 \text{ or } \frac{(T - S)}{T} \times 100.$$

Example 20.2 Calculation of CEC and Base Saturation

A soil has CEC = 8.1 meq/100 g; having exchangeable cations (cmol (P^+)/kg of soil) concentration of, Ca = 5.2, Mg = 2, K = 0.3 and Na = 0.4. What is the % of base saturation?

Solution
7.9/8.1 × 100 = 97.5% (ans)

(B). *Principle*
This method is particularly suited to arid land soils, including those containing carbonates, gypsum and zeolites. Exchange capacity is preferably determined in 1N ammonium acetate extract.

Example 20.3 Calculation of CEC and Base Saturation

Calculate the % of base saturation if soil has the following concentration of cations:

Cations	meq/100 g soil
Na	0.03
K	0.28
Mg	0.12
Ca	1.0
CEC	3.83

Solution
Total bases = Na + K + Mg + Ca
= 1.43 meq/100 g soil
% of base
saturation = (1.43/3.83) × 100
= 37% (ans)

(C). Apparatus
1. Flame photometer
2. Centrifuge; centrifuge tubes of 50 mL
3. Mechanical shaker, reciprocating

(D). Chemicals and Reagents
1. Sodium acetate trihydrate $(NaOAc.3H_2O)$—136 g/L
2. Ammonium acetate (NH_4OAc)—77 g/L
3. Ethanol (C_2H_5OH)—95%

Reagents
1. *Sodium acetate solution (1N)*: Dissolve 136 g of sodium acetate trihydrate in distilled water, and dilute to a volume of 1 L. The pH of the solution should be adjusted to approximately 8.2 either by adding more acetic acid or sodium hydroxide.
2. *Ammonium acetate solution 1(N)*: Dissolve 77-g ammonium acetate in distilled water, and dilute to 800 mL. Adjust pH to 7 with dilute ammonium hydroxide or acetic acid whichever is required, and make 1 L. Alternatively, dissolve 70 mL of reagent grade concentrated NH_4OH to 57 mL of glacial acetic acid (99.5%), and dilute to 1 L. Adjust the pH 7.0 with either addition of NH_4OH or glacial acetic acid.
3. *Standard stock solution of NaCl (1,000 ppm)*: Dry about 5 g of sodium chloride in a hot air oven at 105 oC for 2 h, cool in a desiccator and store in a tightly stoppered bottle. Dissolve 2.5418 g of dry NaCl in distilled water and make the volume up to 1,000 mL. This solution contains 1,000 ppm of Na (stock solution). Prepare a series of standard solution from the stock solution as follows: Dilute 2-, 4-, 6-, 8-, 10- and 15-mL stock solution to 100-mL final volume by adding 1N ammonium acetate solution. These solutions contain 20, 40, 60, 80, 100 and 150 ppm of Na.

(E). Procedure
Take adjacently *5 g of air-dried soil* (correct to overdry moisture content as determined using a separate subsample) and place it in centrifuge tube.

Step 1
1. Add *25 mL of 1-N sodium acetate solution*, stopper the tube and shake for 5 min.
2. Unstopper the tube and centrifuge the tube at 2,000 rpm approximately for 5 min or until the supernatant liquid is clear. Decant the supernatant liquid completely and discard.
3. Treat the sample in this with 25-mL portion of 1N sodium acetate four times, discarding the supernatant liquid each time. (After addition of fresh sodium acetate solution, shake for 30 s then centrifuge and discard the supernatant.)

Step 2
1. Add *25 mL of 95% ethanol* to the tube, stopper, shake for 5 min, unstopper and centrifuge until the supernatant liquid is clear. Decant and discard the supernatant liquid.
2. Wash the sample with 25-mL portion of ethanol three times. The electrical conductivity (EC) of the supernatant liquid from the third washing should be less than 400 µS/cm.

Step 3
Add *25 mL of 0.1N ammonium acetate solution* to the tube, stopper and shake for 5 min. Remove stopper and centrifuge at 2,000 RPM until the supernatant liquid is clear. Decant the supernatant liquid as much as possible as in 100-mL volumetric flask. *Repeat the extraction three times*, decanting into the same flask. By this process, ammonium ions will replace sodium ions, which will come in the supernatant liquid.
Dilute the supernatant liquid up to 100 mL in a volumetric flask and determine sodium concentration by flame photometer.

Observation
- Prepare the calibration curve for exchangeable sodium determination.

- Measure exchangeable sodium concentration in unknown sample (in supernatant) by a Flame photometer at 767 nm wavelength.

(F). Calculation

If the total volume of extract is not 100 mL, then CEC in meq/100 g

$$= \frac{\text{Na con of extrat in me/L} \times 100}{\text{wt of soil in g}}$$

$$\times \frac{\text{Vol of extract (mL)}}{1000}$$

If volume of extract is 100 mL, then

$$\text{CEC in meq/100g} =$$

$$\frac{\text{Na con in extractant in meq/L} \times 10}{\text{wt of soil in g}}$$

(G). Comments

1. CEC is an indicator of the nutrient storage capacity of soil.
2. In the reclamation of sodic (alkali) soil, the principle is to treat them with gypsum which is the sum of Ca^{2+}, so that a major portion of exchangeable Na is replaced with Ca^{2+}.
3. When potential pollutants, such as Pb (Pb^{++}) or cadmium (Cd^{++}), etc., are disposed on soil surface, instead of leaching easily, they are absorbed to the cation exchange sites.

(H). Exercise on CEC Estimation

1. What is CEC?
2. In there any relationship between CEC and % base saturation?
3. Which soil has high CEC sandy soils or clay soils—why?
4. List the different soil cations.
5. CEC is an important indicator of storage capacity of soils.

20.15 Plant-Available Sulphur

Sulphur (S) is the 13th most abundant elements in the earth crust, averaging between 0.06 and 0.10%. Sulphur is absorbed by roots as $SO_4^=$ ions. In soil containing 5 ppm or more $SO_4^=$, all of the requirement of most crop can be supplied by mass flow. $SO_4^=$ concentration of 3–5 ppm in soil solution is adequate for growth of plants. In S-deficient soils, the level of soluble $SO_4^=$ is usually 5–10 ppm. As $SO_4^=$ is anionic in nature, and solubility of sulphate salt is like NO_3, it can be easily leached from surface soil.

Form of S in soil: S is present in soil in both organic and inorganic form, although nearly 90% of the total S in most non-calcareous surface soil exists in organic form. The inorganic forms are $SO_4^=$ in solution, adsorbed $SO_4^=$, insoluble $SO_4^=$ and reduced inorganic S compounds.

(A). Sulphur Testing as $SO_4^=$

Two extractants are generally used for the extraction of $SO_4^=$ from soil:

- *Phosphate extractable $SO_4^=$* (extractant: KH_2PO_4 – 2.195 g/L or $Ca(H_2PO_4)_2.H_2O$- 1.888 g/L) Turbidimetric procedure (Ghosh 1983)
- *$CaCl_2$-extractable $SO_4^=$* (extractant: 0.15% $CaCl_2$. $2H_2O$) colorimetric procedure

(B). Phosphate Extractable $SO_4^=$ (Ghosh 1983)

1. Take 20 g of air-dried soil in a 250-mL conical flask.
2. Add 100 mL of 500-ppm P solution prepared from KH_2PO_4 – 2.195 g/L or $Ca(H_2PO_4)_2$. H_2O- 1.888 g/L into the conical flask.
3. Shake for 30 min.
4. Filter the suspension through Whatman no. 42 filter paper.
5. Determine the sulphate from a suitable amount of aliquot by turbidimetric procedure.

20.15.1 Turbidimetric method for sulphate estimation

(A). Reagents

1. *Nitric acid 25%* (approx.) prepared from AR grade HNO_3.
2. *Acetic-phosphoric acid*: Add 900 mL of AR glacial acetic acid with 300 mL of AR orthophosphoric acid.

3. *Gum acacia-acetic acid solution*: 5 g of chemically pure gum acacia is dissolved in 500 mL of hot water and the hot solution filtered through Whatman no.42 filter paper. The filtrate is cooled and then dilute to 1 L with dilute acetic acid.

4. *Barium sulphate seed suspension*: 18 g of AR grade barium chloride is dissolved in 44 mL of hot water and 0.5 mL of the concentrated standard sulphate solution (given below), is added brought to boiling and cooled quickly. Then 4 mL of the gum acacia-acetic acid solution is put in. The seed suspension is pared a fresh every day before use.

5. *Barium chloride crystals*: $BaCl_2$ analytical grade regent ground to pass through 1-mm sieve.

6. *Concentrated standard sulphate solution (2 mg/mL)*: Dissolve 1.080 g of oven-dried AR potassium sulphate dissolve in distilled water and dilute to 100 mL.

7. *Working-standard sulphate solution (10 µg/mL)*: 5 mL of the concentrated standard solution is dissolved in 1 L with distilled water.

(B). Procedure

1. A suitable aliquot of the extract (not more than 15 mL and containing less than 120 µg of sulphate sulphur) is taken in a 25-mL volumetric flask to which 2.5 mL of 25% nitric acid and 2 mL of acetic-phosphoric acid are added and diluted to about 22 mL.

2. The flask is stoppered and shaken.

3. Then 0.5 mL of barium sulphate seed suspension (which must be shaken before use), and 0.2 g of barium chloride crystals are added successively.

4. The flask is stoppered and inverted three times.

5. After 10 min it is inverted ten times and after 5 min another five times.

6. Allowing for another 5 min, 1 mL of gum acacia-acetic acid solution is put in and diluted to volume and inverted three times and set aside for 1 h and 30 min.

7. Then the flask is again inverted ten times and reading taken in a spectrophotometer (approx.

440 nm). The concentration of sulphur is found out from the standard curve.

(C). Preparation of Standard Curve for $SO_4^=$-Sulphur

Prepare the working standards of $SO_4^=$- sulphur by diluting 0.1, 3, 5, 8, 10 and 12 mL of working-standard sulphate solution in a 25-mL volumetric flask, and make up the volume. Next, proceed as described above.

The standards are to be prepared each time a batch of sample is analysed, and also a blank test with the reagents is to be run side by side.

(D). Exercise on Soil Sulphur Estimation

1. What form of S is found in soil?
2. Which of the S forms is of special importance in plant nutrition?
3. What are the different methods of soil S test?

20.16 Determination of Chloride

Chloride is ubiquitous in the soil environment. It is added to the soil from irrigation water, animal manure, fertilisers, plant residues, rainwater and others. Measurement of Cl^- in soil extract is a routine task in laboratories involved with salinity problems. Other importance of Cl^- estimation is that Cl^- is used in estimating leaching fractions and salt balance. A measurement of CEC of soil is based on the determination of Cl^-, along with NH_4^+ ions in soil extracts.

(A). Principle

Chloride is an essential plant micronutrient; Cl^- can range from as little as 35 ppm in severely deficient tissues to as much as several percent. Chloride is determined in *neutral or slightly alkaline solution* by titration with standard silver nitrate using potassium chromate as an indicator.

Chloride ions get precipitated with the action of $AgNO_3$ to form insoluble white precipitate of AgCl. When all chloride ions are removed, the excess of $AgNO_3$ reacts with potassium chromate forming silver chromate, which is red in colour. The end point is therefore the change in colour from bright yellow (due to K_2CrO_4 indicator) to brick red. The chemical reaction can be shown as:

$$AgNO_3 + Cl^- \rightarrow AgCl \text{ (White ppt.)} \downarrow + NO_3^-$$
$$2AgNO_3 + K_2CrO_4 \rightarrow Ag_2CrO_4 \text{(brick red ppt)}$$
$$\downarrow + 2KNO_3$$

(B). Reagents
1. *Standard sodium chloride, 0.05N (2.935 g/L):* Dissolve 2.935 g NaCl (dried at 140°C for 1 h) and dilute to 1,000 mL with chloride-free water.
2. *Standard silver nitrate titrant, 0.05N (8.494 g/ L):* Dissolve 8.494 g $AgNO_3$ and dilute to 1,000 mL. Standardise against NaCl (0.05N). *Store the solution in a dark glass bottle.*
3. *Potassium chromate indicator solution (50 g/ L):* Dissolve 50 g K_2CrO_4 in 200 mL of distilled water. Add $AgNO_3$ until red precipitate is formed. Allow to stand for 12 h. Filter and dilute to 1,000 mL with distilled water. *Store the solution in a dark-coloured bottle.*

(C). Procedure
I. *Standardisation of $AgNO_3$ Solution*
   1. Pipette out 20 mL of *0.05N* NaCl solution in a 250-mL conical flask.
   2. Add few drops of potassium chromate indicator. The colour turns to bright yellow.
   3. Titrate it against $AgNO_3$ solution. The appearance of brick red colour indicates the end point of titration.
II. *Estimation of Chlorides in Water Sample by Titration Against $AgNO_3$ Solution*
   1. Take 10 g of soil air-dried soil and dissolve it in 100 mL of distilled water.
   2. Filter the suspension and transfer to 250-mL conical flask.
   3. Acidify with 0.02N H_2SO_4 solution, just sufficiently to bring pH to between 6 and 7, checking with a narrow-range pH paper.
   4. Add two drops of saturated potassium chromate indicator solution.
   5. *Preparation of blank set:* Add the same amount to similar amount of indicator solution to a conical flask containing 100-mL distilled water.
   6. Titrate against standard $AgNO_3$ solution till $AgCrO_4$ starts precipitating. A persistent *brick red colour* (or pinkish yellow) will appear which indicates end point of titration.

Precautions
1. The pH must be in a range of 7–8, because Ag^+ is precipitated as AgOH at high pH. At low pH, CrO_4^{2-} is converted to $Cr_2O_7^{2-}$.
2. *Same amount of indicator must be used for the sample and blank titration;* otherwise, Ag_2CrO_7 may form very soon or not soon enough.

(D). Calculation

$$\text{Chloride (mg/L)} = \frac{(A - B) \times N \times 35450}{\text{mL of sample.}}$$

Where,
A = mL of $AgNO_3$ required for sample
B = mL of $AgNO_3$ required for blank
N = normality of $AgNO_3$ used

20.17 Essential Micronutrients and Heavy Metals

Of the seven essential micronutrients, the plant abundance/requirements are as follows: Fe > Mn > B > Zn > Cu > Mo > Cl. These elements if present, greater than the threshold limits, become toxic. The other metals, which are not essential for plant growth, also present in soil or are added by anthropogenic activities such as the following: arsenic (As), cadmium (Cd), Cobalt (Co), Chromium (Cr), Mercury (Hg), Nickel (Ni), Lead (Pb), Selenium (Se), etc. The concentration of heavy metals in soil and plants is shown in Table 20.10.

20.17.1 Lead (Pb)

The average concentration of Pb in lithosphere has been reported as 16 ppm and 15–25 ppm in soil. In the most polluted soil (pollution by

Table 20.10 Concentration of heavy metals in soils and plants (Alloway 1990)

Elements	Normal range in soil (ppm)	Critical soil total concentration[a], (ppm)	Normal range in plants (ppm)	Critical concentration in plants[b] (ppm)
1. As	0.1–40	20–50	0.02–7	5–20
2. Cd	0.01–2	3–8	0.1–2.4	5–30
3. Co	0.5–65	25–50	0.02–1	15–50
4. Cr	5–1,500	75–100	0.03–14	5–30
5. Cu	2–250	60–125	5–20	2–100
6. Hg	0.01–0.5	0.3–5	0.005–0.17	1–3
7. Mn	20–10,000	1,500–3,000	20–1,000	300–500
8. Mo	0.1–40	2–10	0.03–5	10–50
9. Ni	2–750	100	0.02–5	10–100
10. Pb	2–300	100–400	0.2–20	30–300
11. Se	0.1–5	5–10	0.001–2	5–30
12. Zn	1–900	70–400	1–400	100–400

[a]The critical soil concentration in the range of values above which toxicity is considered to be possible
[b]The critical concentration in plants is the level above which toxicity effects are likely

heavy metals), concentration of Pb has been reported to be at 50.9 ppm (10–1,722 ppm), and near Pb/Zn mines, concentration of Pb may be as high as 492 ppm (Gauliulin et al. 1999). The sources of Pb in soil surface are burning of fossil fuels, particularly Pb additives in petrol in 2/3/4 wheelers, and another source is particulate-bound Pb emitted from Pb mining and smelting. Another source of soil pollution is corrosive weathering of Pb-based paints from the outer walls of a building and Pb-based pesticides (lead arsenite).

Accumulation of Pb

As it accumulates in the surface horizon of soil, (0–15 cm) by comparing with subsoil (30–45 cm), an index of surface contamination enhancement (SCE) has been developed. Normally in agriculture soil, SCE ranged from 1.2 to 2.0, while in soil near to a mining and smelting operation, SCE values between 4 to 20 are common.

Lead is not readily lost from the soil profile by leaching. Soil has rather large capacities for the immobilisation of Pb, because organic fraction is largely responsible for immobilisation of Pb. Pb residence time is long; hence, bioavailability is low.

(A). Sampling of Soil

The sampling of soil for Pb analysis should be based on clearly stated objectives. These objectives could be

1. *To measure background concentration of Pb (native Pb)*: For estimation of background concentration of Pb, the sample should be taken from the 45-cm depth. However, researchers reported that at a >5-cm depth, the concentration of Pb was found to be constant in immobility of Pb in soil.

2. *To measure Pb as a pollutant due to atmospheric deposition, vehicle exhaust, mining and smelting of Pb, etc. (Pb pollution)*: For determination of Pb pollution level, the top 5-cm should be taken for Pb analysis. As the top layer contains high organic matter, digestion of the sample should be performed carefully.

3. *To sometimes distinguish between native Pb and Pb pollution*: Take the soil sample at the surface (up to 5-cm depth) and >30–35-cm depth in the same profile.

(B). Total Lead Analysis (HNO_3-$HCLO_4$ Digestion)

It employs prolonged heating of the sample first with HNO_3, then in $HClO_4$ under reflux conditions. It is recommended that the soil to be analysed for inorganic Pb should be dried for 24 h at 65°C. The soil should be ground to pass through a 40-mesh size sieve. The sample should be mixed thoroughly.

Reagents

1. Nitric acid (HNO_3), concentrated
2. Nitric acid (HNO_3), 1%, 1: 100 (vol/vol), HNO_3/H_2O
3. Perchloric acid ($HClO_4$), 70%

Procedure

1. Weigh 2 g of the sample (crushed to pass through a 2-mm sieve) into a 20- by 3.5-cm test tube.
2. Add 10 mL of HNO_3 and insert a funnel condenser. Heat it in a low temperature at 80–90°C overnight.
3. Heat at 125–130°C to dryness.
4. Cool. Add 1 mL of HNO_3 and 4 mL of $HClO_4$.
5. Heat at 200–210°C to dryness.
6. Cool. Add 4 mL of HCl and 50 mL of distilled water. Heat to 70°C for 1 h.
7. Quantitatively transfer contents through a retentive, acid-washed filter paper receiving the filtrate in a 100-mL volumetric flask. Rinse the containers and the filter residue with 1% HNO_3. Make up to a volume with 1% HNO_3.

Experience has shown that 80% or more Pb in most soil sample is extracted by HNO_3–$HClO_4$ digestion.

Comments

The above procedure does not completely dissolve all silicate minerals. The residue on the filter paper can be further decomposed by transferring it into a 50-mL beaker with 1% HNO_3. Evaporate the liquid. Then add 5 mL each of concentrated HNO_3, HF and water and let it evaporate on a hotplate to dryness. Repeat this treatment twice.

Precautions

$HClO_4$ deserves special attention because of its great efficiency in oxidising organic matter at high temperature (72% $HClO_4$, bp. 203°C), and it may lead to explosion of the sample. The danger is generally associated with the incomplete oxidation of organic matter and presence of excessive amounts of fats. These problems can be avoided by extreme first-stage digestion with

HNO_3 (*add concentrated HNO_3 to the sample and heat at 80–90°C overnight*).

H_2SO_4 should not be used for digestion of the sample for Pb analysis, because it precipitates as a Pb sulphate, co-precipitated with other sulphate, and the formation of particles of sulphate in an atomiser leads to excessive light scattering in AAS.

(*C*). *Plant-Available Pb or DTPA: Extractable* (Follow the Procedure Under Sect. 20.18)

20.17.2 Iron (Fe)

Iron in soil exists in both the ferrous (Fe^{2+}) and ferric (Fe^{3+}) states. To determine the total Fe, the sample is decomposed with HF acid in the presence of $HClO_4$, resulting in volatilisation of Si as SiF_4 and oxidation of organic matter. The sample is dissolved using H_2SO_4.

(*A*). *Method for Total Fe*

Apparatus:

Platinum crucible, 30-mL capacity with lid.

Reagents

1. Hydrofluoric acid (HF), 48%
2. Perchloric acid ($HClO_4$), 70–72%
3. Nitric acid (HNO_3), 70%
4. Sulphuric acid (H_2SO_4), 6N

Procedure

1. Digest 0.5 g of oven-dried, 100-mesh soil using HF and $HClO_4$ in a 30-mL Pt crucible.
2. Wet the soil with a few drops of H_2SO_4 and add 5 mL of HNO_3 and 1 mL $HClO_4$.
3. Cool the crucible and then add 5 mL of HF. Cover the lid.
4. Heat the crucible to 200–225°C and evaporate the content to dryness.
5. Cool the crucible and add 2 mL of water and few drops of $HClO_4$. Evaporate the content into dryness.
6. Remove the crucible, and when it is cool, add 5 mL of 6-N HCl and about 5 mL of water.
7. Heat the crucible on a hotplate until the solution boils gently.

8. When the residue is completely dissolved, transfer it to a 100-mL volumetric flask and make up the volume.

(*B*). *Plant-Available Fe or DTPA: Extractable* (Follow the Procedure Under Sect. 20.18)

20.17.3 Copper (Cu)

Total Cu concentration in the lithosphere averages about 55–70 ppm. According to Baker (1984), the average total Cu concentration on the earth crust ranges from 24 to 55 ppm, and average Cu range for soil is 20–30 ppm.

Pollutes Soil
In soil added with fly ash, Cu concentration reported was 14–2,800 ppm. In the area surrounding the smelters (1–3 km), Cu concentration is reported to be at 1,000 ppm. When soil has 25–40 ppm of Cu, and if pH is below 5.5, it may be toxic. If pH is not maintained at a desirable level (6–7), the availability of Cu, Zn, Mn and other elements is more likely to be a problem.

Procedure
Total Cu: Follow HNO_3–$HClO_4$ Digestion Procedure as in that of Pb
Plant-Available Cu: DTPA Method (Sect. 20.18)

20.17.4 Manganese (Mn)

Mn is essential for higher plants. All Mn in soil is derived from parent materials. Average Mn concentration is largely governed by pH. The total Mn concentration in the earth crust averages 1,000 ppm and is found mostly in Fe–Mg rocks. In sedimentary rocks, the concentration range in limestone is about 400–600 ppm, but much lower content is commonly found in sandstone (20–500 ppm). Tolerance level of Mn in plant is 300 ppm. Total Mn concentration in coal mine spoil (Kentucky, USA) has been reported to be at 246 ppm. It is easily translocated in all parts of plant tissue, and maximum concentration is found in leaves (Maiti 1995).

The DTPA-extractable Mn in Indian soil was reported to be as follows: 43 ppm (18–120), Pantnagar, U.P; 29 ppm (8–52), H.P; 20 ppm (2–58), Chikmagaluur, Karnataka and 19 ppm (9.6–34), Gantur, A.P.

Procedure
Total Mn: Follow HNO_3–$HClO_4$ Digestion Procedure as in that of Pb
Plant-Available Mn: DTPA Method (Sect. 20.18)

20.17.5 Zinc (Zn)

The total Zn content of the lithosphere is about 80 ppm and in soils ranged from 10 to 300 ppm (average 50 ppm). Mean Zn content varies from 40 ppm in acid rocks (granite) to 100 ppm in basaltic rock. Kabata-Pendias and Pendias (1984) reports values of 17–125 ppm as background concentration of a large number of surface soils of different countries.

Procedure
Total Zn: Follow HNO_3-$HClO_4$ Digestion Procedure as in that of Pb
Plant-Available Zn: DTPA Method (Sect. 20.18)

20.17.6 Nickel (Ni) and Chromium (Cr)

The average concentration of Ni (total) in the world soil is reported to be at 40, 20 ppm (UK), 93 ppm (USA) and 27 ppm (Scotland). The largest build-up of Ni is in the surface soil and litter layer. A higher concentration of Ni has deleterious effect in soil microbial activity, seed germination and plant growth. Increasing soil pH by addition of lime is an effective and practical means of ameliorating the toxicity of Ni in soil. Increasing the organic matter has also reduced the availability of Ni to plants because of binding of the metals in organic complex.

Chromium: Cr (IV) is more toxic than Cr (III). The average Cr concentration reported for agriculture soil ranged from 5 to 1,500 ppm, but most common values reported were 70–100 ppm. In Scotland and UK agriculture soil, Cr value

Table 20.11 Concentration of Cr and Ni in coal and fly ash (in ppm) (Alloway 1990)

	Cr	Ni
Coal	15	15
Fly ash		
Butiminous	172	11
Sub-butiminous	50	1.8
Lignite	43	13

Table 20.12 Limiting values for concentration of Cr and Ni is soil (ppm) adopted in The Netherlands (Alloway 1990)

Category	A	B	C
Cr	100	250	800
Ni	50	100	500

reported was 62 and 34 ppm. In coal mine soil (Kentucky, USA), the concentration is reported to be at 88 ppm; plant concentration was found to be 10–190 ppm.

Disposal of fly ash on land is the largest source of both Ni and Cr to soil. The average concentration of Ni and Ci in coal and fly ash are reported in Tables 20.11.

The limiting concentration of Cr and Ni as adopted in soil of Netherlands is shown in Table 20.12. The reference value of Cr and Ni is set as level A. If the concentration in soil exceeds B concentration level, further investigation is required, and a concentration greater than the C concentration value, require soil-cleansing operation.

Procedure
Total Ni and Cr: Follow HNO_3-$HClO_4$ Digestion Procedure as that of Pb
Plant-Available Ni and Cr: DTPA Method as Discussed in Sect. 20.18

20.17.7 Cadmium (Cd)

Cadmium CONCENTRATION ranged from 0.005 to 2.4 ppm, with a mean of 0.27 ppm and median of 0.2 ppm (USA). In UK soil, mean Cd concentration reported was 1.2 ppm with a median of 0.9 ppm.

Sources of Cd in Soil Environment
Cd used as protective plating on steels; pigments, stabilisers for plastics; Ni–Cd dry cell batteries; mining and smelting operation of Cd and Zn; atmospheric pollution from metallurgical industries; incineration of plastic container and batteries; burning of fossil fuels etc.

Environmental problems
A large portion of Cd is accumulated in edible portion of plants, thereby causing health hazards to human. Cd (together with Mn, Zn, B, Mo and Se) are recognised as easily translocated to plants tops and absorption through roots. For examples- In Soybean, 2% is accumulated in leaves and 8% in seeds.

Procedure
Total Cd: Follow HNO_3–$HClO_4$ Digestion Procedure as that of Pb
Plant-Available Cd: DTPA Method as Discussed in Sect. 20.18

20.18 Determination of Plant-Available Pb, Zn, Cu, Fe, Mn, Ni and Cd (DTPA-Extractable)

The standards of available trace elements are shown in Table 20.13. The method describes employs DTPA as a chelating agent. The extraction is done at pH 7.3.

(A). Chemicals
1. Calcium chloride, $CaCl_2$. $2H_2O$—*1.47 g/L*
2. DTPA (**D**iethylene triamine **p**entaacetic acid)—*1.967 g/L.*
3. TEA (**T**riethanolamine)—*14.92 g/L*
4. Dilute HCl (AR HCl diluted with double distilled water 1:5)—use for pH adjustment

(B). Reagents
DTPA (Diethylene triamine pentaacetic acid) extracting solution—0.005 M DTPA, 0.01 M $CaCl_2$ and 0.1 M TEA adjusted to pH 7.30 with HCl: take 1-L volumetric flask. Dissolve *1.967 g DTPA, 14.92 g TEA* and *1.47 g $CaCl_2.2H_2O$*

Table 20.13 DTPA-extractable Fe, Zn, Cu and Mn for deficient, marginal and sufficient soils (in ppm) (Maiti 2003)

Category	Fe	Zn	Mn	Cu
Low (deficient)	0–2.5	0–0.5	<1.0	<0.2
Marginal	2.6–4.5	0.6–1.0	—	—
Highly sufficient	>4.5	>1.0	>1.0	>0.2
Indian value	4.5	0.63	2.0	0.20
Average	16–18	0.90	10.0	2.2–2.3

in approximately 200 mL of double distilled water. Allow sufficient time to dissolve DTPA and add 800-mL double distilled water. The pH of the solution is adjusted to 7.3 with dilute HCl while stirring. This solution is stable for several months.

(DTPA is acting as a chelating agent; the pH of 7.3 buffered with TEA is used to prevent excess dissolution of trace metals, which is highly pH dependent. In the presence of 0.01-M $CaCl_2$, enable the extract to attain equilibrium with $CaCl_2$, which minimises the dissolution of $CaCO_3$ from calcareous soils.)

(C). Extraction Procedure (1:2 Soil/Extractant Ratio)

Take *10 g* of air-dried soil in a 100-mL conical flask, add *20-mL DTPA* extracting solution and shake for *2 h*. After exactly 2 h of shaking, filter the suspension by gravity through Whatman filter no. 42. The time of shaking is very important because extraction is not complete before 2 h.

Determine Pb, Cu, Zn, Mn, Ni, Cr, Cd and Fe in ASS.

(D). Calculation

Concentration of DTPA-extractable elements are as follows: (in ppm)

$$= \frac{20 \times (\text{ppm of elements in sample solution} - \text{ppm in blank solution})}{\text{wt of air} - \text{dried sample, g}}.$$

If the results among laboratories are to be compared, the soil weight/extractant volume ratio must be constant.

Comments

A sample preservative is not necessary if analysis can be completed within a day or two. Otherwise, refrigeration may be required to retard microbial growth. This procedure has been widely used on a large range of soil types since initially reported (Lindsay and Norvell 1978).

20.19 Soil nutrient as an index of soil fertility

Generally, the soil testing laboratories use organic carbon as an index of available N, Olsen's and Bray's method for available P and neutral normal ammonium acetate for K. Available nutrient status in the soils is generally classified as low, medium and high which are generally followed at the National level (Table 20.14).

Some states like West Bengal, Maharashtra and some others are followed 6 classes of the nutrient status is given as Table 20.15. In some states 5 levels of soil nutrient ratings are followed such as very low, low, medium, high and very high.

Suitable soil extractants have been developed for various micronutrients to predict the available forms of elements in soils. In India, Zn is the most widely reported deficient nutrient element. Other micronutrients like Cu, Mn, Fe, B, Mo and secondary nutrients like sulphur are also becoming deficient. Suitable testing methods are being standardized under the All India Co-ordinated Research Project on Micronutrients (Anon 2011). Generally accepted critical limits in soils and plants and the soil test methods are given in Table 20.16.

(E). Exercise on Trace Metals

1. List the essential micronutrients required for plant growth.
2. List the important toxic metals added to the soil by anthropogenic activities.

Table 20.14 Soil fertility rating viz. available nutrient status (Anon 2011)

		Soil Fertility Ratings		
Sl No	Soil Nutrients	Low	Medium	High
1	Organic carbon as a measure of available Nitrogen (%)	<0.5	0.5–0.75	>0.75
2	Available N as per alkaline permanganate method (kg/ha)	<280	280-560	>560
3	Available P by Olsen's method (kg/ha) in Alkaline soil	<10	10-24.6	>24.6
	Available K by Neutral N, ammonia acetate method (kg/ha)	<108	108–280	>280

Table 20.15 Classification of fertility status of soil for West Bengal, India (Anon 2011)

Soil Fertility Level	Organic Carbon (%)	Available P_2O_5 (Kg/ha)	Available K_2O (kg/ha)
Very high	>1.0	>115	>360
High	0.81–1.0	93–115	301–360
Medium	0.61–0.80	71–92	241–300
Medium Low	0.41–0.60	46–70	181–240
Low	0.21–0.40	23–45	121–180
Very Low	<0.21	<23	<121

Table 20.16 Critical concentration of nutrients in soil and plant (Anon 2011)

Element	Soil Test Method	Critical level in soil	Critical level in plant
Sulphur	Hot water, $CaCl_2$ or phosphate	Usual 10 ppm (8–30 ppm)	$<0.15–0.2\%$
Calcium	Ammonium acetate	$<25\%$ of CEC or <1.5 meq Ca/100 g	$<0.2\%$
Magnesium	Ammonium acetate	$<4\%$ of CEC or <1 meq Mg/100g	$<0.1–0.2\%$
Zinc	DTPA	0.6 ppm (0.4–1.2 ppm)	$<15–20$ ppm
Manganese	DTPA	2 ppm	<20 ppm (10–30 ppm)
Copper	DTPA or Ammonium acetate	0.2 pm	<4 ppm (3–10 ppm)
Iron	DTPA or Ammonium acetate	2.5–4.5 ppm 2 ppm	<50 ppm (25–80 ppm)
Boron	Hot water	0.5 ppm	<20 ppm
Molybdenum	Ammonium oxalate	0.2 ppm	<0.1 ppm

3. 'For Pb pollution analysis, surface contamination enhancement (SCE) factor is important'—why?

4. How are native Pb and Pb pollution distinguished from each other?

5. Why should H2SO4 not be use for total Pb analysis?

6. Why is DTPA-Fe analysis more important than total Fe analysis?

7. For total Fe analysis, HF is used—why?

8. Where will Cu pollution in soil be at a maximum?

9. 'Translocation of heavy metals to the plants is governed by soil pH'—why?

10. In which soil horizon is Ni accumulation found to be at a maximum?

11. 'Increasing of organic matter in soil helps immobilisation of Ni'—how?

12. Name the important sources of Cr and Ni in soil.

13. What are the sources of Cd soil?

14. Why is Cd considered to be a serious soil pollutant?

15. What is the mechanism of DTPA extraction of metals?

References

Alloway BJ (1990) Heavy metals in soils. Wiley, New York

Anon (2011) Soil testing in India - Methods Manual. Department of Agriculture & Cooperation, Min of Agriculture, Gov of India, New Delhi. agricoop.nic.in/dacdivision/MMSOIL280311.pdf

Baker DE (1990) Copper. In: Alloway BJ (ed) Heavy metals in Soils. Wiley, New York, p 151

Biswas TD, Mukherjee SK (1994) Textbook of soil science, 2nd edn. Tata McGraw Hill, New Delhi

Bray RH, Kurtz LT (1945) Determination of total, organic and available forms of phosphorous in soils. Soil Sci 59:39–45. http://garfield.library.upenn.edu/classics1987/A1987J041400001.pdf

Galiulin RV, Bashkin VN, Birch P, Kucharski R (1999) Influence of phytoextraction on the ferment activity of heavy metal polluted soil. Land Contam Reclam 7(2):133–141

Ghosh AB, Bajaj JC, Hasan R, Singh D (1983) Soils and water testing methods. IARI, New Delhi, p 66

Hendershot et al (1993) Chapter-16: Soil reaction and exchangeable acidity. In: Carter MR (ed) Soil sampling and methods of analysis. Lewis Pub. Boca Raton

Jackson ML (1973) Soil chemical analysis. PHI Pvt Ltd. New Delhi

Kabata-Pendias A, Pendias H (1984) Trace elements in soils and plants. CRC Press, Boca Raton

Lindsay WL, Norvell WA (1978) Development of DTPA soil test for zinc, manganese and copper. Soil Sci Soc Am J 42:421–428

Maiti SK (1995) Some experimental studies on ecological aspects of reclamation in Jharia coalfield. Ph.D. dissertation, Indian School of Mines, Dhanbad

Maiti SK (2003) Handbook of methods in environmental studies, vol 2. ABD Publications, Jaipur

Nelson DW, Sommers LE (1996) Total carbon, organic carbon, and organic matter. In: Page AL et al (eds) Methods of soil analysis, part 2, 2nd edn. American Society of Agronomy Inc., Madison, p 961

Olsen SR, Sommers LE (1982) Phosphorus. In: Page AL, et al (eds), Methods of Soil Analysis, part 2, 2nd edn. Agron Monogr 9. ASA and ASSA, Madison WI, pp 403–430

Shoemaker HE, Mclean EO, Pratt PF (1962) Buffer method for determination of lime requirements of soils with appreciable amount of exchangeable aluminium. Soil Sci Soc Am Proc 15:274–277

Subbiah BV, Asija GL (1956) A rapid procedure for the determination of available nitrogen in soils. Curr Sci 25:259–260

Tisdale SL, Nelson WL, Beaton JD, Havlin JL (1985) Soil fertility and fertiliser, 5th edn. Mcmillan Publishing Co., New York

Walkley A, Black IA (1934) An examination of Degtjareff method for determining soil organic matter and a proposed modification of the chromic acid titration method. Soil Sci 37:29–37

Contents

The soil *is a tremendous biological laboratory.* The roles of microorganisms are as follows: responsible for decomposition of litter, mineralisation process, nutrient cycling, accumulation of organic matter and formation of humus. There are three important groups of microbes: (a) bacteria, (b) fungi and (c) actinomycetes.

21.1 Collection/Processing of Samples

21.1.1 Materials

1. Shovel or auger
2. Plastic bags
3. Knife or scissors
4. Marking pen
5. Scalable plastic containers

21.1.2 Sample Collection

Clearly mention the location of sample, that is, rhizosphere or non-rhizosphere zone.

Rhizosphere: The term rhizosphere was introduced to designate that portion of the soil that is under the immediate influence of the plant root (1–2 cm from root surface). Generally, microbes are found in greater numbers and diversity in rhizosphere compared with non-rhizosphere locations. The density of bacteria, fungi and actinomycetes found in rhizosphere and non-rhizosphere are given in Table 21.1.

S.K. Maiti, *Ecorestoration of the Coalmine Degraded Lands*,
DOI 10.1007/978-81-322-0851-8_21, © Springer India 2013

Table 21.1 Number of bacteria, fungi and actinomycetes (Biswas and Mukherjee 1994)

Soil microorganisms	Location	Number/g of soil
Bacteria	Rhizosphere	$6{,}778 \times 10^6$
	Non-rhizosphere	25×10^6
Fungi	Rhizosphere	918×10^3
	Non-rhizosphere	68×10^3
Actinomycetes	Rhizosphere	83×10^6
	Non-rhizosphere	1×10^6

21.2   Enumeration of Bacteria and Actinomycetes (by Plate Count–*Spread Plate Method*)

Bacteria are the most dominant group of micro-organisms. Population decreases as the depth of soil increases. The values for viable heterotropic plate count in a fertile soil have been reported as 10^8–10^{10} for bacteria and 10^5–10^8 for actinomycetes.

Most common medium used for enumeration of *total aerobic heterotropic bacteria* are:
- Soil extract agar
- Nutrient agar
- Peptone–yeast agar (Plate count agar)

Detailed media composition is given in Table 21.2.

(A) Preparation of Media for Bacteria
Nutrient Agar
Heat until agar and peptone dissolve. Adjust pH to 6.6–7.0, using bromocresol blue as an indicator. Sterilise the medium by autoclaving.

Soil Extract Agar
To prepare 1 L of soil extract, weigh out 500 g of soil and dissolve into 1 L of tap water. Mix well to disperse soil. Sterilise the soil suspension in the autoclave for 1 h at 121°C. A small amount of $CaCO_3$ is added, and the whole is filtered through a double filter paper. Dissolve the agar, K_2HPO_4, glucose in 900 mL of water and 100 mL of soil extract and sterilised in autoclave.

Peptone–Yeast Agar
In 1 L deionised water, add 5 g peptone, 3 g yeast extract 1 g glucose and 15 g agar. After autoclaving at 15 lb pressure at 15 min and cooled to 45°C, add 10 mL of 0.1 M $CaCl_2$. (Note: adding $CaCl_2$ to a hot solution causes flocculation). Adjust the pH to 7.0 with concentrated HCl after autoclaving.

(B) Preparation of Media for Actinomycetes
Actinomycetes were considered to be related to both bacteria and fungi but now recognised as prokaryotic organisms close to the bacteria. They may be defined as Gram-positive bacteria that form branching hyphae (usually 0.5–1-μm diameter) that may develop into a mycelium. They grow nearly neutral pH and more drought resistant. Therefore, drying the soil before performing heterotropic count can generally reduce the bacterial interference (drying of soil at 45°C in sterile Petri dish for 4 h). Antifungal agent such as cycloheximide/or nystatin is added to control fungi.

Actinomycetes are usually isolated and counted using standard dilution procedure. Dilution aliquots are either added to the cooled molten medium (45°C) before pouring the plate

Table 21.2 Chemical composition of bacteriological media

Soil extract agar	Nutrient agar	Plate count agar
1. Glucose 1 g	Peptone 5 g	Tryptone 5 g
2. K_2HPO_4 0.5 g	Beef extract 3 g	Yeast extract 2.5 g
3. Agar 15 g	Yeast extract 1 g	Glucose 1 g
4. Soil extract 100 mL	Agar 15 g	Agar 15 g
5. Tap water 900 mL	Distilled water 1 L	Distilled water 1 L

Table 21.3   Chemical composition of starch–casein agar for actinomycetes (After Williams and Willington 1982)

Sl. no.	Constituents	Amounts
1	Starch (soluble)	10.0 g
2	Casein (vitamin free)	0.3 g
3	Sodium chloride (NaCl)	2.0 g
4	Potassium nitrate (KNO_3)	2.0 g
5	Potassium dihydrogen phosphate (K_2HPO_4)	2.0 g
6	Magnesium sulphate ($MgSO_4. 7 H_2O$)	0.05 g
7	Calcium carbonate ($CaCO_3$)	0.02 g
8	Ferrous sulphate ($FeSO_4. 7H_2O$)	0.01 g
9	Agar	18 g
10	Cyclohexamide (heat stable)	50 mg
	Distilled water	1,000 mL

Adjust the pH to 7 with 0.1 N HCl after autoclaving

(i.e. by *pour plate method*) or spread over the surface of solidified medium (i.e. by *spread plate method*).

Detailed media composition of starch–casein agar is given in Table 21.3 for actinomycetes counts.

(C) Procedure

The details of spread plate method used for heterotropic bacteria and actinomycetes are given in Fig. 21.1.

Stage I: Preparation of Dilution Series

1. Take 10 g of soil and dissolved in 95 mL of dilution water in a dilution bottle. As 10 g of soil occupies approximately 5 mL volume, only 95 mL dilution water is added, which gives dilution ratio 1:10 (w/v; soil:water). It is recommended shaking by hand about 200 times or on a mechanical shaker for 10 min in a horizontal position will be sufficient.

2. Next by using a sterile pipette, 1 mL of suspension is removed from the bottle and added to a tube containing 9 mL dilution water. The dilution series is continued to the highest desire dilution. The three most diluted suspensions are plated. (*Note: You may need to increase the number of dilution depending on the type of soil you are analysing*).

Stage II: Preparation of Plates

Prepare three replicates for each dilution. So a total of nine plates are required. For example, use peptone–yeast agar plate bacteriological enumeration and starch–casein agar for actinomycetes (Pepper et al. 1998).

Stage III: Inoculation of the Plates

1. Prepare three spread plates for each dilution 10^{-3}, 10^{-4} and 10^{-5} for bacteria as shown in Fig. 21.1. Inoculation is to be carried inside laminar flow chamber. Put 0.1 mL of each dilution (this will increase your effective dilution by a factor of 10) to three separate, labelled peptone–yeast agar plates. Inoculate not more than 3 plates at a time before spreading, as standing will allow too much liquid to be absorbed into the agar in one spot.

2. Take a glass-hockey stick spreader, dip it in ethanol and flame the spreader in a Bunsen burner just long enough to ignite the ethanol. After cooling of the spreader, spread the drop of inoculum around the surface of the agar until all traces of free liquid disappear.

3. Replace the lid, reflame the spreader and repeat with the next plates.

4. Incubate the bacteria plates (keep in inverted position) at room temperature for 1 week.

5. For actinomycetes, use the dilutions of 10^{-2}, 10^{-3} and 10^{-4}.

6. Spread 0.1 mL of diluted suspension on starch–casein plates and make three replicates for each dilution.

7. Incubate the actinomycetes plates (inverted) at room temperature for 2 weeks.

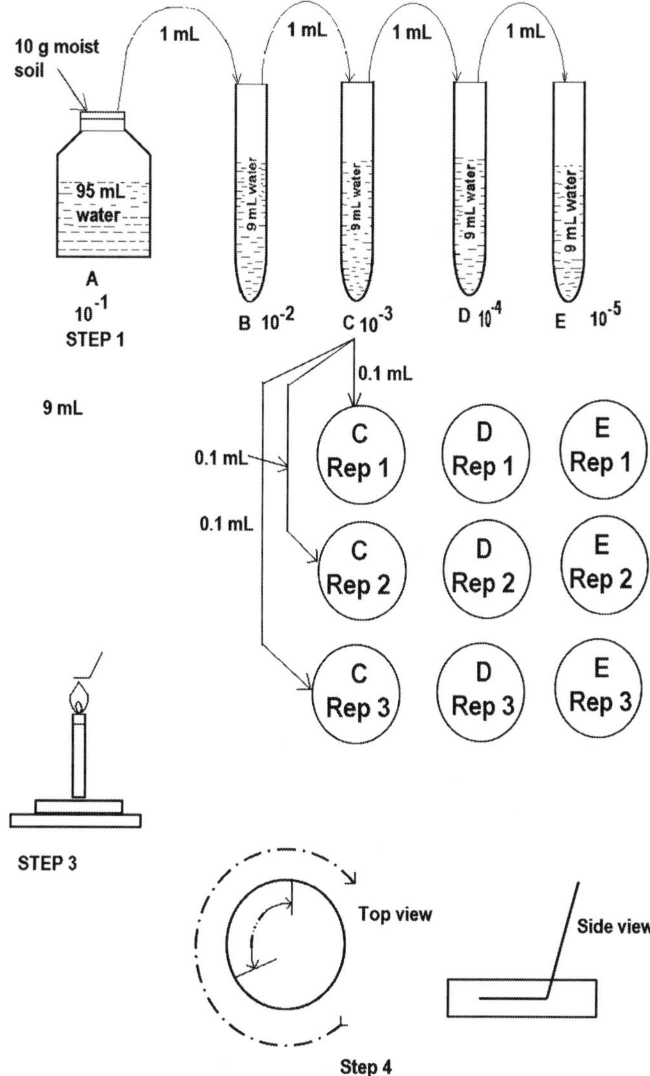

Fig. 21.1 Schematic showing the procedure for viable heterotropic counts of bacteria and actinomycetes (spread place method)

Second Period (Counting)

For Bacteria

1. Examine the bacterial plates carefully. Note the differences in colony size and shape.
2. Count the total number of bacterial colonies (CFUs) for each plate including any actinomycetes. Count only those plates of a dilution that are countable.
3. Calculate CFU/g of dry soil (similarly as discussed under fungal enumeration).

For Actinomycetes

Follow the same procedure like bacterial counting.

21.3 Filamentous Fungi (*Pour Plate Method*)

Values for viable heterotropic plate count for a fertile soil has been reported as 2,000 to 10^6 fungal propagules (spores, hyphae or hyphal figments) per gram of dry soil (Alexander 1977).

Soil sampling: Soil must be collected without allowing the soil to dry. For reliable counts, soil should be collected from at least 3 different places in the field.

Table 21.4 Chemical composition of fungal media (Czapex Dox agar and rose bengal agar)

Czapex Dox agar (yeast)	Rose bengal agar[a] (only fungi)
1. $NaNO_3$ 3 g	1. Peptone 5 g
2. KH_2PO_4 1 g	2. KH_2PO_4 1 g
3. $MgSO_4.7H_2O$ 0.5 g	3. $MgSO_4.7H_2O$ 0.5 g
4. KCl 0.5 g	4. Streptomycin[b] 30 mg
5. $FeSO_4.7H_2O$ 0.1 g	5. Rose bengal 35 mg
6. Sucrose 30 g	6. Glucose 10 g
7. Agar 15 g	7. Agar 15 g
8. Distilled water 1 L	8. Distilled water 1 L
One gram of yeast extract per litre may be added. Dissolve inorganic constituents separately and add $FeSO_4$ last. Add sucrose just before sterilisation	The antibiotic is sterilised separately and added aseptically to the sterilised medium

[a]The medium is especially recommended for the isolation of fungi in the presence of large numbers of bacteria
[b]Auromycine (35–2,000 μg) may be substitute for streptomycin

(A) Isolation of Fungi
1. Acidification of isolation media to pH 5 is frequently used for the isolation of fungi.
2. More recently, the antibacterial antibiotics are being used. Generally, auromycin at 30 mg/L is widely used with or without the addition of streptomycin at 30 mg/L of medium.
3. Crystal violet, rose bengal and oxgall have also been used to restrict bacterial development on isolation media.
4. Care must be taken to add the appropriate antibiotic or antibiotics after autoclaving and cooling.

The media composition of fungi is given in Table 21.4.

(B) Preparation of Plates
Prepare three replicates for each dilution; therefore, total nine plates are require. The medium commonly used is rose-bengal-agar. Dissolved the ingredients without streptomycin in 1 L of deionised water. Sterilise the medium at 15 lb pressure and 121°C for 20 min. Take out the medium from autoclave and cool to about 45°C and then amend the agar with a sterile streptomycin solution (30 mg/L). Add approximately 15 mL of agar to each plate. Pour plate method used for estimation of density of filamentous fungi is shown in Fig. 21.2.
1. Gently swirl the plate to distribute the medium. Do not let flying medium onto the lid or sides of the Petri dishes.
2. Transfer 1 mL of suspension in each plate. After pouring one plate, replace the lid on the Petri dishes and gently swirl the agar to mix in the inoculum.
3. Allow the agar to solidify for at least 10 min then incubate all plates (inverted) at room temperature for 1 week (7 days).

Second Period: After 7 Days of Incubation
1. Count the colony by colony counter for only one dilution of each soil. Note and describe the culture characteristics of colonies.
2. Morphological studies under microscope.
   - Put a drop of lactophenol stain at the centre of a clean glass slide.
   - Take out a small fraction of colony.
   - Put cove slide.
   - Examine under the compound microscope.
   - Describe and illustrate the hyphae and fruit bodies (sporangia, spores, etc.) of fungi. Draw the sketches in your notebook or take photograph.

(C) Calculation
Initial, say 10 g soil has moisture content of 10%.

Dry soil $= 9g$.

Therefore, weight of soil/mL in each dilution bottle is calculated as follows:

$$Bottle\ A = \frac{9g}{100\ mL\ of\ dilution\ water} = 0.09g\ soil/mL$$

$$Bottle\ B = \frac{0.09g}{1\ mL} \times \frac{1\ mL}{(9\ mL + 1\ mL)} = 9 \times 10^{-3}g\ soil/mL$$

$$Bottle\ C = 9 \times 10^{-4}g\ soil/mL.$$

$$Bottle\ D = 9 \times 10^{-5}g\ soil/mL.$$

$$Bottle\ E = 9 \times 10^{-6}g\ soil/mL.$$

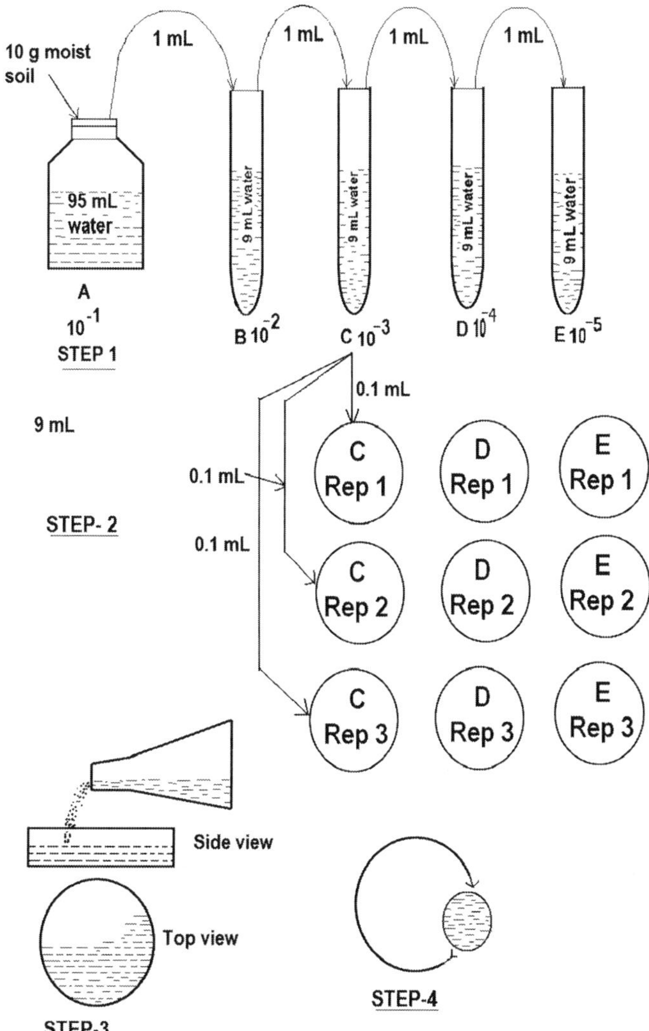

Fig. 21.2 Pour plate method showing the procedure for plate counts of filamentous fungi

Say the count was 24, 26 and 30 colonies on the replicates of dilution D, and then average number of CFU/plate is 26.67.

(*D*) *Calculate the CFU per Gram of Oven-Dried Soil*

$$For\ dilution\ D = \frac{26.67CFUs}{9 \times 10^{-5}g\ dry\ soil}$$

$$= 2.96 \times 10^{5}/g\ of\ oven\ dry\ soil.$$

Final results must be reported as mean and standard deviation with a coefficient of variation.

21.4 Study of Vesicular-Arbuscular Mycorrhiza Fungi (VAMF)

Mycorrhiza (*plural mycorrhizae, literally meaning fungus root*) is formed by association between a plant root and a fungus, and by far, majority of vascular plants are involved in this association. Endomycorrhizas are known as VAM (*vesicular-arbuscular mycorrhiza*) fungi which are largely *Zygomycotina* and *Ascomycotina* group, which do not form sheath. They form *vesicles and arbuscules* within the cells (intracellular) of the root cortex. They appear to serve as *organ of storage and transfer* of carbon compounds and mineral nutrients between the fungal hyphae and host plant. VAM fungi are the most widespread and important root symbionts of all mycorrhizal association. The member of Cruciferae and Chenopodiaceae are devoid of VAM infection. About 80% of all land plants form this type of mycorrhizae. Hosts include most families of angiosperms and gymnosperm including Rosaceae, Gramineae and Leguminosae and also found in Pteridophyta and Bryophyta. About 150 species of VAM are recognised; all are *zygomycetes*. Taxonomy of VAM is purely based on spore morphology. Four genera of VAM are recognised:

1. *Glomus* spp.
2. *Gigaspora* spp.
3. *Sclerocystis* spp.
4. *Aculospora* spp.

A complete analysis of VAMF in soil and root sample consists of four important aspects:

- Collection and storage of soil and root samples
- Extraction of quantification of VAMF spores in soil (density of VAMF spores/g of soil)
- Estimation of VAMF infection in root (% of root colonisation)
- Evaluation of VAMF infection potential (or VAM infectivity)

21.4.1 Collection and Storage of Soil and Root Samples

VAMF inhabit both cortical cells of root and surrounding soil (rhizosphere). The harvest of VAMF association consists of gently excavating and preserving of feeder root (fine root) of chosen plant. About 500 g of rhizosphere soil is collected in a polythene bag. Feeder roots also collected at the site. Feeder roots are generally of .2–.3-mm diameter and found with 0–15-cm depth. Even though some roots are extended up to 4–5-m depth, most of the fine feeder roots are distributed throughout the upper surface within 30–40 cm in soil.

(A) *Materials*
1. Plastic bags and sticker
2. Spade or device to dig soils sample
3. Sampling bottle (filled with 50 mL of fixative solution) for collection of feeder roots
4. Fixative solution: Formaldehyde 37%, glacial acetic acid and ethanol 70% in the ratio of 5:5: 90 (mix 13 mL formalin, 5 mL glacial acetic acid in 200 mL 50% ethyl alcohol)
5. Refrigerator

(B) *Methods*
1. Select the collecting site or special habitat or plant species from where the feeder roots and soil samples will be collected. There are some non-mycorrhizal angiosperms families like Chenopodiaceae, Amaranthaceae, Cistaceae and most Ranunculaceae and Cruciferae.
2. Remove first few cm of soil surface to eliminate plant debris.
3. Dig top 25–30 cm of soil where plant roots will be collected by using a spade.
4. Put feeder roots in a sampling bottle.
5. Put soil sample together with roots in a plastic bag.
6. Number the plastic bags and sampling bottle and note the complete information in field notebook (i.e. collecting number, date of collection, habitat, plant species and types of soil etc.).
7. Store sample in a refrigerator (4°C) until analyzed.

Note: Take atleast 3–4 samples from one site to increase precision.

Spore number and root colonisation are decreases with depth. Spore numbers are found to increase as roots senesced. Fine-textured soils sometimes contain fewer VAM propagules due to the low level of soil aeration.

Plant roots are collected generally during the winter months in rhizosphere of the species. Feeder roots are collected by tracing root system from main stem until young, unthickened roots are encountered.

21.4.2 Procedure for Assessment of VAM Infection in Root

The VAM fungi do not cause obvious morphological changes to the roots; however, they produce arbuscules and vesicles in the roots. The level of root colonization is estimated by the frequency of root colonization and with respect to presence of fungal mycelia, arbuscles or vesicles or all. Colonisation of VAM infection is restricted to the cortical region of root and usually most prevalent in the young feeder roots. The intra-cortical (fungus within cortex of root) morphology of VAMF consists of a network of intracellular coenocytic hyphae, running through the layer of cortical cells; usually bearing intra- or extracellular vesicles, and differentiating intracellular arbuscules. Quantification of the root colonisation can be done by:

The *slide method indicates* the length of colonised roots. In this method, large numbers (50–100) of 1-cm root sections are mounted on slides. The length of the colonised root tissue is measured and compared to the total length of root observed. Express the result as percent as follows:

$$Colonization\ (\%) = \frac{No.\ of\ infected\ segments}{Total\ no\ of\ segments} \times 100$$

(A) Chemicals and Reagents
1. KOH solution (10% w/v)—for root clearing.
2. H_2O_2 solution (3%) or 20% aqueous solution of NH_4OH—*for root bleaching.*
3. Trypan blue solution—dissolve 0.5 g trypan blue in 500 mL Glecerol and add 450 mL distilled water and 50 mL 1% HCl solution—*for root coloration.*
4. *Destaining solution*: Lactophenol (20 g phenol, 20 g lactic acid, 40 g glycerin).
5. *Mounting medium*: Containing acetic acid and glycerol (1:1, v:v).

(B) Methods
1. Collect the fine feeder roots carefully from plants and wash. Put immediately in FAA (13 mL formalin, 5 mL glacial acetic acid and 200 mL 50% ethyl alcohol) solution.
2. *Washing of roots*: Wash the root pieces thoroughly with water to remove attached soil particles and soil debris.

3. *Segmentation*: Cut the washed roots into 1-cm piece.
4. *Root clearing and softening*: Put these roots in 10% KOH solution and boil (90°C) for 1–2 h depending upon the thickness of the root *or* put in autoclave at 15 lb pressure for 15 min at 121°C.
5. *Bleaching of roots*: If the roots are coloured, bleach with H_2O_2 or NH_4OH solution for 10–45 min at room temperature.
6. *Washing*: After cooling, wash the root with water twice or thrice.
7. *Acidification*: The non-coloured roots are acidified with 5 N HCl for a few minutes and then rinsed thoroughly with distilled water for twice or thrice (this is done to nullify the effect of KOH).
8. *Staining:* The root segments are stained with 0.05% trypan blue or cotton blue, and the excess stain is removed in clear lactophenol solution (20 g phenol, 20 g lactic acid, 40 g glycerin).
9. *Mounting*: The stained root segments are mounted on slides containing acetic acid and glycerol (1:1, v:v).
10. *Examination under microscope*: Examine at 100–400 × magnification under compound microscope.

A simple staining technique for VAM fungi is given in Box 21.1 and assessment of root length colonization is given in Box 21.2.

Box 21.1 Simple Staining Technique to Study Root Infection by VAM Fungi (Maiti 1997)

Roots are cleared by boiling in 10% KOH. Boiling time is differed according to the type of plant roots—bean, soybean and maize for 5 min and cucumber, wheat, barley and ryegrass roots for 3 min.
1. The roots are then rinsed several times with tap water.
2. Cleared roots are boiled for 3 min in a 5% ink–vinegar solution with pure white household vinegar (5% acetic acid).

(continued)

Box 21.1 (continued)

3. Roots are destained by rinsing in tap water (acidified with few drops of vinegar) or rinsing in pure vinegar. Destaining time differed according to the type of ink (for blue or black inks, destaining time is 20 min with acidified water, and for red ink, destaining time is 10 min with vinegar).
4. The inks tested by the authors were purple, green, red, blue and black.

Box 21.2 Assessment of Root length colonized by AM fungi (Biermann and Linderman 1981)

Mycorrhizal colonization is assessed using Biermann and Linderman (1981) method (frequency distribution method) in which the colonization is assessed as proportion of root length colonized by mycorrhizal fungi using a compound microscope.

(A). Equipments
1. Plastic plate with grids for measuring root length
2. Beakers
3. Fine forceps
4. Microscopic glass slides, cover slips and lactoglycerol
5. Compound microscope
6. Stereomicroscope
7. Stage and ocular micrometer
8. Petri dish (1 cm grid)

(B). Methods
1. A randomly selected aliquot of stained root segments (1 cm in length) suspended in lactoglycerol are spread in a petri dish.
2. Calibrate the ocular micrometer with the stage micrometer by placing it on the eyepiece of compound microscope.
3. Mount the root pieces on the glass slides (5-10 pieces) and calibrate the ocular micrometer with the stage micrometer at the particular x of compound microscope and observe the root pieces.

(continued)

Box 21.2 (continued)

4. The proportion of the length of each root segment consisting vesicles, arbuscules or hyphae are estimated to the nearest 10%.
5. Data are recorded frequency distributions from samples containing 25, 50, 100 root segments. The percentage of the root length with mycorrhizal fungi in the sample is calculated from the frequency distribution.

Calculation of the percentage of the root length with mycorrhizal colonization in a sample of 25 root segments (1 cm) from a frequency distribution of the percentage of segment lengths with mycorrhizal colonization. An example of calculation of root length colonization is shown below.

Percentage of segments lengths colonized	Frequency	Frequency × distribution	Computed percentage of root length colonized
0	3	0	
10	4	40	
20	1	20	
30	2	60	
40	3	120	
50	3	150	
60	2	120	1140/25 = 45.60%
70	1	70	
80	1	80	
90	2	180	
100	3	300	
	25	1140	

21.4.3 Estimation of VAMF Spores

Spores and sporocarp counts have been widely used for determining the intensity of mycorrhizal infection. The spores are used for identification of VAM species, study of species diversity and the isolation of spore inocula for VAM propagation or pure culture synthesis. Spores numbers can be evaluated by direct counts by using a counting slide. Wet-sieving followed by flotation-bubbling and/or flotation-centrifugation techniques are generally used for estimation of spore density.

Table 21.5 Sieve no. and sieve size opening used for mycorrhiza spore study

Sl. no.	Sieve no.	Sieve size (μm)
1.	No.8	2,000
2.	No. 16	1,000
3.	No. 30	500
4.	No. 40	425
5.	No. 60	250
6.	No. 72	212
7.	No. 100	150
8.	No. 150	100
9.	No. 200	75
10.	No. 240	62
11.	No. 300	50
12.	No. 325	45
13.	No. 400	38

21.4.4 Wet-Sieving and Decanting Method

(Gerdeman and Nicolson 1963)

(A) Material
1. Soil sample and balance
2. Flasks or beaker (500 mL , 1 L),
3. Oven (90°C)
4. Standard sieves: Commonly available sieves in most of the laboratories are given below. Select sieves of 500-, 250-, 150-, 75- and 45-μm sieves (Table 21.5)

(B) Methods
1. Determine soil moisture by weighing 100 g of fresh soil after oven drying at 90°C for overnight.
2. Take 100 g of rhizosphere soil sample.
3. Put in a beaker containing 500 mL of water and thoroughly stir agitate to detach soil debris and aggregates and provide uniform mixing.
4. The suspension is allowed to settle for 1 h.
5. Then decant through a 1,000-μm sieve to remove large debris and root segments.
6. Suspension then passes through 500-, 250-, 150-, 75- and 45-μm sieves.
7. The residues from 250-, 150-, 75- and 45-μm sieves are collected in separate beaker to measure sporocarp, large spores, and small spores.

8. This process may be repeated twice or thrice to ensure that all the spores are extracted from the soil.
9. The residue may be filtered through Whatman filter #1, where spores are retained. Spread the filter paper in a Petri plate and observe directly under stereomicroscope. The yellowish or brownish round-shaped spores may be visible which should be counted.
10. Alternatively, the residue is stained in lactophenol, mounted on slides, and examined under compound microscope.
11. The total number of spores can be expressed in per-gram-soil basis.

21.4.5 Flotation-Centrifugation Technique

(Allen et al. 1974)

(A) Material
1. Fractions of sieved soils
2. Centrifuge apparatus and 50 mL centrifuge tubes
3. Extraction solution: 2 M sucrose with 2% Calgon® or 50% sucrose
4. Separatory funnel
5. Vacuum flask and pump
6. Whatman no.2 filter papers

(B) Methods
1. Make a soil suspension (water to soil (2:1, v/v) with a soil fraction obtained from the wet-sieving technique.
2. Centrifuge at 2,000 RPM for 10 min to remove small organic debris.
3. Pour off supernatant.
4. Resuspend soil in an extraction solution (sucrose solution) and agitate.
5. Centrifuge at 200 RPM for 10 min.
6. Decant supernatant in a flask; repeat steps 4–6 two or three times.
7. Filter the spore suspension in a filter paper by vacuum filtration.
8. Rinse out spores and any debris adhering to the filter paper with distilled water or ringer solution. Rehydrate spores to avoid plasmolysis.

9. Store spore suspension in refrigerator for preservation or vacuum filter on filter papers to isolate them.

(C) Counting Under Stereomicroscope
1. Examine the filter paper under stereomicroscope (10–50X) for counting spores.
2. Use microtip forces or micro-pipettes to hand-picked spores or sporocarps if necessary.

Other popular method as suggested by Singh and Tiwary (2001) is given in Box 21.3.

Box 21.3 Wet Sieving and Decanting Techniques for Enhanced Recovery of VAM Spores

1. About 150 mL tap water was added to 10–20 g of fresh soil in a 250-mL conical flask (glass or plastic).
2. The flask was shaken in a rotary shaker for 10–15 min at 1,000 rpm, and the soil suspension was immediately transferred into a series of sieves (1 mm, 800, 450, 300, 200, 100 and 45 μm).
3. The flask was washed twice or thrice with water into the sieve series. The 1-mm, 800- and 450-μm sieves were removed after completing the washing with a fine jet of water.
4. The 300-μm sieve was also washed with a fine jet of water and checked for spores in the residues in the sieve. If spores were present, the residue was transferred into a 250-mL beaker using little water and fine brass; otherwise, it was discarded.
5. The 200-, 100- and 45-μm sieves were washed again and the residues transferred carefully in the 250-mL beaker. The water in the beaker was increased up to 150 mL and the beaker shaken with the wrist for 30–60 s on the surface of a wooden table and kept standing for 30 s. The surface layer containing spores and organic residues was decanted three to five times into 5- or 100-mL centrifuge tube with the help of a funnel and little water.

(continued)

Box 21.3 (continued)

6. Other steps of centrifugation were same in the standard technique performed in the other set of experiment. Spore counting was done on Whatman no.1 filter paper (Gaur and Adholeya 1994) with rectangular grid lines of 0.5 cm^2 drawn on the filter paper which fits exactly one microscopic field under a stereomicroscope (WF 10× eye piece and 2× or 4× objectives).

21.5 Soil Respiration (CO$_2$ Evolution Method)

Microbial respiration is defined as the uptake of O$_2$ or release of CO$_2$ by bacterial, fungal, algal and protozoan cells in the soil. The soil respiration can be measure either in

- Undisturbed soil sample in Field (in situ condition)
- Disturbed soil sample in laboratory controlled conditions by using soil core. The sieved soils are not used.

(A) Field Measurement of CO$_2$
Field measurements of CO$_2$ depend on moisture conditions and temperature. Several studies have reported that soil temperature and soil moisture plays an important role in affecting the soil respiration. There are four types of soil moisture conditions that influence CO$_2$ evolution.

- Soil can be dry.
- Soil can be well drained and moist.

In well-drained or dry soil, respiration will be mainly aerobic are will involve free exchange of both O$_2$ and CO$_2$ by the microflora, microfauna, macrofauna and plant roots.

(B) Laboratory CO$_2$ Measurement
It measured soil respiration of a disturbed soil, and depending on the degree of disturbance, CO$_2$ evolution will differ even in the same soil sample. Soil core is the least disturbed soil samples, and those that have been sieved and mixed are considered the mostly highly disturbed.

Fig. 21.3 Methods for long-term measurement of CO_2 evolution from the soil surface. A metal cylinder is closed at one end is used to confine the CO_2 evolving from the soil unit and can react with alkali. The quantity of CO_2 trapped in the alkali is determined by titration

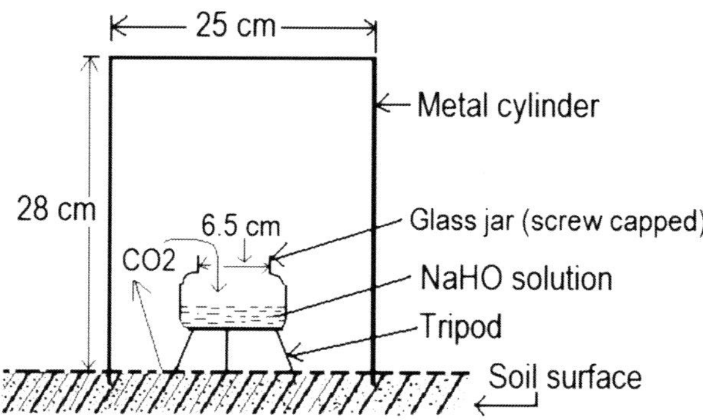

Disadvantage of sieved soil: highly disturbed, lost original structure and lack of macrofauna and plant parts (roots). They are well suited for studying soil microbiota.

21.5.1 Measurement of CO_2 Evolution Rate in Field Conditions *(In Situ)*

(A) Principles

One of the oldest and simplest methods for estimating the rates of CO_2 evolution from undisturbed soils is alkali trap method. For measurement of CO_2 evolution, alkali of a defined concentration is placed in an open jar above the soil surface and covered with a metal cylinder that is closed at the upper end.

As CO_2 evolves from the soil surface, it is trapped in the cylinder and is confined until it can diffuse to and be absorbed by the alkali. After a measured period of time, the alkali is removed and the unreacted portion is determined by titration. By means of subtraction, the amount of CO_2 that combines with the alkali can be determined.

(B) Apparatus

1. *Metal cylinder with one sealed end:* The open end should be at least 25 cm in diameter, and the cylinder should be at least 30 cm high.
2. *Screw-capped glass jars*: The opening should be at least 6.5 cm in diameter, and the jar should be at least 7 cm high.
3. *Tripods made from heavy-gauged wire or plastic*: These should be constructed to hold the base of the jars about 2 cm above the surface of the soil.

Details of field CO_2 measurement apparatus is shown in Fig. 21.3.

(C) Reagents

1. Sodium hydroxide (NaOH) solution, 1 N: Dissolve 40 g of NaOH in 1 L distilled water.
2. Barium chloride ($BaCl_2$), 3 N.
3. Hydrochloric acid (HCl), 1 N.
4. Phenolphthalein indicator: Dissolve 1 g of phenolphthalein in 100 mL of 95% ethyl alcohol.

(D) Procedure

- Collection of CO_2 in field
- Titration at laboratory

Collection of CO_2 in Field

Take 20 mL of 1 N NaOH solution into the glass jar. (The glass jar should be labelled and numbered and filled up with NaOH solution in laboratory itself. All jars must be tightly capped).

1. Select an appropriate site.
2. Put a tripod stand at the centre of the selected site.
3. Open the lid and put the glass jar above the tripod.
4. Immediately place the metal cylinder and press the edge about 2 cm into the surface of the soil.
5. The cylinder should be shield from direct sunlight (either cover appropriate size of wooden box or a weighted piece of aluminium foil).
6. After exposure of the alkali for 24 h, remove the glass jar, cover with lid (airtight seal) and take to the laboratory for analysis.

7. Simultaneously run a control set. Put the same alkali trap in a two-sided closed metal cylinder.

(E) Laboratory Analysis of CO_2
1. Titrate the NaOH solution of the glass jar of both control and exposed to soil air to determine the quantity of NaOH that has not reacted with CO_2.
2. For this purpose, add excess $BaCl_2$ to the NaOH solution to precipitate the carbonate as insoluble $BaCO_3$.
3. Add a few drops of phenolphthalein as an indicator and titrate the unreacted NaOH with HCl directly in the jar.
4. Add HCl slowly to avoid contact with and possible dissolution of the precipitated $BaCO_3$.
5. Note the volume of acid needed to titration.

(F) Calculation
The following formula can be used to calculate the amount of CO_2 evolve from the soil during the exposure to alkali solution:

$$mg\ of\ CO_2 = (B - V) \times N \times E.$$

where
B = HCl needed to titrate the NaOH in control glass jar, mL
V = HCl needed to titrate the NaOH in glass jar exposed to soil, mL
N = normality of HCl
E = equivalent weight, If the data are expressed in terms of C, then E = 6; if expressed as CO_2, then E = 22.

Once the mg of CO_2 has been determined, the result is conveniently expressed as mg of $CO_2/m^2/h$.

(G) Comments
1. Historically, this is one of the oldest method used to investigate the rates of CO_2 evolution from in situ soils.
2. In terms of apparatus needed, it is one of the simplest.
3. One group of investigators contends that the method underestimate the amount of CO_2 evolved from the soil, whereas a second group contends that the method overestimate the amount of CO_2 evolved from the soil. So

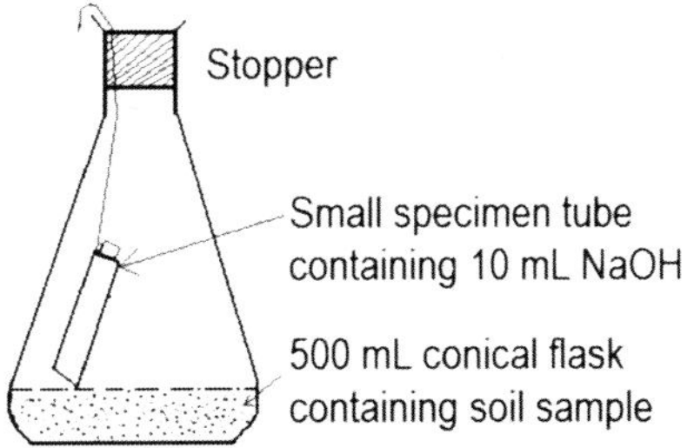

Fig. 21.4 The set of the respirometer

there is a general consensus that, if the method is carefully used, it allows quite accurate measurement of CO_2 evolution from in situ soils.
4. Used as described about, the edge of the metal cylinder is pushed about 2 cm into the soil. Other researchers pushed the edge of cylinder into 5-cm depth.
5. The quantities of CO_2 trapped by alkali solution in related with (a) strength of solution, (b) length of exposure, (c) temperature and (d) surface area. The most efficient trapping occurred when the surface area of alkali solution was about 25% that of surface covered by gas collector.
6. Alkali concentration in glass jar: For stronger concentration (e.g., $\geq$ 0.5 N KOH), the absorption rate remained at >90% of that which was theoretically possible if the alkali was at least 80% unreacted.

21.5.2 Measurement of CO_2 Evolution (Laboratory Method)

(A) Materials
1. Undisturbed soil core
2. Volumetric flask, 500- or 1,000-mL capacity with rubber stopper
3. Test tubes, 30-mL capacity filled with 1 N NaOH solution

(B) Apparatus
A simple respirometer used for laboratory measurement of CO_2 evolution is shown in Fig. 21.4.

(C) Procedure
Place 200–300 g of sample moist soil into a 500-mL conical flask. CO_2 evolved during incubation period is tapped in 10 mL of 1 N NaOH

contained in a small specimen tube inside the flask connected by a thin cotton thread. The conical flask is tightly capped with a rubber stopper and incubated a light/dark cycle of 10/8 h at room temperature (note the room temperate).

CO_2 trapped in NaOH is removed from the flask in the desire interval of times (say 3, 5, 7, 10, 21, etc.). In regard to time, it has been recommended to take reading every 24 h. However, with very active soils, shorter intervals must be used.

The alkali trapped is diluted to 25 mL with CO_2-free deionised distilled water (prepare by boiling the deionised distilled water to get rid of all CO_2).

1. Add excess $BaCl_2$ to the NaOH solution to precipitate the carbonate as insoluble $BaCO_3$.
2. Add a few drops of phenolphthalein as an indicator, and titrate the unreacted NaOH with HCl directly in the specimen tube.
3. Add HCl slowly to avoid contact with and possible dissolution of the precipitated $BaCO_3$.
4. Note the volume of acid needed to titration.

(D) Calculation
The following formula can be used to calculate the amount of CO_2 evolve from the soil during the exposure to alkali solution:

$$\mu g \ of \ CO_2 \ per \ g \ of \ soil = (B - V) \times N \times E.$$

where

B = HCl needed to titrate the NaOH in control glass jar, mL
V = HCl needed to titrate the NaOH in glass jar exposed to soil, mL
N = normality of HCl
E = equivalent weight; If the data are expressed in terms of C, then E = 6; if expressed as CO_2, then E = 22.

21.6 Soil Enzyme

Soil is a living system where all biochemical activities proceed through enzymatic processes. Enzyme accumulated in soils are present as free enzyme, such as exoenzymes released from living cells, endoenzymes released from disinte-grating cells and enzyme bound to cell constituents such as disintegrating cells, in cell fragments and in viable but nonproliferation cells. There is no report on extraction of enzyme from soil or enzyme purified from soil. Most studies measure the rate of enzyme-catalysed reaction of soil samples. Under normal conditions, the activity of a suitably chosen enzyme can give an indication of the total activity.

21.6.1 Dehydrogenase Activities in Soil

Biological oxidation of organic compound is generally a dehydrogenation process. The dehydrogenase enzyme systems apparently fulfil a significant role in the oxidation of soil organic matter as they transfer hydrogen from substrate to acceptor. Therefore, the result of assay of dehydrogenase activity would show the average activity of the active population. The overall process of dehydrogenation may be represented as follows:

$$XH_2 + A \rightarrow X + AH_2$$

where XH_2 is an organic compound (hydrogen donor) and A is a hydrogen acceptor.

In fertile soil, no nutrient amendment is needed, and in fact, one of the advantages of the dehydrogenase assay over other enzyme assays is that it does not require any amendments and not preferentially stimulate any group of microorganisms. However, addition of organic amendment (i.e, 0.5-mL glucose solution after the addition of 3% TTC except blank) is required for low fertility soil, such as mine soils.

(A) Principle
This method involves colorimetric determination of 2,3,5-triphenyl formazone (TPF) produced by the reduction of 2,3,5-triphenyltetrazollium chloride (TTC) by soil microbes (Casida 1977). Dehydrogenase activity is highly correlated with CO_2 release, proteolytic activity and nitrification potential. The highest activity was on the top

3 cm in an arid soil, but again, there was no correlation with microbial numbers.

2, 3, 5 TTC (pale yellow colour) $\xrightarrow[\text{H}^+ \text{ released from microbial respiration}]{\text{Dehydrogenase}}$ 2, 3, 5 TTE (deep red precipitation)

(B) Apparatus
1. Spectrophotometer (485 nm)

(C) Materials
1. Soil—20 g (<2-mm size)
2. Screw-cap test tubes, 16 by 125 mm or plastic vial with a tightly fitting lids—4 nos. (1 for blank and 3 for each soil sample)
3. Bench-top balance (0.01 g)
4. Pipette—1, 5 and 10 mL
5. Whatman filter paper no. 42
6. Filtration funnels for each plastic vial
7. Volumetric flask—100 mL (7 nos.)
8. Spatula, distilled water and glass rods

(D) Reagents
1. *Calcium carbonate* ($CaCO_3$).
2. *Triphenyltetrazollium chloride* (TTC), 3%: Dissolve 3 g of TTC in 80 mL of distilled water and adjust the volume to 100 mL. Not all of the TTC will dissolve, so the solution should be filtered through filter paper. This compound is photosensitive to short wave, UV, which will lead to reduction to TPF, thus avoid exposure to excessive light.
3. *Methanol* (AR).
4. *Triphenyl formazone* (TPF) standard solution. Dissolve 100 mg of TPF in about 80 mL of methanol and adjust the volume to 100 mL with methanol. Mix thoroughly.

(E) Procedure
1st period: Incubation of Soil Sample Saturated with TTC Solution

1. Mix thoroughly 20 g of air-dried soil (<2 mm) and add 200 mg of $CaCO_3$ and place 6 g of this mixture in each of three screw-cap test tube.
2. To each tube, add 1 mL of 3% TTC and 2.5 mL of distilled water. This amount of liquid should be sufficient that a small amount of liquid should be sufficient that a small amount

Table 21.6 Standard curve data of TPF

TPF in µg/mL	Absorbance at 485 nm
0.00 (Blank)	0.000
5.00	0.214
10.00	0.423
15.00	0.642
20.00	0.833
25.00	1.051
30.00	1.244

of free liquid appears at the surface of the soil after mixing. If the soil is low in organic carbon, 0.5 mL of 1% w/v glucose solution is added. Otherwise, microbes take a longer time to reduce the TTC.
3. Mix the contents of each tube with a glass rod, and stopper the tube and incubate it at 37°C for 24 h (*the tube needs to be sealed properly against oxygen from the air*).

2nd period: Analysis of TPF Produced by Spectrophotometer at 485 nm
1. After 24 h, remove the stopper, add 10 mL of methanol and again stopper the tube and shake it for 1 min.
2. Unstop the tube and filter the suspension through a glass funnel plugged with absorbent cotton (or Whatman filter no. 42) into a 100-mL volumetric flask. Wash the soil thrice with additional methanol (in 10-mL portions). Dilute the filtrate to 100 mL with methanol.
3. Measure the intensity of reddish colour by spectrophotometer at a wavelength at 485 nm using 1-cm cuvette with methanol as a blank.
4. Calculate the amount of TPF produced by reference to a calibration graph prepared from TPF standards.

(F) Preparation of TPF Calibration Curve
1. Dilute 10 mL of the TPF standard solution to 100 mL with methanol (100 µg of TPF/mL).
2. Pipette 5-, 10-, 15- and 20-mL aliquots of this solution into 100-mL volumetric flasks, make up the volume with methanol and mix well (500, 1,000, 1,500 and 2,000 µg of TPF/ 100 mL). Measure the intensity of the red colour of TPF at 485 nm wavelength (Table 21.6).

3. Plot the absorbance readings against the amount of TPF in the 100-mL standard solutions.
4. Express the results as *g TPF/g dry soil*. Report the average of the three duplicates.

21.6.2 Invertase, Amylase and Cellulase Activity of Soil

Amylase is widely distributed in soils. The enzyme β-amylase predominates in soil and responsible for hydrolysis of *starch to maltose and glucose*. Among the known amylase, there are two broad groups: α-and β-amylases. β-amylase predominates in soil. The enzyme activity decreases down the profile. The α-amylase hydrolyses starch, glycogen and related polysaccharides and is predominantly distributed in fungi, bacteria and animals.

Like amylase and invertase, cellulase is also an important enzyme in soil, which is responsible for the rate and course of decomposition of plant litter mainly by bacteria *Acetobactor*. The enzymes involved in the random hydrolysis of cellulose to cellobiose have not yet been characterised but have been grouped under the generic term cellulase.

$$Cellulose \rightarrow Cellodextrins \rightarrow Cellobiose \rightarrow Glucose$$

(A) Materials
1. Conical flask with stopper, 150-mL capacity
2. Centrifuge
3. Spectrophotometer

(B) Chemicals and Reagents
Chemicals
1. $Na_2HPO_4.2H_2O$
2. KH_2PO_4
3. Toluene (AR)
4. Rochelle salt (Na K tartrate)
5. 3,5-Dinitro salicylic acid
6. NaOH
7. Soluble starch (for amylase analysis)
8. Carboxymethyl cellulose (for cellulase analysis)
9. Sucrose (for invertase analysis)

Reagents
1. *Sorensen's buffer—pH* 5.5, 0.06 M: Dissolve 10.6896 g Na_2HPO_4. $2H_2O$ in 1 L of distilled water (solution A). Dissolve 8.1654 g KH_2PO_4 in 1 L of distilled water (solution B). Mix 5 mL of solution (A) and 95 mL of solution (B), and the pH is adjusted (if necessary) in pH metre using the above solution.
2. *Toluene* (AR).
3. *Substrate*: 1% soluble starch for amylase (dissolve 1 g soluble starch in 100 mL distilled water), 3% carboxymethyl cellulose for cellulase and 5% sucrose for invertase.
4. *Colour development reagent: 3,5-dinitro salicylic acid solution.* Dissolve 1 g of 3,5-dinitro salicylic acid in 20 mL of 2 N NaOH and dilute to 50 mL with water. Then add 30 g of Rochelle salt (Na K tartrate) and make up to 100 mL with water. Protect the solution from CO_2.

(C) Procedure
1. Fresh soil is screened through a 2-mm sieve.
2. Take a 150-mL flask and put 3 g of soil and 0.2 mL toluene; mix and allow to stand in room temperature for 15 min.
3. Add 6-mL Sorenson's buffer and 6 mL substrate into the flask.
4. Close the flask with stopper and place in the incubator at 30°C
5. Run a control set. Instead of substrate, put 6 mL distilled water.
6. After 24 h, contents of the flask are centrifuged.
7. *Colour development*: Take 1 mL of clear supernatant and add 2-mL 3,5-dinitro salicylic acid solution. The colour reagent also stops enzyme activity. 3,5-dinitro salicylic acid when boiled in an alkaline medium, the acid is reduced by the sugar to 3-amino, 5-nitrosalicyclic acid which is yellow orange in colour.
8. Measure the pink colour at 540 nm.
9. The standard graph is prepared taking fructose as standard. The colour develops after keeping the solution for 5 min in boiling water bath.

21.7 Determination of Microbial Biomass (Carbon and Nitrogen)

(A). Introduction
The microbial biomass accounts for only 1–3% of soil organic C but it is the eye of the needle through which all organic material that enters the soil must pass (Jenkinson 1977). Soil microbial biomass (SMB), which can be either a source or sink of available nutrients, plays a critical role in nutrient transformation in terrestrial ecosystems. Any change in the microbial biomass may affect soil organic matter turnover. Thus, the soil microbial activity has a direct influence on ecosystem stability and fertility. In general, SMB can be used for assessing soil quality of different types of vegetation as well as for evaluating soil perturbation and restoration). Martens (1995) reviewed methods of microbial biomass carbon (MBC) determination and opined that the fumigation-extraction method is good. However, the Kc factor applied to calculate MBC from the organic carbon additionally made extractable by the fumigation is still controversial. A much faster method called chloroform fumigation extraction (CFE), derived from chloroform fumigation incubation (CFI) method, was later proposed by Vance et al. (1987).

(B). Principle
Probably the most commonly used classical technique for determining the microbial biomass size is that of *chloroform fumigation extraction (CFE)* method. Fumigation ruptures microbial cells and releases cell walls and cellular contents into the soil. If incubation follows the fumigation, a "flush of decomposition" of the soil organic matter occurs that is due to the decomposition of microorganisms killed during fumigation. The carbon released from the lysed cell is estimated by submitting the soil samples to a chemical extraction with K_2SO_4 0.5 mol/L, ratio soil:extractant equal 1:4 (Vance et al., 1987). When the fumigation and subsequent extraction procedures are conducted under a strict set of conditions, the size of the biomass pool of Carbon (MBC) can be determined by using the equation as below:

$$MBC = Ec/Kc$$

where
MBC = Concentration of Microbial Biomass Carbon (MBC) in the soil,
  Ec = (organic carbon extracted from fumigated soils) – (organic carbon extracted from non-fumigated soils)
  Kc = Extraction efficiency (0.45) (Vance et al 1987)

Assumptions
Following assumptions are implicit in the calculation of biomass C associated with chloroform fumigation and extraction:
1. The kill caused by the fumigation is essentially complete.
2. The fraction of microorganisms that die in the unfumigated control is negligible compared to those killed in the fumigated sample.
3. Fumigation has no effect on the soil other than the killing of the microbial biomass.

The difference organic carbon is due to the carbon released from the microbial biomass. An extraction efficiency (Kc factor) of 0.45 (or a close similar value) is often used to calculate the microbial biomass carbon value using the fumigation-extraction method.

(C) Materials and Methods
1. Chloroform ($CHCl_3$): Reagent grade, $CHCl_3$ contains a stabilizing agent, ethanol, which must be removed just prior to its use. The purification process consists of double distillation of $CHCl_3$ at 55°C, followed by washing tree times, first with 18 M H_2SO_4 and then with deionized water. Purified $CHCl_3$ is kept in a brown bottle containing anhydrous K_2CO_3 and used within 6 h of preparation. [Ethanol is removed by washing 100 ml chloroform (reagent grade) with 100 ml of 5% H_2SO_4 using a separating funnel. Thereafter, the chloroform was washed 3 times with 100 ml deionized water Christian et al 2000]

2. Large desiccator (vacuum type): Desiccator should be inert to $CHCl_3$ vapour, be of dry seal type; a large desiccator may hold about 20–25 samples simultaneously.
3. Fume hood
4. Boiling chips
5. Two 100-mL glass containers with lids/ assay.
6. One weighing container/ assay for determination of moisture content.
7. K_2SO_4 extraction solution (0.5M).
8. Two Whatman No # 42 filter papers/ assay for filtration.
9. Two 50-mL vials/assay for filtrate.

(D) Procedure
I. Preparation of soil
1. Pass the soil samples through a 2-mm mesh sieve and mix thoroughly.
2. Weigh out three portion of the soil, 15–50g each, one portion into a weighing container for moisture determination. Other two portion into 100-mL glass bottles, one sample to be fumigated for 24 h and then extracted, and one control sample to be extracted immediately.
3. Dry soil in the weighing container in an air oven at 105°C for at least 24 h or until a constant oven-dry weight is achieved. Cool in a desiccator, reweigh, and determine the moisture content of the soil sample.

II. Fumigation treatment
1. The desiccator should be lined with freshly moistened filter paper just prior to fumigation treatment. For fumigation treatment, place the glass sample bottle containing the soil into a desiccator together with a 100-mL beaker containing 50 mL $CHCl_3$, and a few boiling chips. Seal and evacuate the desiccator, taking care to vent the fumes released by the vacuum pump, until the $CHCl_3$ boils vigorously, continue evacuating for 1–2 min. Seal the desiccator under vacuum and place in the dark at 250°C for 24 h.
2. After the fumigation treatment, release the vacuum, open the desiccator, and remove the beaker $CHCl_3$ and moistened filter paper.
3. Remove residual CHCl3 vapor from the soil samples by repeated evacuation, usually 3 to 6 times using vacuum pump.

III. Extraction of microbial biomass carbon
1. Add 40 mL 0.5 M K_2SO_4 to the bottles containing the unfumigated control and fumigated subsamples, the equivalent oven-dry soil weight (g): extractant volume ratio of ¼ and shaken for 1 h on a reciprocal shaker.
2. After shaking, pass the soil suspension through Whatman filter # 42.
3. Measurement of organic carbon and total nitrogen can require 5 to 30 mL of extract, depending on the methods of analysis used. Preliminary tests showed that a small contamination of the soil extract with C derived from the filter paper could be avoided by discarding the first 2 ml of each K_2SO_4 extract (Christian et al 2000).

IV. Determination of organic carbon (OC) and Total nitrogen
Organic carbon dissolved in the K_2SO_4 extracts can be determined by Potassium dichromate method (Nelson and Sommers 1992) (either i or ii) ; and total nitrogen by Kheldahl digestation method.
(i) For dichromate digestion method, an 8-mL aliquot of the K_2SO_4 extract is added to a mixture of 0.2 M $K_2Cr_2O_7$ (2 mL), 18 M H_2SO_4 (10 mL), 14.7 M H_3PO_4 (5 mL), carefully mixed and digested at 150°C for 30 min. The excess dichromat remaining is determined with 0.017 M ferrous ammonium sulfate using ferroin as an indicator. OR,
(ii) Soil extracts (5 ml) were pipetted into 75-ml digestion tubes and treated with 5 ml of 0.07N $K_2Cr_2O_7$, 10 ml of 98% H_2SO_4, and 5 ml of 88% H_3PO_4. Samples were carefully mixed and digested at 150 7C for 30 min using a digestion block. After cooling, samples were titrated with a solution of 0.01N $Fe(NH_4)_2(SO_4)_2$ $6H_2O$ in 0.4 M H_2SO_4 (Christian et al 2000). *[For details calculation of OC refer section 20.4]*
(iii) For organic nitrogen determination, soil extract (5-25 mL) can be converted to ammonical nitrogen by by Kjeldahl digestion for 4 h at 360°C. Each sample is digested in an mixture containing 10 mL concentrated H_2SO_4, 3 g powdered K_2SO_4: $CuSO_4$. $5H_2O$ (30:1 by weight), and a selenium catalyst in the form of the catalyst

mixture. After digestion is over, distillation is carried out by the addition of 40% (w/v) NaOH (50 mL) and distillate is collected into 4% (w/v) boric acid cum- mixed indicator solution in a 250 mL conical flask (pink color solution). After absorption of ammonia colour turns to green. Titrate the distillate with 0.01N HCl to its original colour (i.e. green to pink colour indicates the end point of titration). Run distilled water blank in the same manner. [*For details procedure refer section 20.6*]

(E). Calculation microbial biomass C and N
(i) Soil moisture content

Moisture content (MC%)

$$= \frac{\text{wt of soil (g) - wt of oven dry soil (g)}}{\text{wt of oven dry soil (g)}} \times 100$$

(ii) Weight of soil sample oven dry weight equivalent taken for microbial biomass estimation (MS):

$$MS = \frac{\text{wt of soil (g)} \times 100}{(100 + MC \text{ in } \%)}$$

(iii) Total volume or solution in the extract soil (VS):

VS (mL)

= wet soil weight (g) − oven dry weight soil (g)

+ extractant volume (mL)

(iv) Total weight of extractable C and N in the fumigated (O_F) and unfumigated (O_{UF}) soil samples:

$$OC_F, OC_{UF} \ (\mu g/g \ \text{soil}) = \text{extractable C} \left(\frac{\mu g}{mL}\right) \times \frac{Vs \ (mL)}{Ms \ (g)}$$

$$ON_F, ON_{UF} \ (\mu g/g \ \text{soil}) = \text{extractable C} \left(\frac{\mu g}{mL}\right) \times \frac{Vs \ (mL)}{Ms \ (g)}$$

(v) Microbial biomass C and N in the soil (MBC; MBN)

$$MBC \ (\mu g/g \ soil) = \frac{(OC_F - OC_{UF})}{K_C}$$

$$MBN \ (\mu g/g \ soil) = \frac{(ON_F - ON_{UF})}{K_n}$$

Where, Kc = 0.45 and represents the efficacy of extraction of microbial biomass carbon (Vance et al 1987); Kn = 0.18 represents the efficiency of extraction of microbial biomass nitrogen.

21.8 Exercise on Soil Microbiology

1. Name the different types of microbes study in soil microbiology?
2. Why will there be a variation in microbial density between rhizosphere and non-rhizosphere zone in soil?
3. List the different types of bacteriological media used in laboratory.
4. How to prepare dilution series up to 10^{-9}.
5. What are the difference between pour plate method and spread plate method?
6. Why antibiotics (streptomycin, auromycin, etc.) are used in fungal enumeration. When it is added to the medium?
7. What are mycorrhiza fungi? How to study? Discuss the steps of preparation of mycorrhiza slides.
8. How are mycorrhiza spores (VAM) extracted from a soil sample?
9. Define microbial respiration?
10. "Field measurement of CO_2 depends on the moisture condition of soil"—why?
11. List the important precautions to be taken during the measurement of CO_2 in field.
12. How is microbial respiration measured in the laboratory?
13. What is respirometer?
14. What is dehydrogenase enzyme?
15. How did glucose influence the results of dehydrogenase analysis? What would be the results of using another amendments, for example, sawdust instead of glucose?
16. How did the soil type influence the dehydrogenase estimation?
17. What are some of the errors that could have occurred in dehydrogenase estimation?
18. How dehydrogenase activity is expressed?
19. What is common in invertase, amylase and cellulose enzyme?

References

Alexander K (1977) Introduction to soil microbiology, 2nd edn. Wiley, New York

Allen SE, Grimshaw HW, Parkinson JA, Quarmby C (1974) Chemical analysis of ecological materials. Wiley, New York, p 565

Biermann B, Linderman R G (1981). Quantifying vesicular-arbuscular mycorrhizae: a proposed method toward standardization. New Phytologist 87, 63-67. http://mycorrhizae.org.in/index.php?option=com_content&task=view&id=38&Itemid=71

Biswas TD, Mukherjee SK (1994) Textbook of soil science, 2nd edn. Tata McGraw Hill, New Delhi

Casida LE Jr (1977) Microbial metabolic activity in soil as measured by dehydrogenase determinations. App Environ Microbiol 34(6):630–636. http://www.ncbi.nlm.nih.gov/pubmed/339829

Christian W et al (2000) A rapid chloroform-fumigation extraction method for measuring soil microbial biomass carbon and nitrogen in flooded rice soils. Biol Fertil Soils, 30:510–519. http://xa.yimg.com/kq/groups/13354653/1103835104/name/fulltext.pdf

Gaur A, Adholeya A (1994) Estimation of VAM spores in the soil: a modified method. Mycorrhiza News 6 (1):10–11

Gerdeman WH, Nicolson TH (1963) Spores of Mycorrhizal Endogone species extracted from soil by wet sieving and decanting. Trans Br Mycol Soc 46:235–244

Jenkinson DS (1977) The soil biomass. NZ Soil News, 25:213–218

Maiti SK (1997) Importance of VAM fungi in coalmine overburden reclamation & factors effecting the establishment of VAM Fungi on overburden dumps. Environ Ecol 15(3):602–608

Martens R (1995) Current methods for measuring microbial biomass C in soil: Potentials and limitations. Biol Fertil Soils.19:87–99. http://www.planta.cn/forum/files_planta/0823295488biomass_c_in_soil_111.pdf

Pepper IL, Gerba CP, Brendecke JN (1998) Environmental microbiology- a laboratory manual. Academic press Limited, London

Singh SS, Tiwary SC (2001) Modified wet sieving and decanting techniques for enhanced recovery of vasicular-arbuscular mycorrhizal fungi from forest soil. Mycorrhiza News 12(4):12–13

Vance ED, Brookes PC, Jenkinson DS (1987) An extraction method for measuring soil microbial biomass C. Soil Biol Biochem.19:703-707. http://www.sciencedirect.com/science/article/pii/0038071787900526 OR http://www.scribd.com/doc/58919399/An-Extraction-Method-For-Measuring-Soil-Microbial-Biomass-C

Williams ST, Willington EMH (1982) Actinomycetes. In: Page AL et al (eds) Methods of soil analysis, Part 2: chemical and microbiological properties, 2nd edn. Am Soc Agro Inc, Madison, pp 969–987

Contents

22.1 Plant Material Analysis

Analysis of plant tissue (foliar analysis) can be used as a method of detecting mine soil nutrient deficiencies, the assumption being that sufficiency or deficiency of plant available nutrients in the soil will be directly reflected in their tissue concentration. The technique is subjected to many factors such as time of year, position of leaf tissue on the plant (especially for trees), type of plant, specific plant part, growth stage, level of available soil nutrients and environmental factors.

22.2 Plant Sampling

Standardisation of sample collection procedure and interpretation of results by experts, related in addition to healthy growing of trees in the area, may use foliar analysis, a useful diagnostic technique for determining why stands of trees are growing poorly or not. Leaves are most commonly chosen. Seeds are rarely used for analysis, except for assessing of B toxicity and Zn and P deficiency in certain grain crops. When leaves are sampled, recently mature ones are taken; both new and old leaves are generally avoided. However, young emerging leaves are sampled for diagnosing iron chlorosis by determining ferrous (Fe^{2+}) content of fresh leaves. Damaged or diseased leaves are excluded, and plant should not be sampled when they are under moisture or

S.K. Maiti, *Ecorestoration of the Coalmine Degraded Lands*,
DOI 10.1007/978-81-322-0851-8_22, © Springer India 2013

temperature stress. A representative selection of leaves (10–25 nos) should be collected from the lower part of the crown.

Fresh collected material must be clean to remove dust by washing with deionised water. Air-dried at 65°C for a maximum of 24 h and grind mechanically to produce material suitable for analysis, usually pass through 60-mesh sieve and stored in airtight container. An acid digestion with concentrated sulphuric acid and hydrogen peroxide is suitable for determination of N, P, K, Ca, Mg, Zn, Cu, Al, Fe and Mn.

22.3   Bioassay and Growth Trials

The approach can be two folds, pot experiment to test a wide range of variations and combination of treatments and field trials to test a narrower range of treatment combinations, but it is advantageous because the test is conducted under proper field conditions and can be performed over a longer period of time.

22.4   Pot Experiments

These have the advantages over field trails because pot experiments that are set up are cheaper and can cover a wide range of treatments and combinations, though only cover a short timescale (up to one growing seasons; maximum two). They are particularly valuable for testing the efficacy of amendments, fertiliser doses, different combinations of manures, topsoil and fertiliser. A useful pot size is the standard 12-cm diameter, which has a convenient surface area of $0.01\ m^2$. For longer experiments, the 18-cm diameter is better (Coppin and Bradshaw 1982).

22.4.1   Field Trials

Field trials can be set up in the field on any scale, using any size of plot from 0.5×0.5 m to 10×10 m per treatment. The size will depend on both the space available and whether it will be necessary to superimpose other treatments on the plot at a later stage. A $1\ m^2$ plot is the most useful for ground cover vegetation. Much large area can be used for growth trials of tree species.

Following measurements should be done during field trials experiments:

(i) Field experiments should be monitored regularly. The frequency of monitoring will depend on the observations required to meet the project's objectives. Different types of experiments require different types of measurements. In case of restoration experiment, following measurements could be done at the end of growing seasons (December-January) for the grasses. The parameters to be monitored are:

(a) Aerial height (cm) for tree sapling, increase in root and shoot biomass, numbers of leaves (initially), etc.

(ii) Soil moisture measurements may be carried out during post-monsoon period (November- December) or before planting or both. Gravimetric soil moisture sampling needs a core sampler, weigh scale, containers and oven. Decide on depth increments. For example, use increments of 0–15, 15–30, 30–45, 45–60, 60–90 and 90–120 cm for measuring moisture within the typical rooting zone (1.2 m).

(iii) Weather information, especially temperature and rainfall, should be recorded during the growing season.

(iv) Rainfall should be measured at the site.

(v) Temperatures measured at a nearby weather station may be adequate if the station is located near the experiment.

(vi) Photographs taken during the course of the growing season provide excellent records of visual responses. They are also very useful if you are asked to give a presentation about the project at an extension meeting or to prepare research paper.

22.5 Treatments

For screening, plant species or species combinations (mixture) will form one treatment, with the whole trials perhaps grown at two or more amendments.

For amelioration trials, either a simple increasing treatment, that is, fertiliser is used or a factorial design, by which the effects of combinations of treatment can be tested. Table 22.1 depicts an example of NPK (nitrogen, phosphorous and potassium) factorial, where 3 levels (as an example) of each NPK are applied in every combination.

The above treatment combinations are for $3 \times 2 \times 2$ NPK factorial experiment. The number of factors can be anything from two upwards, though more than three gets very complicated.

22.6 Experimental Design and Layout

In order to be scientifically valid, it is necessary for the layout of the pots or plots to conform to certain patters, so that the results can be analysed statistically. The most important pattern is that the layout should be randomised and that each treatment is replicated. The usual level of replication in pot and field experiment is three, and this gives the best compromise between statistical validity and ease to set up, though up to five would be preferable.

Randomisation can be achieved by allocating treatments to the plots on the ground using random number tables. Shuffling the treatment or spreading them out arbitrarily is not satisfactory randomization. One such arrangement is shown in the Fig. 22.1.

(i) Randomized Complete Block Design
The randomized complete block design (RCB), shown in Figure 22.2, is widely used in field experimentation. It is an extension of the paired t-test. This design is appropriate when quantita-

Table 22.1 An example of showing effects of three factors (in this case, level of nitrogen, phosphorus and potassium) on plant growth experiment

Treatment	Factor 1	Factor 2	Factor 3
1	N level 1	P level 1	K level 1
2	N level 2	P level 1	K level 1
3	N level 3	P level 1	K level 1
4	N level 1	P level 2	K level 1
5	N level 2	P level 2	K level 1
6	N level 3	P level 2	K level 1
7	N level 1	P level 1	K level 2
8	N level 2	P level 1	K level 2
9	N level 3	P level 1	K level 2
10	N level 1	P level 2	K level 2
11	N level 2	P level 2	K level 2
12	N level 3	P level 2	K level 2

tive data, such as yield, and you require a rigorous comparison between treatments.

The two cornerstones of the RCB design are *replication* (i.e., *repetition*) and randomization. These allow to accommodate any variability in the local environment and to determine the probability of the differences in results between "treatments" being real or simply due to chance. Replicate each treatment a minimum of three times. Four of five replicates will be better than three.

Each treatment must be included *once* in each block of replications. The treatment locations must be randomly assigned to plots within the block. The purpose of randomizing the locations is to avoid biasing the results. If field plots are basic experimental unit, the individual plots **should be three to five times long as wide** and should be sized to comfortably handle one or two passes of the field equipment being used.

22.7 Assessment of Growth

The simplest and most usual method of assessment is yield, expressed on a dry weight of shoots per unit area (g/m^2) basis. Other criteria such as plant density, aerial height, increase in diameter, increase in shoot and root biomass and biochemical analysis should be used with the combination of yield.

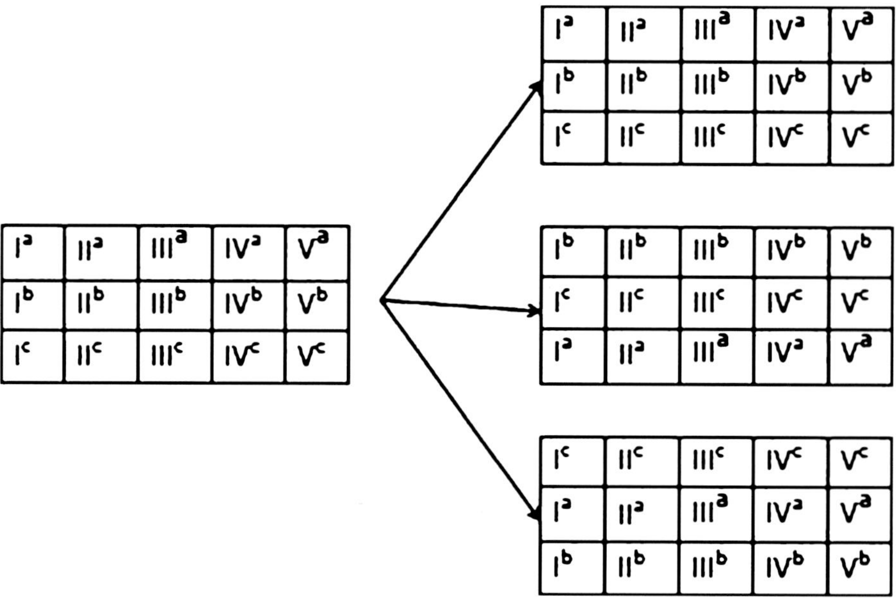

Fig. 22.1 Three alternatives lay out for five treatments

Fig. 22.2 Randomized
Complete Block Design

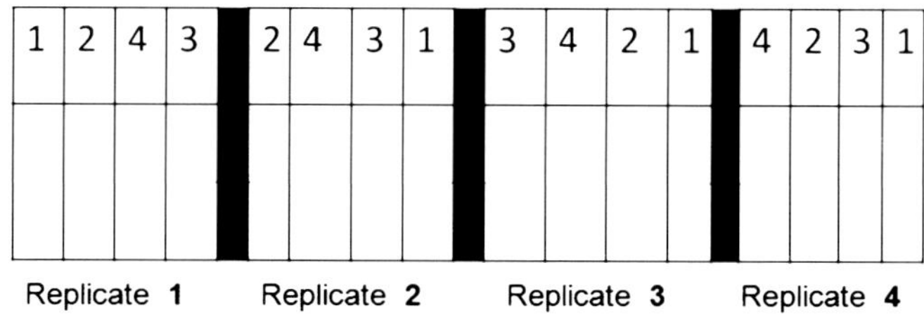

22.8   Vegetation Survey

Although an ecological training will normally be necessary in order to perform a detailed assessment of vegetation, the main requirements are outlined here for general guidance. There are two situations where it is necessary to know precisely the vegetation of the area, without relying on general impressions:

- Description and monitoring of reclaimed areas to assess the progress and condition of the developing vegetation
- Description and analysis of existing vegetation (pre-mining condition or from a control area as baseline study) in order to identify the degree of impacts on native species

The usual layouts of vegetation cover are studied by grid, regular, random or transect methods. Very often the area can be visually subdivided into distinct subareas and sampled. Transects sampled at regular intervals are useful for detecting gradual changes in vegetation or soil characteristics through a transition. The density of sampling will depend on the structure of the vegetation, and variable areas will require a higher density of sampling points, with perhaps transects to define the scale of variation.

22.8.1 Quadrate Sampling

The usual sampling technique is to use a quadrate, which is a square frame that is placed on ground to define an area of vegetation. Usually a square quadrate is used, though rectangular and circular quadrates are also useful in some situations like the one in forests. Round quadrate can be most accurate because they have the smallest perimeter for a given area. Round quadrates are also simple to define in the field, requiring only a center stake and a tape measure (Cox, 1990).

It is necessary to first determine the size of the quadrate for each type of community, and then the minimum number of samples that would be required for the study of that community is decided. This is done by the species–accumulation method. However, the most accurate way to determine the appropriate quadrate size for sampling any vegetation is by establishing and analyzing species accumulation curves. It has been found in several studies that a 1×1 m quadrat is enough for herbaceous communities, while in forests, the area of the quadrat ranges from 100 to 256 m^2. Figures 22.3 and 22.4 shows the set-up of field quadrate during sampling. The data from a series of nested quadrate is plotted on a species vs. area curve (Figure 22.5) to determine the smallest area within which the species of the community are adequately represented (the minimal area). In this method, the presence of additional species in each larger quadrate is recorded and plotted against the size of quadrate. A point is reached where increasing quadrate size does not significantly increase the number of species encountered. The minimal sample area can then be determined from the species-area curve. *Observe the following within the quadrate*:

- Presence or absence of all the plant species
- Number of individuals of each species per unit area
- An estimate of the proportion of ground covered by each species, often using the Braun-Blanquet scale (1–5):

5	75–100 % cover
4	50–75 % cover
3	25–50 % cover
2	5–25 % cover
1	<5 % cover

22.8.2 Dominance

The relative importance of a species in the community is expressed by dominance. The characters like frequency, density, cover (crown and basal) or biomass cannot be used singly to show the relative position of a species. A single tree with a large cover and biomass is of little importance in a grassland community and likewise, grasses although numerous have little importance in a forest community.

The species importance value index (IVI) is calculated, which is the sum of the relative values of the three quantitative characters. The relative values are expressed as percentage of the total values for all species in a community, as given below:

Relative frequency
$$= \frac{\text{Frequency of a species} \times 100}{\text{Sum of frequency of all the species}}$$
Relative Density
$$= \frac{\text{Density of a species} \times 100}{\text{Sum total of densities of all the species}}$$
Relative dominance
$$= \frac{\text{Cover of a species} \times 100}{\text{Sum total of cover of all the species}}$$

Thus, the total of IVIs of all the species in a community equals only 300.

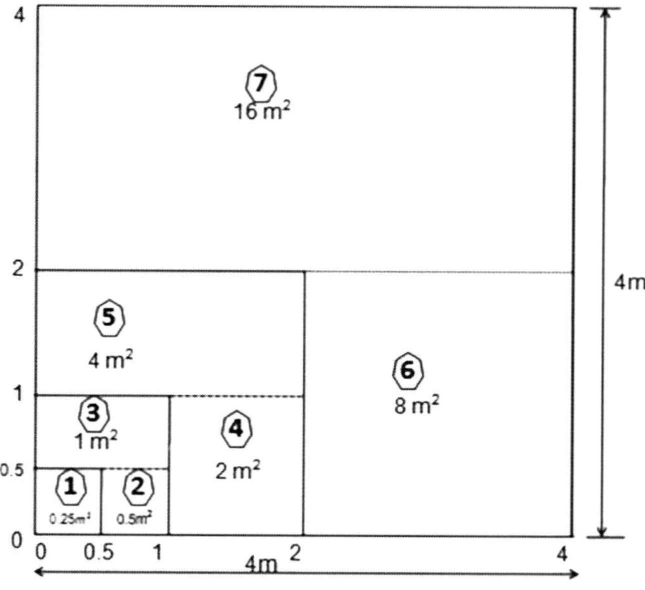

Fig. 22.3 The set-up of quadrate shapes and sizes for establishing a species accumulation curve (by progressively doubling the sampling area)

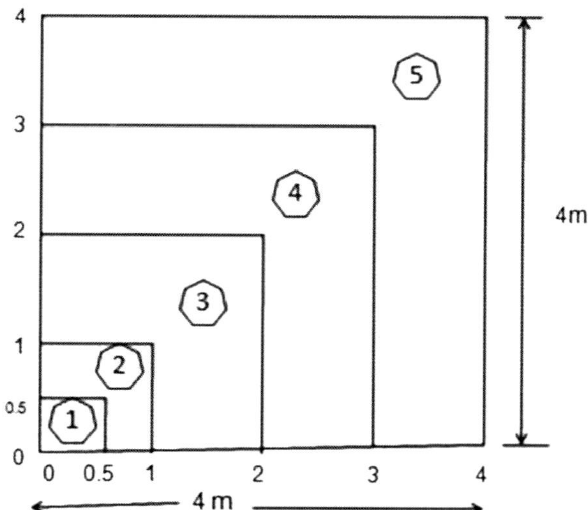

Fig. 22.4 The set-up of quadrate shapes and sizes for establishing a species accumulation curve

22.8.3 Transect Method

Transect is a straight line or belt. It is useful in the study of stratification and in analysing community structure where some of the habitat features show a steep gradient. In hills, on sea shores or margins of a water body, it is important to study the variations in the community along the gradient. A transect line can be made using a nylon rope marked and numbered at

Fig. 22.5 Determination
of minimum quadrate size

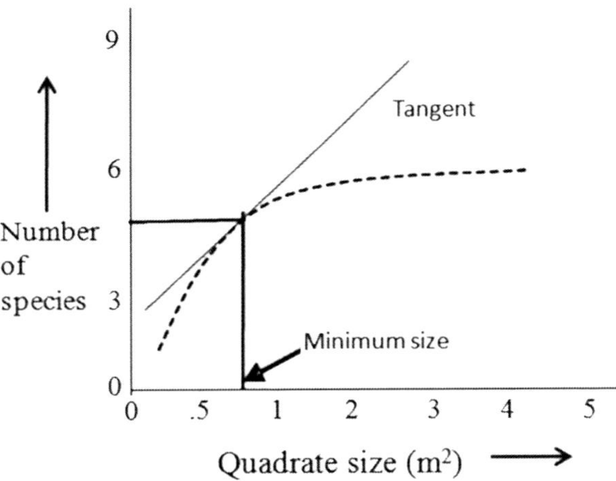

0.5 m, or 1m intervals, all the way along its length. This is laid across the area of study. The position of the transect line is very important and it depends on the direction of the environmental gradient, like change vegetation composition, change in soil characteristics (moisture content, stoniness) etc.

During stratification study, a transect is laid in the community, and all the plants falling along the line or in the belt are marked on a graph paper to a reduced scale. The height, branching and horizontal spreads are properly noted and charted. In forests, the trees and shrubs are distantly placed, and hence, a belt transect is more useful. In such situations, along with the vertical distribution of plants and their parts, the positions of the individuals in the belt are also noted.

In studying the community structure, the transect is divided into small segments, and data for the plants falling along the transect are recorded. Each segment of the transect is treated like one quadrat, and the data are accordingly analysed.

22.8.4 Plotless Method

The plotless method is applied in communities where the determination of the minimum size of the sample plot is either not possible or is very difficult (e.g., in tall glasslands). This method has been developed by the Wisconsin school.

Point-centred quarter method is one of the several variations of the plotless method. The method involves the use of a thin-pointed rod (pin). It is thrown randomly in the field, and the area around the pin is divided into four quarters. The plants nearest to the pin in the four quarters are recorded as shown in Fig. 22.6. The observation of each quadrate is recorded in a data-sheet. A sample data-sheet used during field sampling is given in Table 22.2. From the observation sheets from parameters are calculated:

Frequency of a species (A)

$$= \frac{\text{Number of sampling points with species (A) is present}}{\text{Total number of sampling points observed}} \times 100$$

Average distance

$$= \frac{\text{Distances of all species recorded}}{\text{Number of points } \times 4}$$

$$\text{Average area} = (\text{Average distance})^2$$

Absolute density of all species

$$= \frac{100 \times 100}{\text{Average area, cm}^2}$$

Relative density of the species (A)

$$= \frac{\text{Number of individuals at all points}}{\text{Number of points } \times 4} \times 100$$

Absolute density of the species

$$= \frac{\text{Relative density of the species (A)}}{\text{Absolute density of all species}} \times 100$$

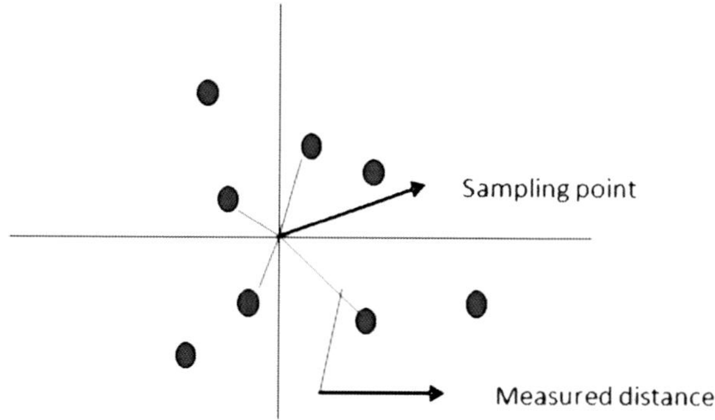

Fig. 22.6 Point centred quarter method of sampling vegetation

Table 22.2 Observation sheet in Point Centred quarter method of sampling vegetation

		Quadrant 1	Quadrant 2	Quadrant 3	Quadrant 4
Sample point #	Name of Species:				
	Distance (m)				
	Circumference:				

The basal cover is determined from the average basal area of the individuals of a species and the absolute density of that species. Importance value indices are calculated as in other methods. The data for the observations are recorded and important community is calculated.

References

Cox GW (1990) Laboratory Manual of General Ecology, (6th ed). William C. Brown Pub, Dubuque Iowa, USA

Coppin NJ, Bradshaw AD (1982) Quarry reclamation. Mining Journal Books, London

Soil Analysis

List of the chemicals required for soil analysis laboratory (price updated up to 2012/2013)

Name of chemicals	Availability (min)	Price (Rs.)
1. Acetic acid, conc. (glacial)	500 mL	235 = 00
2. Ammonium fluoride, NH_4F	500 g	500 = 00
3. Ammonium acetate, CH_3COONH_4	500 g	297 = 00
4. Ammonium hydroxide, NH_4OH	500 ml	250 = 00
5. Ammonium molybdate	100 g	1630 = 00
6. Ammonium metavanadate (NH_4VO_3)	100 g	941 = 00
7. Ammonium fluoride, NH_4F	500 g	500 = 00
8. Boric acid, H_3BO_3	250 g	307 = 00
9. Buffer tablets, pH 4.0, 7, 9.0	10 tablets	135 = 00/ for each pH
10. Bromocresol green	5 g	559 = 00
11. Calcium acetate	500 g	792 = 00
12. Cupric sulphate	500 g	567 = 00
13. Calcium chloride, $CaCl_2.2H_2O$	500 g	756 = 00
14. Copper sulphate, $CuSO_4$	500 g	567 = 00
15. Diphenylamine indicator, $(C_6H_5)_2NH$		
16. DTPA (Diethylenetriaminepentaacetic acid)		
17. Ethyl alcohol	500 mL	575 = 00
18. Ferrous ammonium sulphate	500 g	266 = 00
19. Hydrochloric acid, HCl	500 mL	128 = 00
	2.5 L	500 = 00
20. Hydrogen peroxide, 30% (w/v)	500 mL	118 = 00
21. Hydrofluoric acid	500 ml	684 = 00
22. Methyl orange indicator	125 ml	100 = 00
23. Methyl red indicator	125 ml	100 = 00
24. Mercuric chloride	100 g	2440 = 00
25. Magnesium nitrate	500 g	171 = 00
26. Nitrophenol	500 g	941 = 00
27. Perchloric acid	500 mL	864 = 00
28. Phenolphthalein indicator—powder solution	100 g	154 = 00
	125 mL	50 = 00
29. Potassium chromate, K_2CrO_4	500 g	716 = 00
30. Potassium permanganate, $KMnO_4$	500 g	567 = 00
31. Potassium chloride, KCl	500 g	217 = 00

(continued)

S.K. Maiti, *Ecorestoration of the Coalmine Degraded Lands*,
DOI 10.1007/978-81-322-0851-8, © Springer India 2013

(continued)

Name of chemicals	Availability (min)	Price (Rs.)
32. Potassium dichromate, $K_2Cr_2O_7$	500 g	711 = 00
33. Potassium dihydrogen phosphate, KH_2PO_4	500 g	532 = 00
34. Potassium per sulphate, $K_2S_2O_8$	500 g	595 = 00
35. Potassium hydroxide (pellets), KOH	500 g	338 = 00
36. Phosphoric acid, H_3PO_4	500 mL	437 = 00
37. Selenium powder	25 g	1540 = 00
38. Silver sulphate	25 g	6940 = 00
39. Sodium bicarbonate, $NaHCO_3$	500 g	180 = 00
40. Sodium carbonate, Na_2CO_3	500 g	244 = 00
41. Sodium chloride, NaCl	500 g	140 = 00
42. Sodium hexametaphosphate	500 g	347 = 00
43. Sodium hydroxide, pellets, purified	500 g	180 = 00
44. Sodium acetate	500 g	244 = 00
45. Sodium sulphate, Na_2SO_4	500 g	198 = 00
46. Stannous chloride, $SnCl_2$. AR	100 g	680 = 00
47. Sulphuric acid	500 mL	100 = 00
	2.5 L	500 = 00
48. Triethanolamine (TEA)	500 ml	532 = 00

Apparatus/Instruments Required for Soil Analysis Laboratory

Name of apparatus	Remarks
1. Soil sampling tools: soil sampling cores and augers, Spade, pick axe	Collection of sample
2. Soil sample preparation apparatus: mortar and pestle, brasses, spatula	Preparation of sample
3. Sieves: various sizes	Sieving of soil samples
4. Buchner funnel	Filtration purpose
5. Suction pump	Filtration purpose
6. Weighing bottle	Miscellaneous
7. Hot air oven	Drying of soil sample at 100°C
8. Hot plate	Digestion of soil sample
9. Desiccator: containing anhydrous silica gel	Miscellaneous. Particularly for moisture analysis
10. Mechanical analysis pipette	Texture analysis
11. Top-pan balance	Miscellaneous
12. Keen boxes (perforated circular soil box)	Water holding capacity
13. Infiltrometer (double ring)	Infiltration test
14. Stopwatch	Infiltration test
15. Driving plate	Infiltration test
16. pH metre	pH determination
17. Conductivity metre	EC measurement
18. Porcelain/platinum crucible	Organic matter analysis
19. Muffle furnace	Organic matter analysis

(continued)

(continued)

Name of apparatus	Remarks
20. Kjeldahl distillation apparatus	TKN, ammonia estimation
21. Spectrophotometer	Miscellaneous
22. Flame photometer	Na, K estimation
23. Centrifuge	Miscellaneous
24. Shaker	Miscellaneous
25. Petri dishes	Miscellaneous
26. Filter paper	Miscellaneous
27. Metal container/ plastic container	Wilting point determination
28. Dwarf sunflower seed (wilting point determination)	Wilting point determination

List of Trees Commonly Found in Mining Areas and Reclaimed Areas

Sl. No.	Botanical name (Family)	Local name	Growth characteristics, leaf canopy characteristics	Remarks
1.	*Acacia auriculiformis* A. Cunn (Mimosaeae, Leguminosae)	Australian Wattle; Phyllode Acacia (E), Akashmoni (B)	Quick growing, drought resistant, hardy, nitrogen fixer. Not grazed by cattle; oblong and dense, height 16 m; flowering June–January; regenerated by seed, sapling	Common for ecorestoration
2.	*Acacia nilotica* (L.) Willd. Ex. Delile. (syn *Acacia arabica* (Lam.) Willd. (Mimosaeae, Leguminosae)	Babul (B); Wattles (E)	Other species are: *A. decurrens* (green wattles); *A. dealbata* (silver wattles); *A. meamii* (black wattle)	Common tree for ecorestoration
3.	*Acacia catechu* Willd (Mimosaeae, Leguminosae)	Cutch tree (E), Khair (H), Khayer (B)	Small or medium-sized, thorny tree up to 15 m tall; bark dark grey or greyish-brown, occurs naturally in mixed deciduous forests, common in the drier regions, wood is excellent firewood, seeds viability is maintained for 9 months in open storage at room temperature, contain 15,000–40,000 seeds/kg	Common tree for ecorestoration
4.	*Acacia mangium* Willd (Mimosaeae, Leguminosae)	Mangium acacia	Medium size, fast growing, robust, propagated from seeds	Common tree for ecorestoration
5.	*Adina cordifolia* Hook (Rubiaceae)	Keli-kadam (B), Haldu (H)	Normally 15–20 m, slow growing; oblong, leaves are acute at tip and heart-shaped at base; propagated by seeds	Not used for ecorestoration work
6.	*Aegle mermelos* (L.) Correa. (Rutaceae)	Wood apple (E), Sriphal (H), Bel (B)	Slow growing, xerophytic, hardy. Grow well on OB dumps. Oblong, leaves with three or five leaflets	Very common
7.	*Ailanthus excelsa* Roxb. (Simaroubaceae)	The tree of Heaven (E)	Fast growing, 12–25 m, shade tree	Common in Block – II, BCCL

(continued)

(continued)

Sl. No.	Botanical name (Family)	Local name	Growth characteristics, leaf canopy characteristics	Remarks
8.	*Alstonia scholaris* (Linn) R. Br. (Apocynaceae)	Devil tree (E), Chatim (H, B)	Quick growing, hardy, 4–7 leaves in a whorl; produces latex; round (umbrella shaped); oblong; height 10–15 m; flowering, December–March	Not very common, few could be seen in the reclaimed dumps of Jharia coalfield but growth is poor.
9.	*Albizzia lebbeck*, Benth. (Mimosaeae, Leguminosae)	Silk flower or Parrot tree (E); Siris (B, H)	Moderate growing; grown in dumps; drought resistant. Round/spreading; regenerated by seeds. Other two species *A. odoratissima* (Black siris) and *A. procera* also common in ISM campus	Common. Self-regenerated; have great potential for ecorestoration
10.	*Anthocephalus cadamba* Miq. (=*A. indica* A. Reich) (Rubiaceae)	The Kadamba (E), Kadamba (H, B).	Quick growing, large; spreading; grow rapidly in first 6–8 years; produce golden ball of flowers	Not common
11.	*Anacardium occidentale* L. (Anacardiaceae)	Cashewnut (E), Kaju (B, H)	Slow growing, evergreen; round	Common in Neyveli, trial to be made
12.	*Azadirachta indica* A. Juss. (= *Melia azadirachta* L) (Meliaceae)	Margosa tree (E), Nim (H), Neem (B)	Medium size to large tree; Pinnately compound, semi deciduous; grow slowly on mine dumps; round/oblong, spreading. Blossom in the 1st week of April; good purifier of air. Seeds do not keep long vitality, sown immediately after collection	Common in reclaimed dumps
13.	*Bauhunia purpurea* Linn (Cesalpineae)	The purple bauhinia or Mountain ebony (E); Devkanchan (B)	Medium size tree, deep pink-rosy pink flowers; leaves 10–14 cm long; open branch 9–11 strong nerve radiating from base	Not used in bioreclamation
14.	*Bauhunia variegata* Linn (Cesalpineae)	Variegated Bahunia (E), Raktokchan (B), Kachnar (H)	Small tree, flowers white with light yellow spot, open branch 11–15 strong nerve radiating from base	Not used in bioreclamation
15.	*Butea frondoasa* Roxb, (=*Butea monosperma* L) (Leguminosae)	Flame of forest (E), Dhak (H), Palas (B)	Slow growing; medium size tree; gives flowers onset of Holi. Oblong/ovoid; in March gets covered with scarlet flowers; extremely resistant to draught	Common; grow naturally in reclaimed sites
16.	*Bombax malabaricum* D.C. (= *Salmalia malabarica* Schott; = *Bombax cieba* Linn.) (Bombacaceae)	Silk-cotton tree (E), Shimool (B), Simbal (H)	Tall, wide spreading branches, wood is soft, leaves divided into 5–7 leaflets, flowers are bright crimson or red. Spreading. Often reaches 25 m height, cuttings or seeds, beauty and brightness flower and shadiness of leaves	Common; grow naturally in reclaimed sites
17.	*Cassia siamea* Lamk. (Leguminosae Papilionaceae)	Iron wood tree (E), Chakundi (H, B), Kassod tree	Fast growing, 10–12 m height, dark, evergreen; grow very well in subsided area, OB dumps. Regenerated by seeds/sapling	Very common; planted in reclaimed sites

(continued)

(continued)

Sl. No.	Botanical name (Family)	Local name	Growth characteristics, leaf canopy characteristics	Remarks
18.	*Cassia fistula* L. (Leguminosae, Papilionaceae)	*Indian laburnum* (E), Amaltash (H), Bandar lathi (B)	Slow growing; medium size tree; xerophytic; not eaten by goats or cattle; fruit pods are long 1–2 ft and dark brown and remain hang for next flowing season. Round/oblong; open branched Height 10–12 m; regenerated by seeds. Easily attacked by caterpillar; gets covered with golden yellow flowers in April–May	Good flowering plant; common in reclaimed sites
19.	*Callistemon lanceolatus* Sweet. *C. linearis* L (Myrtaceae)	Lal bottle brush or Bottle brush plant	Handsome evergreen with pendulous branch. Propagate from seed	Not found in reclaimed sites
20.	*Casuarina equisitifolia* Frost. (Casuarinaceae)	Australian or Whistling pine (E), Jau (B), Casuarina, Jungli Saru (H)	Tall tree, quick growing, hardy and evergreen. Appearance like coniferous trees. Multiplied by seeds	Common, good for reclamation work
21.	*Delonix regia* Rafin. (= *Poinciana regia* Bojer.) (Caesalpineae)	The GulMohur (E), Krisnachura (B), Gulmohur (H)	Quick growing. Evergreen, middle size tree, 5–10 m tall. Propagated by seeds; spreading/flat topped; open-branched spreading, irregular shaped	Common in reclaimed sites
22.	*Dalbergia sissoo* Roxb. (Leguminosae Papilionaceae)	Sissoo (E), Shishum (H), Sisu (B)	New leaves appear in mid-Feb; yield excellent timber; grow very well in mining areas. Round; height: 10–12 m	One of the most commonly used tree for ecorestoration
23.	*Duranta repens* L (Verbinaceae)	Duranta (B)	Shrub, 2–3 m, quick growing, evergreen	Hedge plant
24.	*Emblica officinalis* Gaertn. (= *Phyllanthus emblica* L) (Euphorbiaceae)	Gooseberry (E), Amla (B, H), Amloki (B)	Medium size deciduous, quick growing, regenerated by seeds, fruit tree cuttings	Common in reclaimed sites where fruit trees are preferred
25.	*Erythrina india* Lam. Var. *parcelli* Papilionaceae (Leguminosae)	The Indian Coral Tree	Flowering tree, *E. blakei.* – other flowering tree	For roadside plantation
26.	*Eucalyptus citrodora* Hook. (Myrtaceae)	Lemon scented gum (Eucalyptus tree)	Quick growing; conical; height: 15 m or more; Reg. –seed, sapling	Common
27.	*Eucalyptus hybrid* (Myrtaceae)	Mysore gum	Quick growing; conical; height: 15 m or more; Reg.- seed, sapling	Common
28.	*Eugenia jambolana* Lamk., (= *Syzygium cuminii* Skeels) (Myrtaceae)	Java Plum (E); Jambu, Jamun (H); Kala jam (B)	Moderate growing; middle size tree, 10–15 m height; oblong; multiplied by seeds, cuttings	Common fruit tree and suitable during ecorestoration
29.	*Ficus benghalensis* L (Moraceae)	Banyan (E), Bot (H,B)	Moderate growing; spreading; height 20 m or sometimes more. Regeneration by stem cutting, seeds	Naturally growing
30.	*Ficus infectoria* Roxb. (Moraceae)	Pakur (B), Pakar (H)	Quick growing, evergreen, thick shady, sending down aerial roots. Round small to big tree	Naturally growing

(continued)

(continued)

Sl. No.	Botanical name (Family)	Local name	Growth characteristics, leaf canopy characteristics	Remarks
31.	*Ficus religiosa* L. (Moraceae)	Peepal tree (E), Pipal (H), Aswatha (B)	Vigorous seedlings can be seen on some older buildings, sacred tree, grows slowly in early stages	Natural regeneration takes place by birds
32.	*Ficus retusa* Linn. (Moraceae)	Chilkan	Large spreading evergreen tree, glossy barks and green leaves, grow 30–35 ft large as banyan tree. Fruits are yellowish when ripe	Naturally growing
33.	*Ficus cunea*, Ham., (Moraceae)	Fig (E), Dumur (B),	Common tree	Naturally growing
34.	*Ficus glomerata* Roxb. (Moraceae)	Fig tree (E), Gular (H), Jagyadumur (B)	Common tree	Naturally growing
35.	*Ficus hispida* (L), F. (Moraceae)	Kala-umbar, Kakdumur (B), Kagsha, Gobla (H)	Tree, 10 m, quick growing, evergreen Spreading	Naturally growing
36.	*Ficus elastica* Roxb. (Moraceae)	Indian rubber tree (E)	Quick growing; evergreen	Not used in ecorestoration
37.	*Feronia elephantum* Correa (Rutaceae)	Elephant apple (E), Koth-bel (B)	Big tree, evergreen	Not commonly used for ecorestoration
38.	*Gliricidia sepium* Walp (Papilionaceae)	Gliricidia (E)	Quick growing, medium size tree	Roadside plantation
39.	*Gmelina arborea* Roxb. (Verbenaceae)	White teak (E), Gamar (H,B)	Moderate growing, wood is valuable, planted in mining areas. Oblong. Regeneration by seeds, saplings; 1,012 m height	Common tree species in ecorestoration (common in Jharia Coalfield)
40.	*Grevillea robusta* A. Cunn. (Proteaceae)	Silvery oak (E, H, B)	Quick growing, hardy (Silvery color in the back of leaves); oblong. Good for dust and noise attenuation. Regeneration by seeds	Suitable for ecorestoration, very common in West Bokaro areas, SECL
41.	*Heterophragma adenophyllum* B& H (Bignoniaceae)	Katsagoan (E, H, B)	Moderate size deciduous tree, propagated from seeds	Growing in the reclaimed dumps of North Karanpura Coalfields, CCL
42.	*Holoptelea integrifolia* (Ulmaceae)	Indian Elm Tree (E), Chilbil, Papri tree (B, H)	Large deciduous tree	Naturally found in reclaimed sites
43.	*Ixora coccinea* L (Rubiaceae)	Rangan (H, B)	Quick growing, evergreen; oblong, 6 m, by cuttings	Common garden plant
44.	*Jacaranda mimosaefolia* D. Don. (Bignoniaceae)	Nil Gulmohar (E, H, B), The Jacaranda	Native of Brazil. Quick growing; short lived. Introduced in some dumps but growth is not so attractive. Open branched, fruit is flat, disc-like contains numerous papery winged seeds. Feathery leaves and blue color flowers are added attraction	Roadside plantation. After 20 years become ugly
45.	*Leucaena leucocephala* L. (Leguminosae Mimosaeae)	Subabool (B,H); Koo-babool	Fast growing; good fodder and ranked first in mining areas; Regenerated by seeds, saplings, 7–10 m height	Very common in ecorestored sites, grow very well in dumps

(continued)

(continued)

Sl. No.	Botanical name (Family)	Local name	Growth characteristics, leaf canopy characteristics	Remarks
46.	*Lagerstroemia flos-reginae* Retz. (syn. *L. speciosa* Pers.) (Lythraceae)	The Queen's flower or Pride of India (E); Jarul (H, B).	Quick growing; medium size tree, ornamental tree, grown as a avenue tree. Shades all leaves in February–March, looks leafless and dull; Fruit – Woody capsule remain throughout the year	Roadside plantation
47.	*Madhuca indica* Gemel (= *Bassia latifolia* Roxb.) (Sapotaceae)	The Butter tree or Mohwa tree (E), Mauha, Maul (B, H).	Moderate growth, oblong/round, spreading crown	Very good in ecorestoration, naturally grown
48.	*Michelia champaca* L *Magnolia stellata* (Magnoliaceae)	The Golden Champa (E), Swarnachampa (B), Champak (H)	Moderate height tree	Roadside plantation
49.	*Mangifera indica* L. (Anacardiaceae)	Mango (E); Am (H, B)	Quick growing after 1 year. Round/oblong; height 10–15 m; regeneration by seeds, stem cutting	Common fruit tree and suitable for ecorestoration
50.	*Mimusops elengi* Linn. (Sapotaceae)	Bakul (E), Maulsari (H), Bakul (B)	Evergreen, medium to large tree, moderate growth, propagated by seeds/saplings	Suitable for roadside plantation, office complexes
51.	*Melia azedarach* L. (Meliaceae)	The Persian lilac (E), Bakayan(H), Ghoranim (B)	Quick growing, medium tree; 20 m; regeneration by seeds/saplings; confused with Neem (margosa tree)	Grow very well in OB dumps and subsided area (very common in Jharia coalfield)
52.	*Millingtonia hortensis* L. (Syn. *Bignonia subersosa* Roxb.) (Bignoniaceae)	The Indian Cork tree (E), akasneem, neem chameli, Tree Jasmine	Tall, straight evergreen tree, deep green foliage	Could be used for ecorestoration
53.	*Murraya exotica* L (= *M. paniculata* Jack) (Rutaceae)	Orange jessamine (E), Kamini (B)	Shrub, 4–5 m, quick growing, evergreen; round; flowering seasons June–October	Plant in office complex
54.	*Murraya Koeningii* Spreng. (Rutaceae)	Curry-leaf plant (E), Curry-pata (B, H)	Small tree	Could be used for ecorestoration
55.	*Nyctanthus arbortristis* Linn (= *Jasminum multiflorum* Andr.) (Oleaceae)	Night jasmine (E), Sheuli (B), Harsinghar (H)	Shrub, 4–5 m, quick growing, oblong by seeds, cuttings; FOS more or less throughout the year	Plant in office complex
56.	*Pithecolobium dulce* Benth. (Syn. = *Inga dulcis* Willd). (Mimosae)	Manila tamarind, Vilayatiimli (H), Jilapi (B)	Small to big tree, fast growing, Regenerated by seeds. Spreading, oblong	Common tree for ecorestoration, attracts avaifauna
57.	*Plumeria acutifolia* Poir (= *P. rubra* Linn) (Apocynaceae)	The temple tree (E), Gorurchampa, Gulancha (B), Golainchi(H)	Small deciduous tree, open, flowering plant, hardy, propagate by stem cuttings	Plant in office complex
58.	*Polyalthia longifolia*, Var. *pendulous* Benth and Hook (Anonaceae)	The Mast tree (E), Ashok (H), Debadaru (B). (another var is *P. longifolia*, erectus)	Quick growing, tall and majestic tree; branches do not spread out. In erectus varieties branches are spread out. Conical (pyramidal) shape, grow 10–15 m height	Very common, provide good shade, avenue tree

(continued)

(continued)

Sl. No.	Botanical name (Family)	Local name	Growth characteristics, leaf canopy characteristics	Remarks
59.	*Prosopis cineraria* L (Mimosaeae)	Khejri (H)	Deep rooted, nitrogen fixing, multipurpose tree, essential component of the agroforestry land use system, leaves are good fodder, wood is ideal for domestic heating	Suitable for ecorestoration, seen in many places of Jharia coalfields
60.	*Pterospermum acerifolium*, Willd. (Sterculiaceae)	Moochkunda (B), Kanthachampa (H),	Produce fragrant flowers; medium to big size tree, dense canopy	Plant in office complex
61.	*Pongamia pinnata* Pierre. (= *Derris indica* (Lam) Bennett; *Pongamia glabra* Vent.) (Leguminosae, Papilionaceae)	Indian beech (E), Karanj (H), Karanja (B)	Quick growing, grow up to 10–15 m; A single tree is said to yield 9–90 kg seed per tree, indicating a yield potential of 900–9000 kg seed/ha, 25% of which might be rendered as oil (assuming 100 trees/ha). In general, Indian mills extract 24–27.5% oil, village crushers, 18–22% oil regenerated by seeds/saplings, high self regeneration capacity	Grow very well in subsided area and OB dumps
62.	*Peltophorum inerme* (Syn. *P. ferrugineum* Benth; *P. pterocarpum* (DC.). Backer) (Cesalpineae, Leguminosae)	The copper Pod; The Rusty Shield-Bearer (E); Radhachura (B)	Deciduous, quick growing; shades all the leaves in January–February, looks totally leafless and dull; road-side plantation. Propagated by seeds/saplings, blooms twice in a year. Yellow flowers March–May and September–November; pods look copper or rusty brown color, remain on the parent plant whole year, even when the tree is leafless	Grow very well in subsided area and OB dumps, very common in the reclaimed dumps of North Karapura area of CCL
63.	*Pithecolobium dulce* Benth. (= *Inga dulcis* Willd.) (Mimosae)	Manila tamarind, Vilayatiimli (H), Jilapi (B)	Medium to big tree, fast growing spreading, regenerated by seeds/saplings	Suitable for ecorestoration
64.	*Putranjiva roxburghii* Wall. (Euphorbiaceae)	Lucky Bean tree (E), Putranjiva (E, B)	Medium size, evergreen with drooping foliage, provides pleasant shade	Suitable for office complexes
65.	*Saraca indica* Linn. (= *Saraca asoka* Roxb.) (Cesalpinaceae, leguminosae)	The Asoka tree (E), Sita Asoka (H), Asoka (B)	Handsome, small (5 m) dense crown, foliage 7 pairs of leaflets +1, produce beautiful flowers, propagated from seeds. Beautiful sheltered plant	Preferably plant in office complex, residential areas
66.	*Samanea saman* Jacq. (Mimosaeae Leguminosae)	Rain tree (E)	Big tree, umbrella shaped canopy. Spreading	Suitable for ecorestoration
67.	*Spathodea campanulata* Beauv. (Bignoniaceae)	Tulip tree (E)	Tall tree, ornamental flowering plant	Suitable for office complexes, residential areas
68.	*Streblus asper* Lour. (Urticaceae)	Siamese rough bush (E), Shaorah (B), Siora (H)	Medium size tree	Grow in waste places

(continued)

(continued)

Sl. No.	Botanical name (Family)	Local name	Growth characteristics, leaf canopy characteristics	Remarks
69.	*Spondias pinnata* (L.f.) Kurz. Syn = *Mangifera pinnata* L.f.; *Spondias mangifera* Willd. (Anacardiaceae)	Indian Hogplum (E), Amra (B)	Big tree	Suitable for ecorestoration, attracts avaifauna
70.	*Swietenia mahogoni* L. (Meliaceae)	Mahogony tree (E, B, H)	Beautiful evergreen plant. Other species are *S. macrophylla* and *S. microphylla*	Suitable for ecorestoration
71.	*Tamarix aphylly* L. (Tamaricaceae)	Athel tree, Laljhar (H)	Fast growing, moderate size evergreen tree grow up to 18 m	Suitable for ecorestoration
72.	*Tamarindus indica* Linn. (Cesalpineae)	The Tamarind tree (E), Imli (H), Tentul (B)	Tree, moderate growing, evergreen	Suitable for ecorestoration, attracts avaifauna
73.	*Tectona grandis* Linn. (Verbenaceae)	Teak tree (E), Segun (B), Sagwan (H)	Large tree, quick growing, 25–30 m tall, propagated by seeds/saplings	Not commonly used for ecorestoration. In some reclaimed dumps, it found growing, but tree is not so healthy
74.	*Terminalia arjuna* (Roxb.) Weight & Arn. (Combretaceae)	The Arjuna (E), Arjun (B, H)	Large evergreen tree with smooth gray barks; quick growing, oblong; regeneration by seeds, cuttings	Common tree for ecorestoration
75.	*Ziziphus jujuba* Lamk. (Rhamnaceae)	Kul (B), Ber (H)	Tree, quick growing (early); evergreen, round to oblong	Common tree for ecorestoration, attracts avaifauna
76.	*Z. oenoplia* Mill (Rhamnaceae)	Siakul (B)	Straggler shrub, 5 m quick growing, evergreen, round. Other sp: *Z. rugosa* Lamk. and *Z. xylopyra* Willd.	Common tree for ecorestoration, attracts avaifauna

List of Other Common Trees

Sl. No.	Botanical name (Family)	Local name	Characteristics
1.	*Amherstia nobilis* Wall (Cesalpineae)	The Nobel Amherst	Good plant
2.	*Acacia dealbata* (Mimosae)	Silver wattle	Evergreen, fast growing, 15 m, tree, by seeds. FS: April–June
3.	*Adenanthera pavonia* (Mimosae)	Coral or red wood	Tree, quick growing, deciduous, spreading crown
4.	*Barringtonia acutangula* (Barringtoniaceae)	Indian oak	Tree, 9–12 m, quick growing, by seeds, evergreen
5.	*Cassia javanica* (Cesalpineae)	Pink Cassia	Tree, 12 m, quick growing, by seeds, deciduous
6.	*Crataeva religiosa* (Capparidaceae)	The Barna	Ornamental trees

(continued)

(continued)

Sl. No.	Botanical name (Family)	Local name	Characteristics
7.	*Ceiba pentandra* (L.) Gaertn (Bombacaceae)	Kapok tree, silk-cotton tree	Very large, deciduous tree grow up to 60 m, umbrella-shaped crown, leaves digitately compound, 5–9 leaflets
8.	*Diospyros melanoxylon*	Tendu (B, H)	
9.	*Garuga pinnata* (Burseraceae)	Garuga	Elegant tree, 10 m, medicinal The Sausage tree
10.	*Grewia elastica* (Tiliaceae)	Dhamni (B), Dhaman (H)	Tree, quick growing, 10 m height
11.	*Holarrhena antidyseninterica* Wall. (Apocynaceae)	Easter tree (E), Kurchi (B), Karchi (H)	Tree, medicinal value
12.	*Kigelia pinnata* (Bignoniaceae)	Jhar phanos	Medium size tree, 6 m height
13.	*Kleinhovia hospita* L. (Sterculiaceae)	Tree antigonon (E), Bola (B)	Medium size beautiful tree
14.	*Kydia calycina* Roxb. (Malvaceae)	Potari (H)	Medium tree
15.	*Memecylon umbellatum* (Melastomataceae)	Iron wood tree	Tree, quick growing 15 m height
16.	*Mallotus philipiensis* (Euphorbiaceae)	Kamala (H, B)	Tree, slow growing, 12 m height
17.	*Morus alba* (Moraceae)	Mulberry (B, H)	Tree, 8 m height, quick growing
18.	*Pterocarpus santalinus* (Fabaceae)	Red Sanders	Tree
19.	*Sapium sebiferum* Roxb. (Euphorbiaceae)	The Chinese Tallow tree or Vilayati Shisham	Medium size deciduous tree
20.	*Spathodea nilotica* (Bignoniaceae)	Tree	Medium size tree
21.	*Sterculia foetida* (Sterculiaceae)	Jangli badam (H, B)	Tree, quick growing, 15 m height
22.	*Strychnos nux-vomica* (Loganiaceae)	Kuchila (B), Kulcha (H).	Tree, 12 m height, quick growing
23.	*Syncarpia glomulifera* (Myrtaceae)	Turpentine tree	Tree, 20 m height, quick growing
24.	*Terminalia belerica* (Combretaceae)	Behra	Good tree
25.	*Tecoma undulata* (Bignoniaceae)	The Lahura	Medium size tree
26.	*Tabebuia speciosa* (Bignoniaceae)	The Mauve Tabebuia	Roadside flowering plant
27.	*Toona ciliata* (Meliaceae)	The Toon tree	Shade tree

Index

S.K. Maiti, *Ecorestoration of the Coalmine Degraded Lands*,
DOI 10.1007/978-81-322-0851-8, © Springer India 2013

Printed by Publishers' Graphics LLC
AMZ20130320.09.06.77